Systems
Analysis, Design, and Computation

Systems
Analysis, Design, and Computation

Bradley W. Dickinson

Department of Electrical Engineering
Princeton University

PRENTICE HALL INFORMATION AND SYSTEM SCIENCES SERIES
Thomas Kailath, Series Editor

PRENTICE HALL, Englewood Cliffs, New Jersey 07632

```
Library of Congress Cataloging-in-Publication Data

Dickinson, Bradley W.
    Systems : analysis, design, and computation / Bradley W.
  Dickinson.
        p.    cm.
    Includes index.
    ISBN 0-13-338047-5
    1. System design.   2. System analysis.    I. Title.
  QA76.9.S88D53   1991
  004.2'1--dc20                                              90-25133
                                                                  CIP
```

Editorial/production supervision and
   interior design: *Rick DeLorenzo*
Cover design: *Karen Stephens*
Logo design: *A. M. Bruckstein*
Manufacturing buyers: *Linda Behrens and Patrice Fraccio*
Acquisitions editor: *Peter Janzow*

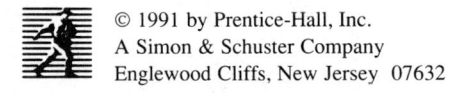

© 1991 by Prentice-Hall, Inc.
A Simon & Schuster Company
Englewood Cliffs, New Jersey 07632

All rights reserved. No part of this book may be
reproduced, in any form or by any means,
without permission in writing from the publisher.

Printed in the United States of America
10 9 8 7 6 5 4 3 2 1

ISBN 0-13-338047-5

Prentice-Hall International (UK) Limited, *London*
Prentice-Hall of Australia Pty. Limited, *Sydney*
Prentice-Hall Canada Inc., *Toronto*
Prentice-Hall Hispanoamericana, S. A., *Mexico City*
Prentice-Hall of India Private Limited, *New Delhi*
Prentice-Hall of Japan, Inc., *Tokyo*
Simon & Schuster Asia Pte. Ltd., *Singapore*
Editora Prentice-Hall do Brasil, Ltda., *Rio de Janeiro*

To Colette and Jim

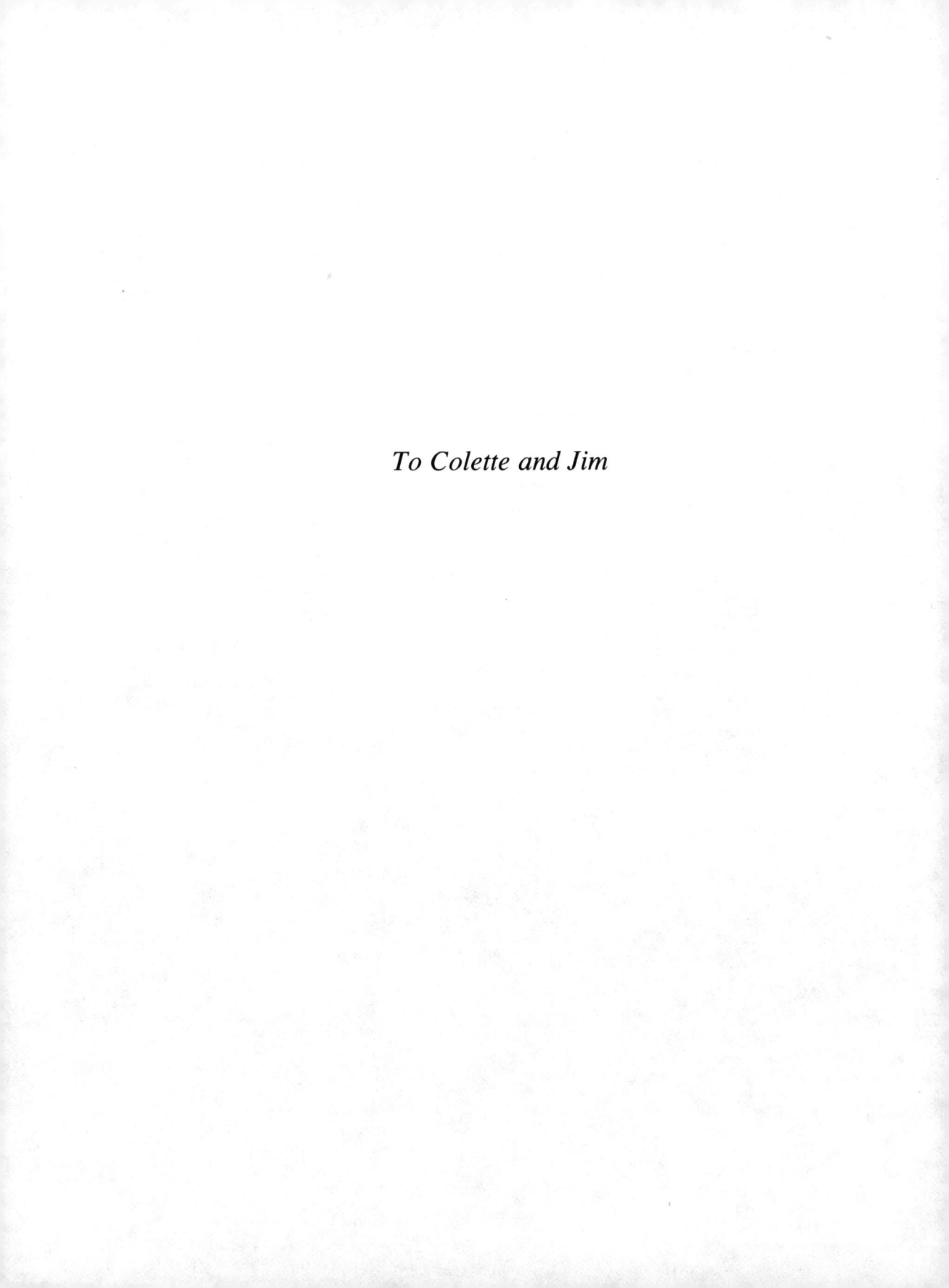

Contents

Preface ix

1 The Mathematical Toolbox: Linear Algebra and Matrices 1

    1.1   Some Topics in Matrix Algebra 1

    1.2   Linear Equations and Matrix Factorizations 14

    1.3   Matrix Calculus 36

    A1.   Some Basic Facts About Matrices and Linear Algebra 42

    1.4   Notes and References 60

2 Linear Systems 63

    2.1   Differential Equations and Linear State Space Systems 70

    2.2   Inputs, Outputs, and States 77

    2.3   Stability 88

    2.4   Frequency-Domain Characteristics of Linear Systems 97

    2.5   Time-Domain Characteristics of Linear Systems 108

    2.6   Discrete-Time Linear Systems 111

    2.7   Notes and References 118

3 Discretization of Continuous-Time Systems — 124

- 3.1 Basic Discretization Methods   125
- 3.2 Analysis of Discretization Techniques   131
- 3.3 Discretization and Digital Filter Design   141
- 3.4 Discretization for Distributed Parameter Systems   145
- 3.5 Notes and References   156

4 Nonlinear Systems — 159

- 4.1 Derivatives and Applications   165
- 4.2 Linearization and Stability   180
- 4.3 Qualitative Behavior of Nonlinear Systems   184
- 4.4 Nonlinear Systems and Neural Networks   202
- 4.5 Input-Output Analysis of Nonlinear Systems   211
- 4.6 Piecewise Linear Systems   226
- 4.7 Notes and References   229

5 Optimization — 234

- 5.1 Parameter Optimization   235
- 5.2 Numerical Optimization Techniques   254
- 5.3 Path Optimization Problems and the Principle of Optimality   267
- 5.4 Notes and References   278

Index — 283

Preface

Computer-aided techniques for analysis and design of systems have come into widespread use in recent years thanks to a proliferation of powerful desktop workstations and the availability of sophisticated software packages that provide easy-to-use "tools" for analysis and design which incorporate powerful numerical and graphical procedures. These technological innovations in computer hardware and software have been rapidly deployed in the commercial world; they have also made a major impact on research and graduate-level teaching in systems, control, and signal processing.

The major premise behind this book is that a valuable and interesting approach to the teaching of undergraduate systems and signals material can be based on a set of topics having clear importance in modern computer-aided system analysis and design. In an electrical engineering curriculum, the "systems and signals" courses have traditionally provided considerable exposure to important applied mathematics material; it seems very appropriate that these courses reflect the evolution of engineering practice, especially with regard to the recent rapid advances in computer-aided design and analysis technology. The fact that computer-aided methods for electronic circuit analysis have already proved to be beneficial for use in courses starting at an introductory level is taken as further evidence supporting the value of a systems and signals course of the kind for which this book was developed. As for its interest, experience suggests that engineering students are attracted to courses that exploit and complement a high level of computer expertise.

This book provides an introduction to analytical and computational methods of fundamental importance for computer-aided systems analysis and design. The aim is to provide an understanding of basics, and the book is written at an introductory level, in the sense that no substantial sophistication in any application is assumed. A single course could not cover the vast range of applications where computer-aided techniques play an important role, but various specialized courses can build on the material presented here. As in a "traditional" systems and signals course, the tools discussed are important and versatile ones. It is essential that students be provided with opportunities (with some training if necessary) for using these tools in practice; this means, typically, an opportunity for gaining familiarity with using one or more software packages while doing analysis and design problems on PC-class workstations.

The book is structured to support several modes of integrating the subject matter into a curriculum. It was developed at Princeton University to support a junior-level Electrical Engineering elective course that sets the stage for a traditional signals and linear systems course which follows. The course is taken concurrently with a course in analog electronic circuits that includes SPICE simulations. The PC-MATLAB software package is used in course assignments. About 80% of the material is covered in one semester. The book could also be used for a course in which a fluency in Laplace and Fourier transforms is assumed as a prerequisite; the material on linear systems can be covered quickly, and more time can be devoted to nonlinear systems and optimization. How best to structure a year long course sequence which combines the material presented in this book with other important analytical methods covered in "conventional" systems and signals textbooks is a challenging problem of curriculum design. It is hoped that this book some will suggest some new possibilities for consideration.

Basic mechanical and electrical systems provide many of the examples and problems which are treated in this book. For example, frequency response and time response characteristics of such systems are two aspects of behavior that are often the subject of design and analysis tasks. Examples from a variety of fields are included, and one goal of the presentation is broad appeal. The mathematical and computational tools presented are of fundamental importance in a wide variety of applied problems arising in fields such as electronic circuits, signal processing, telecommunications, dynamics, automation and control, and econometrics. Thus the ideas and methods developed in this book offer essentially unlimited possibilities for use in studying quantitative models of all kinds of dynamical phenomena.

In addition to a sophomore-level engineering familiarity with electrical and mechanical systems, a mathematical background including linear algebra and differential equations is assumed. Important background material is introduced, reviewed, and motivated by applications. The discussion of a few relatively sophisticated mathematical ideas relies on intuition and analogies rather than on rigorous formalism. Examples are used throughout to reinforce ideas and to demonstrate applications.

The book does not include any discussion of implementation details at the programming language level. Thus the book is not tied to a particular software development environment. At Princeton, a self-guided introduction to PC-MATLAB is assigned at the beginning of the course, and later assignments provide additional suggestions for implementing programs for specific tasks. The justification of this approach is that the book deals with system analysis and design, not development of reliable numerical software; nor is a main purpose of the book to provide a thorough understanding of the use and limitations of particular software tools. However, a significant appreciation of such issued can be conveyed through hands-on experience in using a computer to solve problems such as those included in the book.

There are four major areas covered by the book.

1. *Mathematical Fundamentals.* The coverage of this first topic is intended to illustrate the utility of linear algebra, differential equations, and difference equations in formulating quantitative descriptions of system behavior. Parts of the first chapter will primarily serve as a review of material that students are expected to have seen before. Based on experience with classes of first-semester junior electrical engineering students, covering this material is essential for developing an adequate understanding of topics that provide the foundation for much of what follows in later chapters. Almost all undergraduate readers will find some new material introduced and some new approaches given; instructors will need to tailor the coverage of this material to suit particular audiences. One particular focus of the presentation is to motivate the kinds of numerical techniques that are useful in system analysis and design, e.g. solving systems of linear equations, computing eigenvalues and eigenvectors, etc. The emphasis is on understanding fundamental ideas and their interpretations and applications.

2. *Linear and Nonlinear Systems.* Linear models are widely used because they adequately describe a wide variety of dynamical systems, and they are amenable to detailed analysis, including the use of special techniques such as frequency response methods which very much rely on the superposition principle for justification. Various analytical and computational methods are available to handle linear and nonlinear systems; time-domain methods are the focus of the presentation. A variety of topics involving Fourier and Laplace transforms is covered in a "traditional" introductory (linear) systems and signals course; in this book, these topics have been replaced by some "nontraditional" ones, broadening the scope of the course while keeping the presentation at an introductory level. As one example, a number of topics concerning nonlinear systems are treated in the book. Since nonlinear effects arise in all real systems, (e.g. at extremes of the operating ranges of system variables) and since there are important physical phenomena that require nonlinear models (e.g. Newton's law of gravitation, models for competition in population ecology, electronic device models, etc.), a knowledge of some fundamentals of nonlinear systems is quite important for applications of systems analysis and design. The uses of linearization and other approximations are covered, and other approaches arising from qualitative notions of nonlinear system behavior are discussed.

3. *Discretization, Discrete-Time Models, and Simulation.* Since differential equations are rarely solvable by any kind of "closed form" analytical expression, numerical methods are indispensable for determining properties of solutions. Digital simulations methods require some kind of discretization process and provide information about the solutions on some grid of points rather than for all values of the independent variable. In applications such as audio signal processing and computerized control, it is necessary to interface a discrete-time system (the signal processor or the digital process controller) with the continuous-time variables of the "real-world" (the output of a microphone or the measured values of temperature, pressure, etc.). The coverage of this topic is intended to provide insight and understanding by presenting and analyzing a variety of discretization techniques that are widely used.

4. *Optimization.* System design involves consideration of various performance objectives in view of physical and geometrical constraints and limitations on available resources. In short, a designer tries to achieve best possible performance within the limits posed by technology, economics, the time available for design, etc. By modeling performance criteria and constraints in mathematical form, analytical and numerical methods can be used to obtain optimum designs. Even in cases where system behavior is judged in a qualitative way by the designer or the end user of the system, numerical methods and graphical displays of system response characteristics may be used to guide the designer. This book covers the basic theory behind the principal kinds of optimization problems that have been used in systems applications, and it provides a selected overview of numerical optimization techniques.

ACKNOWLEDGMENTS

I received encouragement, advice, and help from many people during the months I spent writing this book. My wife Colette and son Jim patiently endured through several overly optimistic targeted completion dates; there would be no book without the love and support that they have provided. My Princeton colleagues are an amiable and inspiring group; they have been the source of many influential ideas and insights. Stu Schwartz encouraged and facilitated the course development effort that underlies this book. Valuable technical interactions with Bede Liu, Peter Ramadge, Ken Steiglitz, and Rob Stengel at Princeton and Eduardo Sontag at Rutgers are gratefully acknowledged. I sincerely appreciate the comments of several individuals who provided technical reviews of parts of the manuscript; their efforts led to many improvements in style and substance. I particularly thank Tom Kailath, Series Editor of the Prentice Hall Information and System Sciences Series, for his many helpful suggestions. I would also like to thank the following reviewers for their thoughtful comments: Bernard C. Levy, MIT; J.B. Pearson, Rice University; Marc Bodson, Carnegie-Mellon University; George Cybenko, University of Illinois; and George Verghese, MIT.

Several classes of Princeton electrical engineering students (mostly juniors) have cheerfully offered their comments, corrections, and criticisms of preliminary versions of much of the material in this book distributed as course notes. I also benefitted from the assistance of several graduate students. Doug Gallager, my teaching assistant during the first time the course was taught, provided a great deal of help while I was getting this writing project underway. Joe Serrano, my most recent teaching assistant, read a nearly final draft of most of the text and offered many valuable comments. Sean Carroll and Nick Skiadas also read and critiqued various parts of the manuscript.

As I look over the finished book and think back over the time I spent in writing it, I'm reminded of a quotation attributed to Earl Weaver in 1968:

"It's what you learn after you know it all that counts."[*]

Bradley W. Dickinson

---

[*]*Voices of Baseball*, edited by Bob Cheiger, Atheneum Publishers, New York, 1983.

CHAPTER 1

The Mathematical Toolbox
Linear Algebra and Matrices

It is not uncommon for engineering and science students to finish a mathematics course involving linear algebra without much appreciation for the importance of this subject in a variety of applications. A "systems course" can offer a perspective that provides a basis for an intuitive understanding of the underlying mathematical ideas. Linear algebra is an extremely useful subject for systems analysis, and students will find its mastery to be essential. The utility of the subject arises in two quite different ways. First, its concrete manifestation in the guise of vectors, matrices, and the algebraic operations involving such quantities (e.g., vector-matrix multiplication, matrix inversion and the solution of linear equations, eigenvalues and eigenvectors) provides an important framework for the analysis of systems. The framework is amenable to the development of a set of computational tools well suited to software implementation. Second, "abstract" linear algebra concepts, particularly the notions of linear independence, basis, and linear transformations, are crucial for describing the most basic notions of linearity in systems analysis, such as the superposition principle and coordinate changes. In this book we will have occasion to see both parts of linear algebra in action.

1.1 SOME TOPICS IN MATRICES AND VECTOR SPACES

A thorough familiarity with matrix algebra is an essential prerequisite for a study of dynamic systems. The Appendix covers a selection of basic material of particular importance, and it should be read in conjunction with this first section of this chapter. In this section we will present some specialized and particularly useful topics that will also serve to indicate the notation used and to review some important terminology.

Students should not underestimate the importance of linear algebra. From past experience in teaching undergraduate and beginning graduate courses in systems, an attention to detail in this area will pay off handsomely in subsequent studies of all the basic systems ideas. One argument for this viewpoint is the following. We reason mathematically about systems using matrices and vectors as tools, and it is important for gaining a clear understanding of systems concepts to be fluent in the mechanics of manipulating matrices and vectors.

For example, it is surprising how many problems can be avoided simply by paying attention to the requirement of conformable dimensions that arises in the definition of matrix multiplication. Recall that if $\mathbf{A}$ is an $m \times n$ matrix and $\mathbf{x}$ is an n-vector (by which we mean a column vector with n entries or equivalently an $n \times 1$ matrix), the product $\mathbf{Ax}$ is well defined, whereas the "product" $\mathbf{xA}$ makes no sense (i.e., is not defined at all). The requirement of matrix multiplication is that for the product $\mathbf{BC}$ to make sense the "adjacent" dimensions of $\mathbf{B}$ and $\mathbf{C}$ must be equal (i.e., $\mathbf{B}$ must be $m \times n$ and $\mathbf{C}$ must be $n \times p$). Said another way, in terms of the "shape" of the matrices, to form the matrix product $\mathbf{BC}$, the number of columns of $\mathbf{B}$ must equal the number of rows of $\mathbf{C}$. Another kind of restriction about dimensions is so "natural" and familiar that it usually is not even thought about twice: in order to add two matrices, we require that both matrices have the same number of rows and columns.

The biggest difference between matrix algebra and the familiar algebra of real numbers is the lack of *commutativity* of the matrix multiplication operation. Thus, even if the two products of the matrices $\mathbf{A}$ and $\mathbf{B}$ make sense (when does this happen?), we may find that $\mathbf{AB} \neq \mathbf{BA}$. While this is a real headache, it is one that we must live with, and it necessitates care in simplifying expressions involving matrix products. Fortunately, there are some situations where commutativity holds: the identity matrix $\mathbf{I}$ commutes with all square matrices of the same size, and any invertible square matrix commutes with its inverse, $\mathbf{AA}^{-1} = \mathbf{I} = \mathbf{A}^{-1}\mathbf{A}$. Also note that because a square matrix commutes with itself, we may denote products like $\mathbf{AAA}$ as $\mathbf{A}^3$ without ambiguity. Indeed, an inductive argument may be used to define $\mathbf{A}^n$, the product of n factors of $\mathbf{A}$, as $\mathbf{A}^n = \mathbf{AA}^{n-1} = \mathbf{A}^{n-1}\mathbf{A}$.

Bases and Transformations

Suppose that we have two sets of basis vectors (two bases) for $\mathbb{C}^n$, the set of n-vectors of complex numbers (or for $\mathbb{R}^n$, the set of n-vectors of real numbers). Then there is an invertible matrix that relates any two corresponding basis vectors (see the last section of the Appendix). In particular, if $\{\mathbf{x}_1, \ldots, \mathbf{x}_n\}$ and $\{\mathbf{z}_1, \ldots, \mathbf{z}_n\}$ are two bases for $\mathbb{C}^n$ (or $\mathbb{R}^n$), let $\mathbf{T_x}$ be the $n \times n$ matrix whose columns are the $\mathbf{x}_i$ basis vectors (in natural order) and similarly for $\mathbf{T_z}$. Both $\mathbf{T_x}$ and $\mathbf{T_z}$ are invertible because each has columns that form a linearly independent set of vectors. Now since

$$\begin{bmatrix} \mathbf{z}_1 & \cdots & \mathbf{z}_n \end{bmatrix} = \mathbf{T_z} = \mathbf{T_z}\mathbf{T_x}^{-1}\mathbf{T_x} = (\mathbf{T_z}\mathbf{T_x}^{-1}) \begin{bmatrix} \mathbf{x}_1 & \cdots & \mathbf{x}_n \end{bmatrix} \quad (1.1)$$

we see that each of the $\mathbf{z}_i$ basis vectors is obtained by multiplying the matrix $\mathbf{T} = (\mathbf{T_z}\mathbf{T_x}^{-1})$ times the corresponding $\mathbf{x}_i$ basis vector. Furthermore, the basis

Sec. 1.1 Some Topics in Matrices and Vector Spaces

transformation matrix $\mathbf{T}$ is invertible, with inverse $\mathbf{T}^{-1} = (\mathbf{T_x T_z^{-1}})$. Thus, one important use of inverse matrices arises in the transformation of bases for the vector space $\mathbb{C}^n$ (or $\mathbb{R}^n$).

Examples

Consider the two bases for $\mathbb{R}^2$:

$$B_1 = \{\mathbf{x}_1, \mathbf{x}_2\} = \left\{ \begin{bmatrix} 2 \\ 2 \end{bmatrix}, \begin{bmatrix} -1 \\ 2 \end{bmatrix} \right\}$$

$$B_2 = \{\mathbf{z}_1, \mathbf{z}_2\} = \left\{ \begin{bmatrix} -2 \\ -2 \end{bmatrix}, \begin{bmatrix} 1 \\ 2 \end{bmatrix} \right\}$$

The matrix transforming B_1 into B_2 is

$$\mathbf{T} = \begin{bmatrix} -2 & 1 \\ -2 & 2 \end{bmatrix} \begin{bmatrix} 2 & -1 \\ 2 & 2 \end{bmatrix}^{-1} = \begin{bmatrix} -1 & 0 \\ -4/3 & 1/3 \end{bmatrix}$$

Now consider two different bases for $\mathbb{R}^2$:

$$B_3 = \{\mathbf{x}_1, \mathbf{x}_2\} = \left\{ \begin{bmatrix} \sqrt{3}/2 \\ 1/2 \end{bmatrix}, \begin{bmatrix} -1/2 \\ \sqrt{3}/2 \end{bmatrix} \right\}$$

$$B_4 = \{\mathbf{z}_1, \mathbf{z}_2\} = \left\{ \begin{bmatrix} 1/2 \\ \sqrt{3}/2 \end{bmatrix}, \begin{bmatrix} -\sqrt{3}/2 \\ 1/2 \end{bmatrix} \right\}$$

The matrix transforming B_3 into B_4 is

$$\mathbf{T} = \begin{bmatrix} 1/2 & -\sqrt{3}/2 \\ \sqrt{3}/2 & 1/2 \end{bmatrix} \begin{bmatrix} \sqrt{3}/2 & -1/2 \\ 1/2 & \sqrt{3}/2 \end{bmatrix}^{-1} = \begin{bmatrix} \sqrt{3}/2 & -1/2 \\ 1/2 & \sqrt{3}/2 \end{bmatrix}$$

Notice that multiplying a vector $\mathbf{x}$ by this $\mathbf{T}$ has the effect of rotating $\mathbf{x}$ by $\pi/6$ radians when we view vectors as lying in the Cartesian plane (Figure 1.1). For this case the matrix $\mathbf{T}$ is an *orthogonal matrix*; its two columns are mutually orthogonal and have unit length (see (A.54)).

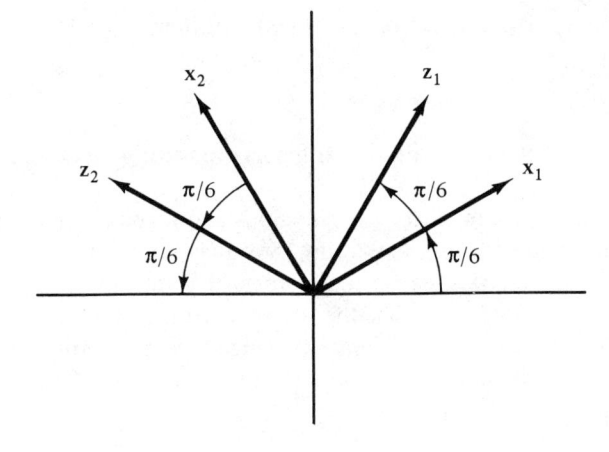

Figure 1.1 Transformation of bases. Vectors in $\mathbb{R}^2$ plotted in the Cartesian plane.

Of course, a transformation of bases is usually carried out for some purpose, and now we will describe a very important use of basis transformations for purposes of system analysis. When the standard basis is used, linear functions defined on the vector space $\mathbb{C}^n$ (and $\mathbb{R}^n$, too, of course) correspond to matrix multiplication (see the last section of the Appendix). A question of obvious importance is how the representation is changed when we perform a transformation of bases. To answer this question, we study the situation with the facts we have available.

Suppose we have two n-vectors, **x** and **y**, related through the matrix equation **y** = **Ax**. Suppose that we transform **x** and **y** using an invertible matrix **T**, so that $\mathbf{x} = \mathbf{T}\mathbf{x}_1$ and $\mathbf{y} = \mathbf{T}\mathbf{y}_1$. How are $\mathbf{x}_1$ and $\mathbf{y}_1$ related? A simple calculation shows that $\mathbf{y}_1 = \mathbf{T}^{-1}\mathbf{A}\mathbf{T}\mathbf{x}_1$:

$$\mathbf{y}_1 = \mathbf{T}^{-1}\mathbf{y} = \mathbf{T}^{-1}\mathbf{A}\mathbf{x} = \mathbf{T}^{-1}\mathbf{A}\mathbf{T}\mathbf{x}_1 \tag{1.2}$$

Some thought will show that the matrix **T** represents the basis change from the "new" basis to the "old" basis, while $\mathbf{T}^{-1}$ represents the change from "old" to "new." (The reader will find this "backward" interpretation for **T** to be traditional in systems textbooks.)

We will denote this special product with its own symbol: $\mathbf{T}^{-1}\mathbf{A}\mathbf{T} = \mathbf{A}_1$. It is customary to say that $\mathbf{A}_1$ is obtained from **A** by the *similarity transformation* **T**. Obviously, in this case **A** is obtained from $\mathbf{A}_1$ by the similarity transformation $\mathbf{T}^{-1}$. It is common to drop the reference to **T** or $\mathbf{T}^{-1}$ altogether and simply say that **A** and $\mathbf{A}_1$ are *similar*.

Eigenvalues, Eigenvectors, and Similarity

In many situations we will try to find a basis transformation matrix **T** to make the form of the transformed matrix equation, as described by $\mathbf{A}_1$, take a particularly simple form. This is a well-studied problem of linear algebra, and a very useful transformation is described in terms of the *eigenvalues and eigenvectors* of a matrix. Because of its importance, we will turn to this topic now.

First recall the definitions:

> **For a square matrix A, if the (real or complex) number λ and nonzero vector u satisfy**
>
> $$\mathbf{A}\mathbf{u} = \lambda \mathbf{u} \tag{1.3}$$
>
> then λ is called an *eigenvalue* of A and u is the corresponding *eigenvector*.

For now we'll ignore the problem of how eigenvalues and eigenvectors might be found, assuming what we need in order to show how they may be used to simplify a matrix by similarity transformation. Assume for the moment that the size of **A** is $n \times n$, that **A** has n eigenvalues $\lambda_1, \ldots, \lambda_n$, and that the set of corresponding eigenvectors $\{\mathbf{u}_1, \ldots, \mathbf{u}_n\}$ form a basis for $\mathbb{C}^n$. In particular, the eigenvectors form a linearly independent set and therefore can be used as the columns of a nonsingular (i.e., invertible) matrix **T**. Consider the following product:

Sec. 1.1 Some Topics in Matrices and Vector Spaces

$$\begin{aligned} \mathbf{AT} &= \mathbf{A} \begin{bmatrix} \mathbf{u}_1 & \mathbf{u}_2 & \cdots & \mathbf{u}_n \end{bmatrix} \\ &= \begin{bmatrix} \mathbf{Au}_1 & \mathbf{Au}_2 & \cdots & \mathbf{Au}_n \end{bmatrix} \\ &= \begin{bmatrix} \lambda_1 \mathbf{u}_1 & \lambda_2 \mathbf{u}_2 & \cdots & \lambda_n \mathbf{u}_n \end{bmatrix} = \mathbf{T}\Lambda \end{aligned} \quad (1.4)$$

where Λ is a diagonal matrix formed from the eigenvalues:

$$\Lambda = \begin{bmatrix} \lambda_1 & 0 & \cdots & 0 \\ 0 & \lambda_2 & & \cdot \\ \cdot & & \cdot & \cdot \\ \cdot & & & 0 \\ 0 & \cdots & 0 & \lambda_n \end{bmatrix} \quad (1.5)$$

By premultiplying equation (1.4) by $\mathbf{T}^{-1}$, we obtain

$$\mathbf{T}^{-1}\mathbf{AT} = \Lambda \quad (1.6)$$

and we have found a change of coordinates that puts $\mathbf{A}_1$ in a very simple form indeed (namely, diagonal).

The assumption that the eigenvectors can be used as the columns of a nonsingular matrix is one that requires some further consideration. We are interested in finding out when (if ever!) the set of eigenvectors of a matrix $\mathbf{A}$, $\{\mathbf{u}_1, \ldots, \mathbf{u}_n\}$ forms a basis for n-dimensional space, $\mathbb{C}^n$, which means that they must be a linearly independent set of vectors. So, the question of when a matrix can be brought to diagonal form by a similarity transformation requires determining which matrices have a set of n linearly independent eigenvectors. As evidence that this is not always the case, consider the 2×2 matrix

$$\mathbf{A} = \begin{bmatrix} a & 1 \\ 0 & a \end{bmatrix} \quad (1.7)$$

which has only the number a as an eigenvalue, and only the vector

$$\mathbf{u} = \begin{bmatrix} u_1 \\ u_2 \end{bmatrix} = \begin{bmatrix} 1 \\ 0 \end{bmatrix} \quad (1.8)$$

(or any nonzero multiple) as an eigenvector. (This is easily checked by determining all possible solutions of the defining equation $\mathbf{Au} = \lambda \mathbf{u}$ (1.3) for this particular matrix $\mathbf{A}$.) This is an example of a matrix that cannot be transformed to diagonal form by any similarity transformation.

The complete characterization of all matrices that can be diagonalized is rather technical and not so important from an applications point of view. It is a nice exercise in using the definitions of eigenvectors and of linear independence to show that if a matrix $\mathbf{A}$ has two eigenvectors, say $\mathbf{u}_1$ and $\mathbf{u}_2$, corresponding to eigenvalues λ_1 and λ_2, $\lambda_1 \neq \lambda_2$, then the two eigenvectors are linearly independent. It then follows that a matrix with n *distinct* eigenvalues has a set of n linearly independent eigenvectors. This case is one that often arises in practical applications. A second case arising in many applications involving "passive" RLC electric networks and "passive" mechanical systems concerns *symmetric* matrices. (The matrix **A** is symmetric if it equals its own transpose, $\mathbf{A} = \mathbf{A}^T$, where $\mathbf{A}^T$, the transpose of **A**, is obtained by interchanging the rows and columns of **A**: in terms of matrix elements, $(\mathbf{A}^T)_{ij} = (\mathbf{A})_{ji}$.) Symmetric matrices, even if they do not have n distinct eigenvalues, always have a set of n linearly independent eigenvectors; in fact, the linearly independent eigenvectors may also be chosen to be orthogonal and scaled to unit length so as to be orthonormal. (In the case of matrices of complex numbers, the Hermitian, or conjugate, symmetric ones have real eigenvalues and n linearly independent eigenvectors; see the Appendix.)

Our discussion has glossed over one practical point: How are eigenvalues and eigenvectors of a matrix **A** determined? For the answer, at least from a theoretical viewpoint, consider the defining equation, (1.3), repeated here for convenience:

$$\mathbf{A}\mathbf{u} = \lambda \mathbf{u} \tag{1.9}$$

Bringing everything to the right-hand side of the equation, we have

$$\mathbf{0} = \lambda \mathbf{u} - \mathbf{A}\mathbf{u} = \left[\lambda \mathbf{I} - \mathbf{A}\right] \mathbf{u} \tag{1.10}$$

Recognizing this as a *homogeneous* set of linear equations (i.e., the "known" vector on the left side is the zero vector, so any solution is defined only up to multiplication by a constant), we note that a nonzero solution vector **u** will exist if and only if the columns of the matrix $[\lambda \mathbf{I} - \mathbf{A}]$ are linearly *dependent*. This is the case if and only if the determinant of this matrix is zero. (A more detailed discussion of linear equations will be given in the next section.) Thus the condition for a number λ_o to be an eigenvalue of **A** is that

$$\det\left[\lambda \mathbf{I} - \mathbf{A}\right] = 0 \text{ for } \lambda = \lambda_o \tag{1.11}$$

Since **A** is $n \times n$, we may look at the form of this determinantal equation and make some important observations:

$$\det\left[\lambda \mathbf{I} - \mathbf{A}\right] = \det \begin{bmatrix} \lambda - a_{11} & -a_{12} & \cdots & -a_{1n} \\ -a_{21} & \lambda - a_{22} & \cdots & -a_{2n} \\ \vdots & \vdots & & \vdots \\ -a_{n1} & -a_{n2} & \cdots & \lambda - a_{nn} \end{bmatrix} \tag{1.12}$$

Using the procedure for evaluating a determinant by Laplace expansion, see (A.21) or (A.22), or using the formula for the determinant as a sum of signed products of elements chosen from each row and column, see (A.20), it is straightforward to show that this determinant takes the form of a monic polynomial in λ having degree exactly n. (Monic means that the coefficient of the highest power of λ, λ^n in this case, is 1.) We will denote this polynomial by $p(\lambda)$, which may be written in terms of its coefficients as

$$p(\lambda) = \lambda^n + p_1 \lambda^{n-1} + \cdots + p_{n-1} \lambda + p_n \tag{1.13}$$

Now the problem of finding eigenvalues is reduced to one of finding the zeros of $p(\lambda)$, a monic polynomial of degree n. The Fundamental Theorem of Algebra states that

Sec. 1.1   Some Topics in Matrices and Vector Spaces 7

such a polynomial (with real or complex coefficients) has exactly n values (that may be real or complex even if the coefficients of the polynomial are all real) at which it vanishes, counting these zeros according to their multiplicities. That is, $p(\lambda)$ can be written in the factored form

$$p(\lambda) = (\lambda - \lambda_1)(\lambda - \lambda_2) \cdots (\lambda - \lambda_n) \tag{1.14}$$

for some set of complex numbers $\lambda_1, \ldots, \lambda_n$, not necessarily distinct. It follows immediately that if λ_o is one of the zeros of $p(\lambda)$ (i.e., $\lambda_o = \lambda_i$, for some i, $1 \le i \le n$), then

$$\lambda_o^n + p_1 \lambda_o^{n-1} + \cdots + p_{n-1} \lambda_o + p_n = 0 \tag{1.15}$$

Having found the eigenvalues of $\mathbf{A}$ by finding the zeros of the polynomial $p(\lambda) = \det[\lambda \mathbf{I} - \mathbf{A}]$, we are sure to find at least one eigenvector corresponding to each distinct eigenvalue, since we are guaranteed that the columns of $[\lambda \mathbf{I} - \mathbf{A}]$ are linearly dependent when λ is a zero of the polynomial, and hence there is at least one nontrivial solution to the equation defining an eigenvector. If λ is a zero of multiplicity 1 (i.e., not repeated), then only a single linearly independent eigenvector can be found. There may be more when an eigenvalue is *repeated*, by which we mean that the polynomial $p(\lambda)$ has a multiple zero at some point. However, as the 2×2 example in (1.7) shows, there may be only one eigenvector in this case as well. It is exactly those matrices having a multiple eigenvalue without a full set of corresponding eigenvectors (i.e., a linearly independent set consisting of as many vectors as the multiplicity of the eigenvalue), often described as being *degenerate*, which make up the class of matrices that cannot be diagonalized by similarity transformation.

Example

For a 2×2 matrix, the familiar formula for solutions of a quadratic equation provides an expression for the eigenvalues. If

$$\mathbf{A} = \begin{bmatrix} a_{11} & a_{12} \\ a_{21} & a_{22} \end{bmatrix}$$

then

$$\det[\lambda \mathbf{I} - \mathbf{A}] = \lambda^2 - (a_{11} + a_{22})\lambda + (a_{11}a_{22} - a_{12}a_{21})$$

so the eigenvalues are

$$\lambda = \frac{a_{11} + a_{22} \pm \sqrt{(a_{11} - a_{22})^2 - 4a_{12}a_{21}}}{2}$$

From this expression it is clear that the eigenvalues of $\mathbf{A}$ are (a) real and distinct, (b) real and repeated, or (c) complex (in fact, a conjugate pair) according to whether the quantity $(a_{11} - a_{22})^2 - 4a_{12}a_{21}$ is (a) positive, (b) zero, or (c) negative.

The Characteristic Polynomial and the Cayley-Hamilton Theorem

Because of its importance, the polynomial $p(\lambda) = \det[\lambda \mathbf{I} - \mathbf{A}]$ is given a name, the *characteristic polynomial* of the matrix $\mathbf{A}$. Not only does the characteristic polynomial of a matrix determine the eigenvalues of the matrix, but a beautiful and useful result known as the Cayley-Hamilton theorem involves the characteristic polynomial. Since we will make use of this result almost immediately, we will now state it; its proof will be discussed at the end of this section.

If A is an $n \times n$ matrix with characteristic polynomial $p(\lambda)$, then the matrix A satisfies the matrix equation

$$\mathbf{A}^n + p_1 \mathbf{A}^{n-1} + \cdots + p_{n-1} \mathbf{A} + p_n \mathbf{I} = \mathbf{0} \tag{1.16}$$

In words, the matrix **A** satisfies exactly the same kind of polynomial equation that its eigenvalues satisfy! (Notice that the right-hand side of equation (1.16) is the $n \times n$ zero matrix.) Written in compact form, equation (1.16) is simply $p(\mathbf{A}) = \mathbf{0}$. This notation suggests that we might be able to define *functions* of a matrix in some useful way. We will return to this point in the last section of this chapter and also in the second chapter in connection with the solution of linear differential equations.

Our immediate interest is in applying the Cayley-Hamilton theorem to obtain a second useful similarity transformation of a matrix, one that can be performed without the need for computing eigenvalues and eigenvectors, and one that gives a matrix with real entries whenever we start out with such a matrix. The matrix **T** is constructed as follows. First choose an arbitrary vector **v** and compute the vectors **Av**, $\mathbf{A}^2 \mathbf{v}, \ldots, \mathbf{A}^{n-1} \mathbf{v}$. If these vectors are linearly independent, then take

$$\mathbf{T} = \begin{bmatrix} \mathbf{v} & \mathbf{A}\mathbf{v} & \mathbf{A}^2 \mathbf{v} & \cdots & \mathbf{A}^{n-1} \mathbf{v} \end{bmatrix} \tag{1.17}$$

(For "almost all" matrices **A**, the vectors will "almost always" be linearly independent when **v** is chosen at random; for instance, as long as **A** is not degenerate, the only "bad" choices for **v** are those vectors that may be expressed as a linear combination of fewer than n of the eigenvectors of **A**. In practical terms, if a particular choice of **v** doesn't work, we try again with another choice; if **T** is singular for several different choices of **v**, it is quite likely that **A** is degenerate.) If we write **T** in terms of its columns as $\mathbf{T} = [\mathbf{t}_1 \, \mathbf{t}_2 \, \cdots \, \mathbf{t}_n]$, the construction of **T** gives the following: $\mathbf{A}\mathbf{t}_1 = \mathbf{t}_2$; $\mathbf{A}\mathbf{t}_2 = \mathbf{t}_3$; $\cdots$; $\mathbf{A}\mathbf{t}_{n-1} = \mathbf{t}_n$. Then the Cayley-Hamilton theorem may be used to find $\mathbf{A}\mathbf{t}_n$, since

$$\mathbf{A}\mathbf{t}_n = \mathbf{A}\mathbf{A}^{n-1}\mathbf{v} = \mathbf{A}^n \mathbf{v} \tag{1.18}$$

and the polynomial equation of the Cayley-Hamilton theorem, (1.16), may be rearranged to give

$$\mathbf{A}^n = -(p_1 \mathbf{A}^{n-1} + \cdots + p_{n-1} \mathbf{A} + p_n \mathbf{I}) \tag{1.19}$$

Multiplying on the right by **v** gives

$$\mathbf{A}\mathbf{t}_n = -(p_1 \mathbf{t}_n + \cdots + p_{n-1} \mathbf{t}_2 + p_n \mathbf{t}_1) \tag{1.20}$$

We may collect these equations into a single matrix equation involving the matrix **T**:

Sec. 1.1 Some Topics in Matrices and Vector Spaces

$$\mathbf{A}\begin{bmatrix} \mathbf{t}_1 & \mathbf{t}_2 & \cdots & \mathbf{t}_n \end{bmatrix} = \begin{bmatrix} \mathbf{t}_1 & \mathbf{t}_2 & \cdots & \mathbf{t}_n \end{bmatrix} \begin{bmatrix} 0 & 0 & \cdots & 0 & -p_n \\ 1 & 0 & \cdots & 0 & -p_{n-1} \\ 0 & 1 & & & \cdot \\ \cdot & \cdot & & \cdot & \cdot \\ \cdot & & & \cdot & \cdot \\ \cdot & & & 1 & 0 & -p_2 \\ 0 & \cdots & 0 & 1 & -p_1 \end{bmatrix} \quad (1.21)$$

so by using the symbol $\mathbf{A}_c$ for the rightmost matrix in this equation we may write

$$\mathbf{AT} = \mathbf{TA}_c \quad (1.22)$$

which gives the desired similarity transformation: $\mathbf{T}^{-1}\mathbf{AT} = \mathbf{A}_c$. Because of the form of $\mathbf{A}_c$, being determined by the coefficients of the polynomial $p(\lambda)$, it is called a *companion matrix* associated with $p(\lambda)$.

Examples

The companion matrix associated with the polynomial $\lambda^2 + 3\lambda - 4$ is

$$\mathbf{A}_c = \begin{bmatrix} 0 & 4 \\ 1 & -3 \end{bmatrix}$$

and the companion matrix for the polynomial $\lambda^3 + 3\lambda^2 + 4\lambda + 2$ is

$$\mathbf{A}_c = \begin{bmatrix} 0 & 0 & -2 \\ 1 & 0 & -4 \\ 0 & 1 & -3 \end{bmatrix}$$

It is an exercise in manipulating determinants to verify directly that $p(\lambda) = \det[\lambda \mathbf{I} - \mathbf{A}_c]$. Moreover, using two familiar properties of determinants, namely $\det(\mathbf{AB}) = \det \mathbf{A} \det \mathbf{B}$ and $\det \mathbf{T}^{-1} = 1/\det \mathbf{T}$, it is easy to show more generally that two matrices related by similarity transformation have the same characteristic polynomial.

A final point to notice is the systematic way of determining both the companion and diagonal forms. We can describe the method in general terms. Given a set of basis vectors, we construct a transformation matrix $\mathbf{T}$ by using the vectors as columns. Then the equation

$$\mathbf{AT} = \mathbf{TX} \quad (1.23)$$

must be solved for the matrix $\mathbf{X}$, which is the desired matrix similar to the matrix $\mathbf{A}$. In both cases we looked at, it was much easier to solve this equation directly than it would be to find the inverse of $\mathbf{T}$ and perform the multiplications to compute $\mathbf{T}^{-1}\mathbf{AT}$.

Example

We summarize and illustrate this discussion with a simple example. Consider the matrix

$$\mathbf{A} = \begin{bmatrix} 0 & -2 & -2 \\ 1 & 0 & 0 \\ 1 & -2 & -3 \end{bmatrix}$$

Computing the characteristic polynomial of **A**, we obtain

$$\det[\lambda \mathbf{I} - \mathbf{A}] = \begin{bmatrix} \lambda & 2 & 2 \\ -1 & \lambda & 0 \\ -1 & 2 & \lambda+3 \end{bmatrix} = \lambda^3 + 3\lambda^2 + 4\lambda + 2$$

which can be written in factored form as $(\lambda+1)(\lambda+1+j)(\lambda+1-j)$. So the eigenvalues of **A** are $\lambda_1 = -1$, $\lambda_2 = -1-j$, and $\lambda_3 = -1+j$. At this point we can give both the diagonal and companion matrices that are similar to **A**:

$$\Lambda = \begin{bmatrix} -1 & 0 & 0 \\ 0 & -1-j & 0 \\ 0 & 0 & -1+j \end{bmatrix}, \quad \mathbf{A}_c = \begin{bmatrix} 0 & 0 & -2 \\ 1 & 0 & -4 \\ 0 & 1 & -3 \end{bmatrix}$$

To check these results we need to find the corresponding transformation matrices. To get the diagonal form, we must determine the eigenvectors of **A**. As an example of such a calculation we have for the third eigenvector

$$\mathbf{0} = \left[\lambda_3 \mathbf{I} - \mathbf{A}\right] \mathbf{u}_3 = \begin{bmatrix} -1+j & 2 & 2 \\ -1 & -1+j & 0 \\ -1 & 2 & 2+j \end{bmatrix} \mathbf{u}_3$$

which may be solved to obtain

$$\mathbf{u}_3 = \begin{bmatrix} -1+j \\ 1 \\ 1-j \end{bmatrix}$$

Similar calculations give

$$\mathbf{u}_1 = \begin{bmatrix} 2 \\ -2 \\ 3 \end{bmatrix}, \quad \mathbf{u}_2 = \begin{bmatrix} -1-j \\ 1 \\ 1+j \end{bmatrix}$$

As for the transformation to companion form, we pick the "random" vector

$$\mathbf{v} = \begin{bmatrix} 1 \\ 0 \\ 0 \end{bmatrix}$$

and we find that

$$\mathbf{A}\mathbf{v} = \begin{bmatrix} 0 \\ 1 \\ 1 \end{bmatrix}, \quad \mathbf{A}^2 \mathbf{v} = \begin{bmatrix} -4 \\ 0 \\ -5 \end{bmatrix}$$

These three vectors are linearly independent and can be used as the columns of the transformation matrix taking **A** to $\mathbf{A}_c$. The remaining details of the example are left for the student to verify.

It is natural to question the relationship between the companion and diagonal forms, and we now turn our attention to this subject. It turns out that a companion matrix can be transformed to diagonal form, and vice versa, if and only if the zeros of the characteristic polynomial are distinct. Of course, given a diagonal matrix Λ, with diagonal elements (hence with eigenvalues) $\lambda_1, \ldots, \lambda_n$, the matrix $[\lambda \mathbf{I} - \Lambda]$ is also a

Sec. 1.1 Some Topics in Matrices and Vector Spaces

diagonal matrix and its determinant, the characteristic polynomial of Λ, is just what we expect: $p(\lambda) = \det[\lambda \mathbf{I} - \Lambda] = (\lambda - \lambda_1) \cdots (\lambda - \lambda_n)$ which can be multiplied out to give the coefficients of $p(\lambda)$. Notice the following interesting and useful relationships: p_1, the coefficient of λ^{n-1}, equals $-\lambda_1 - \lambda_2 - \cdots - \lambda_n$, and p_n, the constant term of $p(\lambda) = p(0)$, equals $(-1)^n \lambda_1 \lambda_2 \cdots \lambda_n$. Even though we know the form of $\mathbf{A}_c$ corresponding to Λ, it may be of interest to know a transformation matrix $\mathbf{T}$ giving $\mathbf{A}_c = \mathbf{T}^{-1} \Lambda \mathbf{T}$. By using the general procedure described above to obtain such a matrix $\mathbf{T}$ we find the following:

$$\mathbf{T} = \begin{bmatrix} \mathbf{v} & \mathbf{A}\mathbf{v} & \mathbf{A}^2\mathbf{v} & \cdots & \mathbf{A}^{n-1}\mathbf{v} \end{bmatrix} = \begin{bmatrix} 1 & \lambda_1 & \lambda_1^2 & \cdots & \lambda_1^{n-1} \\ 1 & \lambda_2 & \lambda_2^2 & & \lambda_2^{n-1} \\ \cdot & \cdot & \cdot & & \cdot \\ \cdot & \cdot & \cdot & & \cdot \\ \cdot & \cdot & \cdot & & \cdot \\ 1 & \lambda_n & \lambda_n^2 & \cdots & \lambda_n^{n-1} \end{bmatrix} \quad (1.24)$$

This matrix is nonsingular precisely in the case that the eigenvalues are distinct, which is the condition for there to be a similarity transformation between the diagonal and companion forms. One way to see this is to carry out the computation to show that

$$\det \mathbf{T} = \prod_{1 \leq i < k \leq n} (\lambda_k - \lambda_i) \quad (1.25)$$

Of course the inverse of this $\mathbf{T}$ matrix provides the transformation back from companion form to diagonal form.

Some Unfinished Business

We will conclude this section with some follow-up discussion of two topics already introduced. The first concerns the importance of an eigenvector basis. As described earlier, when a matrix $\mathbf{A}$ has n linearly independent eigenvectors (which will certainly be the case when $\mathbf{A}$ has distinct eigenvalues or when $\mathbf{A}$ is symmetric), then these eigenvectors form a basis for $\mathbb{C}^n$. In this case, any n-vector may be expressed as a linear combination of eigenvectors, say

$$\mathbf{x} = \sum_{i=1}^{n} \alpha_i \mathbf{u}_i \quad (1.26)$$

where $\mathbf{x}$ is an n-vector and the $\mathbf{u}_i$ vectors are the eigenvectors of $\mathbf{A}$, just as in our earlier notation. We can recognize the right side of equation (1.26) as the componentwise expression for the multiplication of the matrix $\mathbf{T}$, whose columns are the eigenvectors, times a vector $\mathbf{x}_\alpha$, whose components are the α_i coefficients in the linear combination. This expression is called the *spectral decomposition of* $\mathbf{x}$ *corresponding to* $\mathbf{A}$. Given the spectral decomposition of $\mathbf{x}$, (1.26), multiplication by $\mathbf{A}$ is easy, since

$$\mathbf{A}\mathbf{x} = \sum_{i=1}^{n} \alpha_i \mathbf{A}\mathbf{u}_i = \sum_{i=1}^{n} \alpha_i \lambda_i \mathbf{u}_i \quad (1.27)$$

which specifies the spectral decomposition of the result.

Actually, this observation is nothing but a restatement of results obtained as part of our previous discussion of eigenvectors and similarity transformations. From (1.6) we may express **A** in terms of Λ as

$$\mathbf{A} = \mathbf{T}\Lambda\mathbf{T}^{-1} \tag{1.28}$$

where the columns of **T** are the eigenvectors of **A**:

$$\mathbf{T} = \begin{bmatrix} \mathbf{u}_1 & \mathbf{u}_2 & \cdots & \mathbf{u}_n \end{bmatrix} \tag{1.29}$$

Now suppose we write $\mathbf{T}^{-1}$ in terms of its rows as

$$\mathbf{T}^{-1} = \begin{bmatrix} \mathbf{v}_1 \\ \mathbf{v}_2 \\ \vdots \\ \mathbf{v}_n \end{bmatrix} \tag{1.30}$$

Then equation (1.6) for **A** may be rewritten as

$$\mathbf{A} = \sum_{i=1}^{n} \lambda_i \mathbf{u}_i \mathbf{v}_i \tag{1.31}$$

which is called the *spectral representation of* **A**. (Although the equivalence of the two expressions for **A**, (1.28) and (1.31), may not be apparent at first glance, it is easily checked by using the definition of matrix multiplication.) As seen from the derivation, the spectral representation is nothing more than an alternative way of writing the matrix product representation (1.28) for **A**. Using the spectral representation of **A**, we obtain exactly the same expression for the spectral decomposition of the product **Ax** as before in (1.27):

$$\mathbf{Ax} = \sum_{i=1}^{n} \lambda_i \mathbf{u}_i \alpha_i = \sum_{i=1}^{n} \alpha_i \lambda_i \mathbf{u}_i$$

where $\alpha_i = \mathbf{v}_i \mathbf{x}$, which may be seen from (1.30) to be the ith component of the vector $\mathbf{T}^{-1}\mathbf{x}$.

Example

To illustrate these ideas, we will consider a simple example. Let

$$\mathbf{A} = \begin{bmatrix} -4 & 15 \\ -2 & 7 \end{bmatrix}$$

The usual calculations show that the characteristic polynomial is $p(\lambda) = \lambda^2 - 3\lambda + 2$, so the eigenvalues of **A** are $\lambda_1 = 1$ and $\lambda_2 = 2$. Solving for the eigenvectors gives

$$\mathbf{u}_1 = \begin{bmatrix} 3 \\ 1 \end{bmatrix}, \quad \mathbf{u}_2 = \begin{bmatrix} 5 \\ 2 \end{bmatrix}$$

Taking these as the columns of the transformation matrix **T**, we find that

$$\mathbf{T}^{-1} = \begin{bmatrix} 2 & -5 \\ -1 & 3 \end{bmatrix}$$

The first row of this matrix is $\mathbf{v}_1$ and the second row is $\mathbf{v}_2$. The spectral representation of $\mathbf{A}$ is thus given by

$$\mathbf{A} = \sum_{i=1}^{2} \lambda_i \mathbf{u}_i \mathbf{v}_i = 1 \begin{bmatrix} 3 \\ 1 \end{bmatrix} \begin{bmatrix} 2 & -5 \end{bmatrix} + 2 \begin{bmatrix} 5 \\ 2 \end{bmatrix} \begin{bmatrix} -1 & 3 \end{bmatrix}$$

An example of a spectral decomposition is easy to obtain, since the coefficients in the spectral decomposition of a vector $\mathbf{x}$ are given by the products $\alpha_1 = \mathbf{v}_1 \mathbf{x}$ and $\alpha_2 = \mathbf{v}_2 \mathbf{x}$. As an example,

$$\mathbf{x} = \begin{bmatrix} 3 \\ 0 \end{bmatrix} = 6\mathbf{u}_1 - 3\mathbf{u}_2$$

The last point to be discussed in this section is the proof of the Cayley-Hamilton theorem. The results on eigenvalues, eigenvectors, and similarity transformations will be combined to give a proof that is valid for any matrix $\mathbf{A}$ having a linearly independent set of eigenvectors.

We start with the product representation of a matrix $\mathbf{A}$ that can be diagonalized by a similarity transformation, equation (1.28). Subtracting from the identity $\lambda\mathbf{I} = \lambda\mathbf{I}$ and taking determinants leads to

$$p(\lambda) = \det(\lambda\mathbf{I} - \mathbf{A}) = \det(\lambda\mathbf{I} - \mathbf{T}\Lambda\mathbf{T}^{-1}) \tag{1.32}$$

and by using familiar properties of determinants the last expression may be simplified to give

$$\det(\lambda\mathbf{I} - \mathbf{T}\Lambda\mathbf{T}^{-1}) = \det \mathbf{T} \det(\lambda\mathbf{I} - \Lambda) \det \mathbf{T}^{-1}$$
$$= \det(\lambda\mathbf{I} - \Lambda) \tag{1.33}$$

Thus, the two similar matrices $\mathbf{A}$ and Λ have the same characteristic polynomial, $p(\lambda)$.

Because Λ is a diagonal matrix, it is easy to compute $p(\Lambda)$, which turns out to be the diagonal matrix having diagonal elements $p(\lambda_1), p(\lambda_2), \ldots, p(\lambda_n)$. But each one of these diagonal elements is equal to zero because the eigenvalues are zeros of the characteristic polynomial, so $p(\Lambda) = \mathbf{0}$. (This proves the Cayley-Hamilton theorem for diagonal matrices.)

Since

$$\mathbf{A}^k = \mathbf{A}\mathbf{A} \cdots \mathbf{A} \quad (k \text{ factors})$$
$$= \mathbf{T}\Lambda\mathbf{T}^{-1} \mathbf{T}\Lambda\mathbf{T}^{-1} \cdots \mathbf{T}\Lambda\mathbf{T}^{-1} \tag{1.34}$$

it follows that

$$\mathbf{A}^k = \mathbf{T}\Lambda^k\mathbf{T}^{-1} \quad \text{for all } k \geq 0 \tag{1.35}$$

Thus the similarity transformation relating $\mathbf{A}$ and Λ may be applied term by term to polynomials, so that

$$p(\mathbf{A}) = \mathbf{T}p(\Lambda)\mathbf{T}^{-1} \tag{1.36}$$

Hence $p(\mathbf{A}) = \mathbf{0}$, and the Cayley-Hamilton theorem is proved.

1.2 LINEAR EQUATIONS AND MATRIX FACTORIZATIONS

A major part of the history of linear algebra can be traced to the problem of solving systems of simultaneous linear equations. While this problem, and variations such as solving eigenvalue-eigenvector problems, motivated a great deal of beautiful abstract mathematics, the amazingly diverse and important applications of linear equations have kept practical computational aspects of linear algebra in the forefront of research activities, especially since the development of machine computation. The needs of scientific and technological calculation stimulated the development of parallel methods suitable for distributed computation in the environment of mechanical calculators (i.e., a roomful of clerks using adding machines!); the invention of electronic digital computers profoundly altered the course of science and technology, and their use in the solution of linear equations played no small role in this revolution.

Along with its potentials, electronic computation brings its own limitations, mainly the need for numerical computations to be carried out with limited precision arithmetic. Numerical analysis, and more specifically numerical linear algebra, is the branch of mathematics that has developed as a response to various capabilities and limitations inherent to carrying out linear algebraic computations on digital computers. This active research area encompasses a wide range of topics, from error analysis of computational techniques, to studies of special-purpose computer architectures suited for particular linear algebraic computations, to fundamental studies of computational complexity. But for the purposes of applications to computer-aided design and analysis, the topic of this book, perhaps the most profound impact results from a more general development: the "bundling" of reliable numerical linear algebra programs into easy-to-use "toolboxes" that are becoming widely available and integrated into software packages for the current generation of personal computers. And having made their way into the computing technology "of the masses," it can be expected that advanced capabilities for more powerful engineering workstations, as well as for larger machines up to the supercomputer category, will follow promptly on the heels of every advance in hardware technology.

Just as we assume that the reader of this book has a background in linear algebra, we assume some computer "literacy" to go along with it. The discussion of numerical linear algebra will be correspondingly brief, since it is not our intention to cover such material from first principles. Anyway, more and more courses in linear algebra are being taught with some substantial emphasis on computation. This is facilitated by the wide availability of user-friendly software packages that virtually eliminate all the tedium of computational exercises. Linear algebra courses that emphasize topics of importance in numerical linear algebra can also exploit the ability of PC-based software packages to deal with complex problems that would require substantial efforts to handle with a general-purpose programming language such as FORTRAN or Pascal.

Finally, a note about the use of real and complex numbers in this discussion. We follow the convention of treating the case of linear equations involving real quantities. However, in applications, it may be necessary to deal with complex numbers. Calculations with complex numbers may even be required in problems that at first glance seem to involve only real quantities, since the eigenvalues of a real matrix can be complex

Sec. 1.2   Linear Equations and Matrix Factorizations

numbers. For the most part, it suffices to remember that the complex counterpart of matrix transposition is transposition and conjugation. The complex generalization of an orthogonal matrix is a unitary matrix. Symmetry is generalized to Hermitian (conjugate) symmetry.

Linear Equations

Matrices provide a compact notation for describing simultaneous linear equations. If A is an $m \times n$ matrix and x and y are n- and m-vectors, respectively, then the equation

$$Ax = y \qquad (1.37)$$

specifies m equations in the n unknowns (the components of x). From the definition of matrix multiplication, we see that the left side is some linear combination of the n columns of A, the coefficients of the linear combination being the elements of the vector x. Thus for a given vector y, there will be a solution if and only if y lies in the *range space* of A, the subspace of $\mathbb{R}^m$ spanned by the columns of A. And if we are interested in knowing that a solution x can be found for every possible choice of y, it is necessary and sufficient that the range space of A have dimension m, or equivalently that there are m linearly independent columns among the n columns of A.

A first observation is that if there are fewer unknowns than equations, so that $n < m$, then it will "rarely" be the case that a solution can be found for a particular y, in the sense that the dimension of the range space of A is less than m and there are (many) choices of y for which no solution exists. Thus for $n < m$, the equations are *inconsistent*, or *overdetermined*, except for "special" choices of y and A.

Some caution is required to make valid conclusions in the opposite situation, when $n \geq m$. The right intuition is that there will "rarely" be cases when a solution cannot be found; apart from an exceptional set of matrices A, the equations will be consistent for all y. Even among the exceptional set, if y lies in the range space of A, solutions can still be found.

Conditions for uniqueness of a solution may be written in several ways. The geometric condition is simply that the columns of A be linearly independent vectors. Hence in the case $n > m$, which would be the *underdetermined* case, uniqueness never holds. In the case $n = m$, the uniqueness condition coincides with the existence condition: the matrix A must be invertible. (This is equivalent to the condition that the determinant of A be nonzero.)

For $n \leq m$, there is a more general uniqueness test to apply in cases where existence of a solution holds. It is a criterion that may be used to test for the linear independence of the set of columns of any matrix, in particular A. The *Gram matrix* of the columns of A is the (symmetric) matrix $A^T A$; the columns of the matrix A are linearly independent if and only if the corresponding Gram matrix is invertible. Hence, if the Gram matrix $A^T A$ is nonsingular, solutions to $Ax = y$, when they exist, are unique.

Examples

We will write the equations as $Ax = y$, so we only need to specify the matrix A and vector y to describe each different example.

1. A solution exists but is not unique.

$$\mathbf{A} = \begin{bmatrix} 1 & 1 & 2 \\ 0 & 2 & 2 \\ 3 & 0 & 3 \end{bmatrix}, \quad \mathbf{y} = \begin{bmatrix} 4 \\ 4 \\ 6 \end{bmatrix}$$

Notice that **y** can be written as twice the last column of **A** and as the sum of the columns of **A**. So **y** is in the range space of **A** and a solution exists, but evidently it is not unique. This is verified by a check of the Gram matrix of the columns of **A**:

$$\mathbf{A}^T\mathbf{A} = \begin{bmatrix} 10 & 1 & 11 \\ 1 & 5 & 6 \\ 11 & 6 & 17 \end{bmatrix}$$

which is not invertible. (Its determinant is zero since its third row is the sum of its first two rows.) In fact, the general form of solutions to the equations is

$$\mathbf{x} = \begin{bmatrix} 1 \\ 1 \\ 1 \end{bmatrix} + \alpha \begin{bmatrix} 1 \\ 1 \\ -1 \end{bmatrix}$$

where α is any real number. This takes the form of a particular solution plus the general solution to the associated homogeneous system of linear equations $\mathbf{Ax} = \mathbf{0}$.

2. A solution exists and is unique.

$$\mathbf{A} = \begin{bmatrix} 1 & 3 \\ 0 & 2 \\ 5 & 0 \end{bmatrix}, \quad \mathbf{y} = \begin{bmatrix} 4 \\ 2 \\ 5 \end{bmatrix}$$

A solution exists because **y** can be written as the sum of the first and second columns of **A**. Evidently the solution is unique, and the Gram matrix test may be used to verify this. The Gram matrix is

$$\mathbf{A}^T\mathbf{A} = \begin{bmatrix} 11 & 3 \\ 3 & 13 \end{bmatrix}$$

whose determinant is 134, so the Gram matrix is invertible.

3. No solution exists. Using the **A** matrix from either of the two previous examples, and choosing

$$\mathbf{y} = \begin{bmatrix} 0 \\ 0 \\ 1 \end{bmatrix}$$

gives an inconsistent set of equations.

One particular case of importance where uniqueness can fail involves the solution of homogeneous linear equations:

$$\mathbf{Ax} = \mathbf{0} \tag{1.38}$$

where it is often desired to find a "nontrivial" (i.e., nonzero) solution. Existence of a nontrivial solution amounts to requiring that the obvious zero solution be nonunique. From our discussion of existence, it follows that a nontrivial solution may be found if and only if the columns of **A** are not linearly independent. This is equivalent to having the Gram matrix be singular (not invertible); when **A** is a square matrix it is equivalent

Sec. 1.2 Linear Equations and Matrix Factorizations

to the usual determinant condition $\det \mathbf{A} = 0$. Notice that the set of vectors $\{\mathbf{x} : \mathbf{A}\mathbf{x} = \mathbf{0}\}$ is a subspace of $\mathbb{R}^n$ whose dimension is greater than zero if any nontrivial solutions exist. This subspace is called the *nullspace* of $\mathbf{A}$.

The nullspace plays an important role in specifying the entire set of solutions to underdetermined sets of equations. Part 1 of the preceding example serves as a model for the general situation. If $\mathbf{A}\mathbf{x} = \mathbf{y}$ has a solution, say $\mathbf{x}_0$, the set of all solutions can be expressed as

$$S_\mathbf{x} = \left\{ \mathbf{x} \,\middle|\, \mathbf{x} = \mathbf{x}_0 + \sum_{i=1}^{k} \alpha_i \mathbf{x}_i \right\} \tag{1.39}$$

where the α_i are arbitrary scalars and $\{\mathbf{x}_1, \ldots, \mathbf{x}_k\}$ is a basis for the nullspace of $\mathbf{A}$.

We also encountered homogeneous linear equations in our previous discussion about eigenvalues and eigenvectors. Recall that to find an eigenvector of $\mathbf{A}$ corresponding to an eigenvalue λ requires finding a nonzero vector in the nullspace of the matrix $(\lambda \mathbf{I} - \mathbf{A})$.

The notion of inverse matrix is closely related to the solution of n consistent linear equations in n unknowns, and the basic ideas involved in the use of the inverse matrix can be extended to the general case of m equations in n unknowns by introducing the pseudo-inverse matrix. We develop this idea by starting with the basic facts involving inverse matrices.

From the fundamental defining equations $\mathbf{A}\mathbf{A}^{-1} = \mathbf{A}^{-1}\mathbf{A} = \mathbf{I}$, we can deduce some basic facts in connection with the linear equations

$$\mathbf{A}\mathbf{x} = \mathbf{y} \tag{1.40}$$

First $\mathbf{x} = \mathbf{A}^{-1}\mathbf{y}$ provides an explicit expression for $\mathbf{x}$ as a linear function of $\mathbf{y}$. Second, notice that the ith column of $\mathbf{A}^{-1}$ is the solution to the equations formed by taking $\mathbf{y} = \mathbf{e}_i$, the ith unit vector. So computation of $\mathbf{A}^{-1}$ is equivalent to solving n sets of equations. In cases where only one or a few sets of equations are to be solved, an efficient direct method of solution of equations would usually be preferred to a computation of $\mathbf{A}^{-1}$ followed by a matrix-vector multiplication to form $\mathbf{A}^{-1}\mathbf{y}$. One exception would be in a case where $\mathbf{A}^{-1}$ takes a special form that is efficiently determined and that allows the matrix-vector multiplication to be carried out efficiently also.

Now suppose we consider the linear equations

$$\mathbf{A}\mathbf{x} = \mathbf{y} \tag{1.41}$$

for general choices of m and n and think of how we might associate a *unique* vector $\mathbf{x}$ with every possible choice of $\mathbf{y}$. Thus, we need to consider how to select a particular $\mathbf{x}$ in cases when the system of equations is underdetermined, and we need to find a way to select some $\mathbf{x}$ in cases when the system is inconsistent. A geometric perspective is useful in deciding how to resolve these two issues. Moving the right side of the equation to the left, we obtain

$$\mathbf{A}\mathbf{x} - \mathbf{y} = \mathbf{0} \tag{1.42}$$

so we see that solving the equations is equivalent to making the length of the difference vector $\mathbf{A}\mathbf{x} - \mathbf{y}$ equal to zero. If the equations are inconsistent, no choice of $\mathbf{x}$ will make the length of this difference vector zero, so a rather natural thing to do is to choose **x** to make the length as small as possible.

Example

A very simple case will be helpful in illustrating the geometric viewpoint. Consider the two linear equations in a single unknown **x**,

$$\begin{bmatrix} 1 \\ 2 \end{bmatrix} \mathbf{x} = \begin{bmatrix} 3 \\ 4 \end{bmatrix}$$

Clearly, no choice of **x** can make equality hold. The left side of the equation defines a line in the Cartesian plane (Fig. 1.2); it passes through the origin and has slope 2. For each choice of **x**, the difference between the left and right sides of the equation is

$$\mathbf{Ax} - \mathbf{y} = \begin{bmatrix} x-3 \\ 2x-4 \end{bmatrix}$$

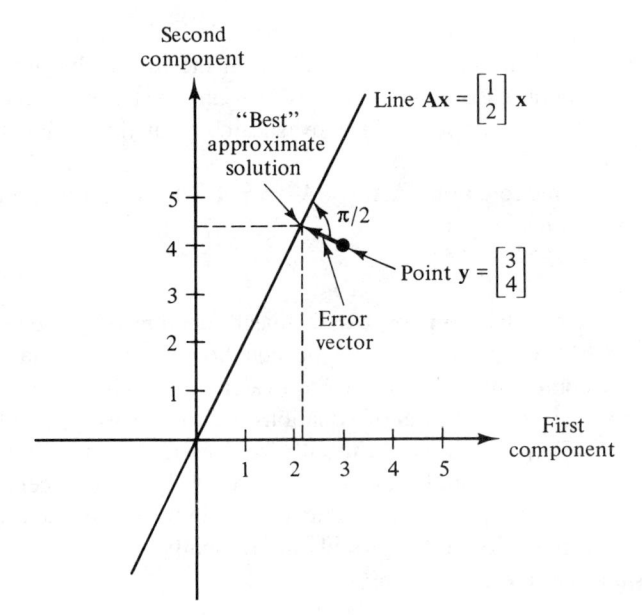

Figure 1.2 Cartesian plane, illustrating inconsistent equations and the best approximate solution; the length of the error vector is minimized.

which defines a vector connecting the point **y** to the point **Ax** on the line, and by elementary geometry the vector of shortest length is the one that is perpendicular to the line. In other words, we want the vectors **Ax** and **Ax** − **y** to be orthogonal, or in terms of an equation $(\mathbf{Ax})^T(\mathbf{Ax} - \mathbf{y}) = 0$. Substituting for the variables and grouping terms gives

$$0 = \begin{bmatrix} x & 2x \end{bmatrix} \begin{bmatrix} x-3 \\ 2x-4 \end{bmatrix} = 5x^2 - 11x$$

The trivial solution is obviously extraneous, leaving the solution $\mathbf{x} = 11/5$ as the one providing the desired minimum-length error.

For the underdetermined case, where many choices of **x** lead to a zero difference vector, it is also quite natural, at least from a mathematical standpoint, to single out the

Sec. 1.2 Linear Equations and Matrix Factorizations

(unique!) choice that is smallest in length. A simple geometric example is helpful in illustrating this idea also.

Example

Consider the following single linear equation involving two unknowns:

$$\mathbf{Ax} = \begin{bmatrix} 2 & 1 \end{bmatrix} \begin{bmatrix} x_1 \\ x_2 \end{bmatrix} = \mathbf{y} = 3$$

Every point in the Cartesian plane lying on the straight line with slope -2 and x_2 intercept 3 is a solution to this equation (Fig. 1.3). In choosing the solution of minimum length, we are choosing the point on the line that is closest to the origin, the point where the vector from the line to the origin is perpendicular to the line. The resulting solution is $x_1 = 6/5$ and $x_2 = 3/5$.

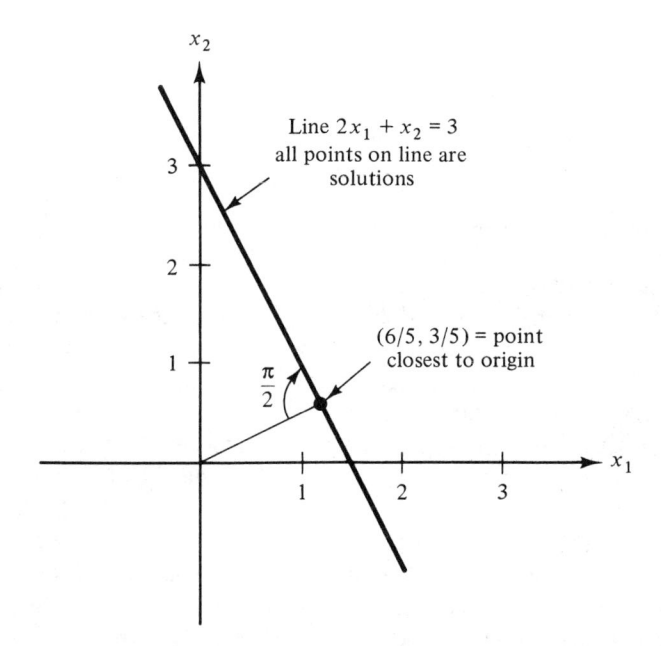

Figure 1.3 Cartesian plane, illustrating underdetermined equations and the minimum-length solution.

We pause for a brief, but important, comment. In resolving the problems of inconsistency and nonuniqueness, length plays the role of a "cost function" or "error measure" introduced to provide a mathematical criterion upon which a choice of solution may be based; selection of an appropriate cost function for any particular application is a matter of judgment. Using the minimum-length criterion is often the first thing to try because of the resulting mathematical simplicity. It is always worth keeping in mind that what is "best" is highly problem dependent, especially when there is some "real world" situation that the solution to the linear equations is attempting to model.

Now we continue with the development. It turns out that the geometric considerations involving minimum length may be turned into corresponding linear algebraic conditions. The end result is that **x** may be expressed in a very simple way, namely

$$\mathbf{x} = \mathbf{A}^\dagger \mathbf{y} \tag{1.43}$$

where the matrix $\mathbf{A}^\dagger$ is called the *pseudo-inverse* of $\mathbf{A}$ (since it coincides with $\mathbf{A}^{-1}$ when $\mathbf{A}$ is square and nonsingular). Remarkably, our choices of how to resolve overdetermined and inconsistent problems lead to a formula for $\mathbf{x}$ that involves matrix multiplication by a suitable matrix, namely $\mathbf{A}^\dagger$. The result is a clear generalization of the relationship that holds between $\mathbf{y}$ and $\mathbf{x}$ in the special case of n consistent equations in n unknowns discussed before.

The computation of $\mathbf{A}^\dagger$ is often done with the help of the Singular Value Decomposition of $\mathbf{A}$, to be described later; in the general case there is no simple "formula" to apply, but it is known that it provides the unique solution to the following set of matrix equations:

$$\mathbf{A}\mathbf{A}^\dagger\mathbf{A} = \mathbf{A} \tag{1.44}$$

$$\mathbf{A}^\dagger\mathbf{A}\mathbf{A}^\dagger = \mathbf{A}^\dagger \tag{1.45}$$

$$(\mathbf{A}\mathbf{A}^\dagger) \text{ is symmetric} \tag{1.46}$$

$$(\mathbf{A}^\dagger\mathbf{A}) \text{ is symmetric} \tag{1.47}$$

You might check to see that when $\mathbf{A}$ is square and invertible, then $\mathbf{A}^{-1}$ satisfies these equations. If $\mathbf{A}$ is rectangular, with linearly independent columns, then the Gram matrix $\mathbf{A}^T\mathbf{A}$ is invertible and $\mathbf{A}^\dagger = (\mathbf{A}^T\mathbf{A})^{-1}\mathbf{A}^T$. And if $\mathbf{A}$ is rectangular and $\mathbf{A}^T$ has linearly independent columns, then the Gram matrix associated with $\mathbf{A}^T$, $\mathbf{A}\mathbf{A}^T$ is invertible and $\mathbf{A}^\dagger = \mathbf{A}^T(\mathbf{A}\mathbf{A}^T)^{-1}$. When neither $\mathbf{A}$ nor $\mathbf{A}^T$ has linearly independent columns, a simple expression for $\mathbf{A}^\dagger$ is not available. However, as we will see in the following section, there are certain matrix factorizations that can be used to express $\mathbf{A}^\dagger$ for any matrix, and these factorizations are the preferred numerical methods for computing $\mathbf{A}^\dagger$. And just as in the case when $\mathbf{A}$ is invertible, it is often preferable to use a computational method for solving equations directly rather than a method that involves finding $\mathbf{A}^\dagger$ first.

As a demonstration that the equations defining $\mathbf{A}^\dagger$ really do reflect certain geometric properties, consider the matrix $(\mathbf{A}\mathbf{A}^\dagger)$. The third defining equation shows that it is symmetric. Multiplying the first defining equation on the right by $\mathbf{A}^\dagger$ shows that $(\mathbf{A}\mathbf{A}^\dagger)^2 = (\mathbf{A}\mathbf{A}^\dagger)$. This means that the matrix $(\mathbf{A}\mathbf{A}^\dagger)$ is an *orthogonal projection*, where we use terminology that will now be made clear. Notice that for any vector $\mathbf{v}$, the vector $(\mathbf{A}\mathbf{A}^\dagger)\mathbf{v}$ is a linear combination of the columns of $\mathbf{A}$. Thus, the range space of $(\mathbf{A}\mathbf{A}^\dagger)$ is contained in the range space of $\mathbf{A}$. Furthermore, for any vector $\mathbf{y}$ that can be written as a linear combination of the columns of $\mathbf{A}$ (i.e., for which $\mathbf{y} = \mathbf{A}\mathbf{x}$ for some $\mathbf{x}$), the first defining equation gives $(\mathbf{A}\mathbf{A}^\dagger)\mathbf{y} = \mathbf{y}$. Hence the range spaces of $\mathbf{A}$ and $(\mathbf{A}\mathbf{A}^\dagger)$ coincide, and the effect of multiplying a vector by $(\mathbf{A}\mathbf{A}^\dagger)$ is to project the vector onto the range space of $\mathbf{A}$. The identity

$$\mathbf{v} = (\mathbf{A}\mathbf{A}^\dagger)\mathbf{v} + (\mathbf{I} - \mathbf{A}\mathbf{A}^\dagger)\mathbf{v} \tag{1.48}$$

expresses a general vector $\mathbf{v}$ as a sum of two vectors (Fig. 1.4); we have shown that the first is in the range space of $\mathbf{A}$. From the second of the defining equations, the vector

Sec. 1.2   Linear Equations and Matrix Factorizations

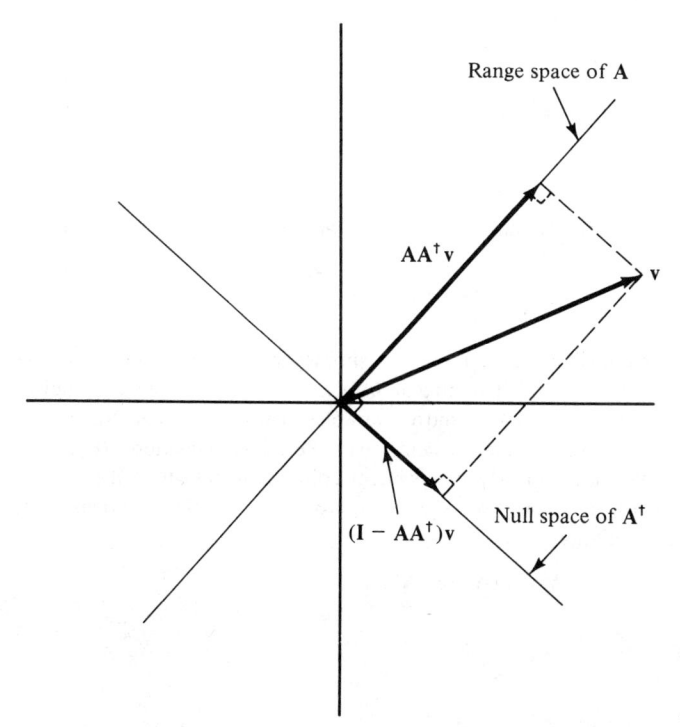

Figure 1.4 Geometric decomposition of vector **v** with orthogonal projections; $\mathbf{v} = \mathbf{AA}^\dagger \mathbf{v} + (\mathbf{I} - \mathbf{AA}^\dagger)\mathbf{v}$.

$(\mathbf{I} - \mathbf{AA}^\dagger)\mathbf{v}$ is in the nullspace of $\mathbf{A}^\dagger$. (The matrix $(\mathbf{I} - \mathbf{AA}^\dagger)$ is called the *complementary orthogonal projection*.) Furthermore, these two vectors are orthogonal since

$$\mathbf{v}^T(\mathbf{AA}^\dagger)^T(\mathbf{I} - \mathbf{AA}^\dagger)\mathbf{v} = \mathbf{v}^T(\mathbf{AA}^\dagger - \mathbf{AA}^\dagger\mathbf{AA}^\dagger)\mathbf{v} = 0 \qquad (1.49)$$

where the symmetry of $(\mathbf{AA}^\dagger)$ was used to obtain the first equality.

Examples

We will give two more examples that illustrate typical situations in which overdetermined and underdetermined linear equations arise.

  1. *Inconsistent, overdetermined case.* Curve fitting is often used to derive an empirical model for a set of observed data. The observations are available as a set of ordered pairs, $\{(x_1,y_1),(x_2,y_2),\ldots,(x_N,y_N)\}$, and based on a qualitative analysis, for example, a plot of the ordered pairs on a Cartesian plane, a model structure with undetermined parameters is selected and a numerical procedure is employed to find "good" values for the parameters. The fitting of a straight-line model, $y = mx + b$, is the most common example of curve fitting, and finding the slope and intercept parameters to fit a set of observations amounts to assuming that the data arise from the following linear equations:

$$\mathbf{Ax} = \mathbf{y}$$

where the components of **y** are the y_i coordinates of the data. The matrix **A** is formed from the x_i coordinates of the data as follows:

$$\mathbf{A} = \begin{bmatrix} x_1 & 1 \\ x_2 & 1 \\ . & . \\ . & . \\ . & . \\ x_N & 1 \end{bmatrix}$$

and the vector of unknowns corresponds to the slope and intercept parameters of the model:

$$\mathbf{x} = \begin{bmatrix} m \\ b \end{bmatrix}$$

In many situations, the straight-line model is only a simple approximation that does not hold exactly, perhaps due to unmodeled nonlinear effects or because there are "noise" effects corrupting the measurements of the x_i and y_i variables. Thus the typical straight-line fitting application results in a set of inconsistent, overdetermined equations (Fig. 1.5). Notice that the $\mathbf{A}$ matrix will have linearly independent columns if not all of the x_i variables are the same. In such a case, the least-squares approximate solution is obtained by using the pseudo-inverse $\mathbf{A}^\dagger$, giving

$$\mathbf{x} = \mathbf{A}^\dagger \mathbf{y} = (\mathbf{A}^T \mathbf{A})^{-1} \mathbf{A}^T \mathbf{y}$$

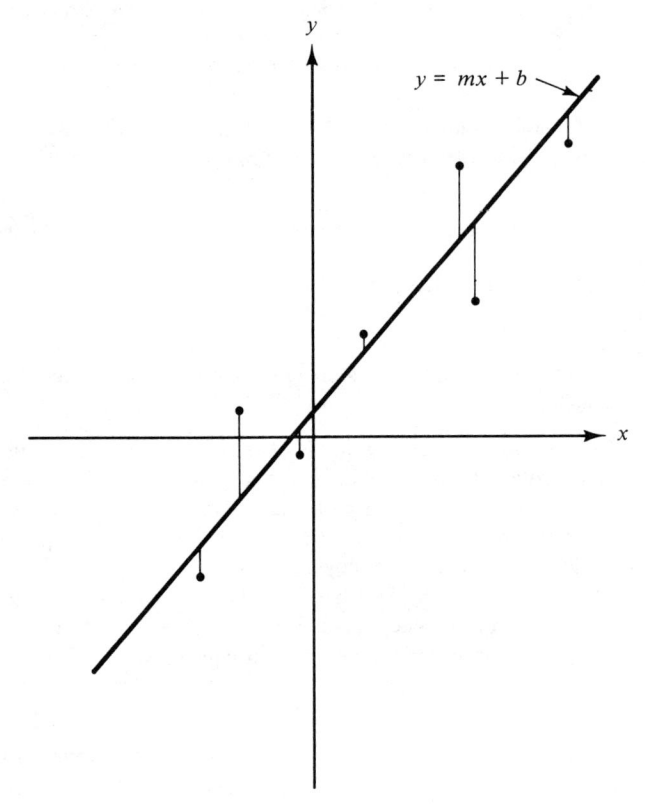

Figure 1.5 Fitting a straight line to $(\mathbf{x}, \mathbf{y})$ data points. The "least-squares fit" minimizes the sum of the lengths of vertical line segments as shown.

Sec. 1.2 Linear Equations and Matrix Factorizations

Using the forms of **A** and **y** we may write down explicit formulas for the least-squares slope and intercept parameters:

$$\begin{bmatrix} m \\ b \end{bmatrix} = \begin{bmatrix} \sum x_i^2 & \sum x_i \\ \sum x_i & N \end{bmatrix}^{-1} \begin{bmatrix} \sum x_i y_i \\ \sum y_i \end{bmatrix}$$

where all the sums run from 1 to N.

2. *Consistent, underdetermined case.* Consider the equation describing the points on a plane in Cartesian 3-space:

$$\mathbf{n}^T(\mathbf{x} - \mathbf{x}_0) = 0$$

where **n** is a *normal vector* to the plane (a vector perpendicular to the plane), $\mathbf{x}_0$ is a point lying on the plane, and **x** is a general point in 3-space. The equation for the points on the plane is a single linear equation in the three unknowns x_1, x_2, and x_3 used to denote the Cartesian coordinates. The minimum-length solution gives the point on the plane closest to the origin of the Cartesian plane. To give a particular example, take the plane passing through the point $(1,3,2)$ and perpendicular to the normal vector

$$\mathbf{n} = \begin{bmatrix} -1 \\ 4 \\ -3 \end{bmatrix}$$

The resulting equation for points on the plane is

$$\mathbf{n}^T(\mathbf{x} - \mathbf{x}_0) = -1(x_1 - 1) + 4(x_2 - 3) - 3(x_3 - 2) = 0$$

Simplifying, this gives the single linear equation

$$-x_1 + 4x_2 - 3x_3 = 5$$

So **A** is the 1×3 matrix $[-1 \ \ 4 \ -3]$, $\mathbf{A}^T$ has one nonzero (hence linearly independent) column, and using the appropriate formula for the pseudo-inverse gives the minimum-length solution

$$\mathbf{x} = \mathbf{A}^T(\mathbf{A}\mathbf{A}^T)^{-1}\mathbf{y} = \begin{bmatrix} -5/26 \\ 10/13 \\ -15/26 \end{bmatrix}$$

This gives the point on the plane closest to the origin.

Some geometric interpretation of linear equations provides useful insights. Using $\mathbf{A}_i$ to denote the ith row of **A**, we may write

$$\mathbf{A}\mathbf{x} = \begin{bmatrix} \mathbf{A}_1 \\ \vdots \\ \mathbf{A}_m \end{bmatrix} \mathbf{x} = \mathbf{y} \tag{1.50}$$

As before, we assume that **x** is an n-vector and **y** is an m-vector; let $\mathbf{y}_i$ denote the ith component of **y**. The row-wise decomposition of **A** makes clear that we are considering a system of m linear equations in n unknowns,

$$\mathbf{A}_i \mathbf{x} = \mathbf{y}_i, \quad 1 \leq i \leq m \tag{1.51}$$

and the solution set of each equation determines a *hyperplane* in n-dimensional Euclidean space (a line when $n = 2$, a plane when $n = 3$, etc.). Thus the condition for existence of a solution **x** to the system of m equations is that the hyperplanes must have a nonempty intersection. If the set of intersection is a point, there is a unique solution. If the set of intersection is a line or other hyperplane, solutions exist but are not unique; as in the last example above, a unique representative solution can be obtained by choosing the solution of minimum length, which is the point closest to the origin on the hyperplane of intersection.

This perspective also provides some intuition about the sensitivity of the solution **x** to variations in **A** and **y**. When only a small change in **A** or a small change in **y** will result in an empty intersection of the perturbed hyperplanes, we say that the linear equations are *ill-conditioned*. Evidently, very small changes in an ill-conditioned system can produce large changes in the solution vector **x**; the *condition number*, denoted $\kappa(\mathbf{A})$ (see (A.48)), provides a useful measure for determining when a system of n equations in n unknowns is ill-conditioned.

Matrix Factorizations and Applications

Matrix factorizations make up an important set of tools for the numerical linear algebra trade. Our intent in this section is to exhibit some of the more important factorizations and to indicate some of their applications. We will not cover the details of the numerical methods for obtaining the factorizations. That is a subject in itself, and it is already covered in a number of well-done books. Still, as users of numerical methods, systems analysts and designers should know some basic facts to satisfy a basic mathematical literacy requirement. Reliable software for these calculations is widely available in software packages such as LINPACK, EISPACK, IMSL, MATLAB, MATRIX-X; FORTRAN subroutines are even available by an electronic mail service known as **netlib**. Thus the need for writing programs to perform the basic factorizations described in the following discussion will rarely arise.

PLU Factorization

Upper and lower triangular square matrices yield systems of easily solved linear equations, the corresponding solutions being readily obtained by a series of substitution operations, solving for the variables in natural order (i.e., starting with the first variable in the lower triangular case) or in reverse order (starting with the last variable in the upper triangular case). Invertibility, the necessary and sufficient condition for existence of a unique solution is also easily checked since the determinant of a triangular matrix is the product of the diagonal elements. It is also easily checked that the inverses of nonsingular upper and lower triangular matrices are also upper and lower triangular, respectively.

If a matrix can be brought to lower triangular form by a permutation of its rows, say $\mathbf{A} = \mathbf{PL}$, then the solution of the corresponding system of linear equations, $\mathbf{Ax} = \mathbf{y}$, can again be easily found by "un"-permuting rows and solving $\mathbf{P}^{-1}\mathbf{Ax} = \mathbf{Lx} = \mathbf{P}^{-1}\mathbf{y}$; the inverse of a permutation matrix is another permutation matrix, so no significant

Sec. 1.2　Linear Equations and Matrix Factorizations

computational work is required by this preliminary reordering. The same procedure works for matrices that are upper triangular after a row permutation.

For other square matrices the situation is more complicated, but it turns out that any square matrix can be written as a product

$$\mathbf{A} = \mathbf{PLU} \tag{1.52}$$

where the terms in the product on the right side are a permutation matrix, a lower triangular matrix, and an upper triangular matrix, respectively. Given this PLU factorization, the solution of a system of n equations in n unknowns, $\mathbf{Ax} = \mathbf{y}$, is reduced to solving a sequence of two triangular problems: Solve $\mathbf{PLz} = \mathbf{y}$ followed by $\mathbf{Ux} = \mathbf{z}$. Notice that the factorization needs to be done only once to enable solutions to be obtained in this way for any number of different vectors of unknowns.

The computation of the PLU factorization really amounts to Gaussian elimination with pivoting, with certain partial results kept along to form the various factors. To see that this is the case (although we leave the details for the interested reader to work out), consider the equation

$$\mathbf{L}^{-1}\mathbf{P}^{-1}\mathbf{A} = \mathbf{U} \tag{1.53}$$

Note that $\mathbf{L}^{-1}$ is a lower triangular matrix, and $\mathbf{P}^{-1}$ is a permutation matrix. The product $\mathbf{P}^{-1}\mathbf{A}$ therefore results in a reordering of the rows of $\mathbf{A}$, which is used to bring numerically large "pivot" entries to the diagonal elements during the Gaussian elimination process. Premultiplication by the lower triangular matrix $\mathbf{L}^{-1}$ amounts to performing row-by-row elimination steps, working from the top row downward, bringing $\mathbf{P}^{-1}\mathbf{A}$ to upper triangular form. Storing the PLU factors enables efficient solution of several systems of linear equations involving the same matrix $\mathbf{A}$ but different vectors $\mathbf{y}$.

A significant benefit of using PLU factorization for solving linear equations instead of finding the matrix inverse (and also better than explicitly inverting to obtain $\mathbf{A}^{-1} = \mathbf{U}^{-1}\mathbf{L}^{-1}\mathbf{P}^{-1}$) is realized in the case of *banded* matrices, which are matrices where the only nonzero entries occur in the vicinity of the main diagonal. For a banded matrix, the elements a_{ij} are all zero when the row and columns indices satisfy $i - j > k$ or $j - i > k$ for some k much less than n, the dimension of $\mathbf{A}$. There are many applications where k is 1 or 2. Taking $\mathbf{P} = \mathbf{I}$ for banded matrices is customary, and with this choice the triangular factors $\mathbf{U}$ and $\mathbf{L}$ are then banded, too. This means that the LU factorization is easier to compute (no need to compute the elements that are known to be zero!) and also the solution of the two triangular sets of linear equations can be carried out more efficiently, since only a few terms (as few as two or three) are involved in each equation to which forward or backward substitution must be applied. These advantages of PLU factorization also apply to certain other classes of *sparse* matrices (ones with relatively few nonzero elements).

Eigenvector-Eigenvalue Factorization

As indicated earlier, when the square matrix $\mathbf{A}$ has a complete set of linearly independent eigenvectors (e.g., when its eigenvalues are distinct or when it is a symmetric matrix), then it may be diagonalized by a similarity transformation

$$\mathbf{T}^{-1}\mathbf{A}\mathbf{T} = \Lambda \tag{1.54}$$

where $\mathbf{T}$ is a matrix whose columns are the eigenvectors of $\mathbf{A}$ and Λ is a diagonal matrix of corresponding eigenvalues. Notice that by taking the terms in $\mathbf{T}$ to the right side of this equation we obtain a factorization of $\mathbf{A}$:

$$\mathbf{A} = \mathbf{T}\Lambda\mathbf{T}^{-1} \tag{1.55}$$

We will call this the *eigenvector-eigenvalue factorization*. Its relationship to coordinate transformations has already been mentioned. In the next chapter this factorization will be shown to be a useful theoretical tool for the study of the dynamics of linear systems described by differential or difference equations. It is also useful for the study of various functions involving matrices, as was seen in our earlier proof of the Cayley-Hamilton theorem. Basically, this is because the eigenvector-eigenvalue factorization of any power of $\mathbf{A}$ is given by

$$\mathbf{A}^k = \mathbf{T}\Lambda^k\mathbf{T}^{-1} \tag{1.56}$$

Similarly, using linearity, for any polynomial $q(\mathbf{A})$,

$$q(\mathbf{A}) = \mathbf{T}q(\Lambda)\mathbf{T}^{-1} \tag{1.57}$$

Another application involves *quadratic forms* (homogeneous quadratic functions of several variables). These functions are, in an intuitive sense, the simplest nonlinear functions, and they have applications in the study of optimization problems as will be seen in a later chapter. For now, we only consider how the eigenvector-eigenvalue factorization leads to a nice representation for quadratic forms.

If we define the function

$$f(x_1, \ldots, x_n) = \sum_{i=1}^{n}\sum_{k=1}^{n} p_{i,j} x_i x_j \tag{1.58}$$

then using $\mathbf{x}$ to denote the column vector whose components are the n variables $x_1, \ldots, x_n$ and $\mathbf{P}$ to denote the $n \times n$ matrix of coefficients, we may write the function in matrix-vector notation as

$$f(\mathbf{x}) = \mathbf{x}^T \mathbf{P} \mathbf{x} \tag{1.59}$$

Noting that $f(\mathbf{x})$ is a scalar, and hence equal to its own transpose, we see that

$$f(\mathbf{x}) = \mathbf{x}^T \mathbf{P}^T \mathbf{x} \tag{1.60}$$

so when defining a quadratic form we may as well assume that $\mathbf{P}$ is symmetric, i.e., that $\mathbf{P}^T = \mathbf{P}$, because any asymmetric part contributes nothing to the value of the function f.

Now we turn to the role of the eigenvector-eigenvalue factorization. The eigenvalues of a symmetric matrix are real numbers. Every symmetric matrix has a full set of linearly independent eigenvectors; what is more, the eigenvectors may be chosen to be orthonormal. These facts mean that in the eigenvector-eigenvalue factorization of a symmetric matrix, $\mathbf{P} = \mathbf{T}\Lambda\mathbf{T}^{-1}$, the matrix $\mathbf{T}$ may be taken to be *orthogonal* (i.e., $\mathbf{T}^{-1} = \mathbf{T}^T$). Furthermore, all of the eigenvalues of a symmetric matrix are real numbers. So for a symmetric matrix, the eigenvector-eigenvalue factorization takes the form

Sec. 1.2   Linear Equations and Matrix Factorizations

$$\mathbf{P} = \mathbf{T}\Lambda\mathbf{T}^T \tag{1.61}$$

where $\mathbf{T}$ is an orthogonal matrix and Λ is a real diagonal matrix. Using the eigenvectors of $\mathbf{P}$ to form an orthogonal transformation of the variables, one that preserves the length of transformed vectors (see (A.55)), we set $\mathbf{Ty} = \mathbf{x}$. This transforms the quadratic function f into a weighted sum of squares, since

$$f(\mathbf{x}) = \mathbf{x}^T \mathbf{P} \mathbf{x} = \mathbf{y}^T \mathbf{T}^T \mathbf{P} \mathbf{T} \mathbf{y} = \mathbf{y}^T \Lambda \mathbf{y} = \sum_{i=1}^n \lambda_i y_i^2 \tag{1.62}$$

The transformation from $\mathbf{x}$ to $\mathbf{y}$ is a systematic way of describing the process of "completing the square," by which a homogeneous quadratic function is represented as a weighted sum of squares of linear functions. If all of the eigenvalues are nonnegative, then the weighting factors can be absorbed into the squared quantities, and the quadratic form may be expressed as a sum of squares:

$$f(\mathbf{x}) = \sum_{i=1}^n (\sqrt{\lambda_i}\, y_i)^2 \tag{1.63}$$

A symmetric matrix whose eigenvalues are nonnegative is said to be *nonnegative definite*; if its eigenvalues are all positive, it is said to be *positive definite*.

Example

For the quadratic function $2x_1^2 + 10x_1 x_2 + 13x_2^2$, the corresponding quadratic form is $\mathbf{x}^T \mathbf{P} \mathbf{x}$, where

$$\mathbf{P} = \begin{bmatrix} 2 & 5 \\ 5 & 13 \end{bmatrix}$$

The eigenvalues of $\mathbf{P}$ are $(15 \pm \sqrt{221})/2$, so the quadratic function may be written as the sum of two squares. The remaining details are left for the reader.

QR Factorization

A factorization of a quite different kind is commonly known as the QR factorization. It is applied most commonly to matrices with at least as many rows as columns (i.e., when $\mathbf{A}$ is $m \times n$ with $m \geq n$). The application to overdetermined systems of linear equations provides an alternative to computation of the pseudo-inverse matrix. Another important use of the QR factorization arises in an algorithm for determining the eigenvalues of a square matrix. Before elaborating on these applications, let us describe the factorization explicitly.

For any $m \times n$ matrix $\mathbf{A}$ with $m \geq n$, we may write

$$\mathbf{A} = \mathbf{QR} \tag{1.64}$$

where $\mathbf{Q}$ is also an $m \times n$ matrix, and $\mathbf{R}$ is a square, $n \times n$, upper triangular matrix. $\mathbf{Q}$ has one additional property: its columns are orthogonal so that $\mathbf{Q}^T \mathbf{Q} = \mathbf{I}$, the $n \times n$ identity matrix.

The QR factorization is simply the matrix expression of a well-known construction called the *Gram-Schmidt procedure*. This is a procedure for replacing a linearly independent sequence of vectors with an orthogonal sequence of vectors that span the same "nested" set of subspaces. This nesting is reflected in the upper triangular form of **R** in the QR factorization: the first column of **A** is a scalar multiple of the first column of **Q**, the first two columns of **A** span the same subspace of $\mathbb{R}^m$ as the first two columns of **Q**, and so on. (See the Appendix for a description of the Gram-Schmidt procedure.)

Example

The QR factorization is rarely computed column by column because better numerical accuracy can be achieved by taking a different approach. For simplicity, assume that **A** is $n \times n$. If a sequence of orthogonal matrices $\mathbf{Q}_1, \mathbf{Q}_2, \ldots, \mathbf{Q}_k$ can be found so that **A** is transformed to an upper triangular matrix,

$$\mathbf{Q}_k \cdots \mathbf{Q}_2 \mathbf{Q}_1 \mathbf{A} = \mathbf{R}$$

then multiplying both sides of this equation by $\mathbf{Q} = \mathbf{Q}_1^T \mathbf{Q}_2^T \cdots \mathbf{Q}_k^T$ gives the QR factorization (1.64). A commonly applied computational procedure involves a sequence of $k = n$ orthogonal matrices known as Householder transformations. A matrix of the form

$$\mathbf{Q}_\mathbf{w} = \mathbf{I} - \frac{2\mathbf{w}\mathbf{w}^T}{\|\mathbf{w}\|^2}$$

is known as the *Householder transformation* determined by the vector **w**. It has the following easily verified properties: (a) $\mathbf{Q}_\mathbf{w}$ is a symmetric, orthogonal matrix; (b) $\mathbf{Q}_\mathbf{w}\mathbf{w} = -\mathbf{w}$; (c) **x** is a vector orthogonal to **w** if and only if $\mathbf{Q}_\mathbf{w}\mathbf{x} = \mathbf{x}$; (d) the ith components of **x** and $\mathbf{Q}_\mathbf{w}\mathbf{x}$ are equal if and only if the ith component of **w** is zero.

To carry out the QR factorization of **A**, take $\mathbf{w}_1 = \mathbf{a}_1 - \|\mathbf{a}_1\|\mathbf{e}_1$, where $\mathbf{a}_1$ denotes the first column of **A** and $\mathbf{e}_1$ is the first standard unit vector (i.e., $\mathbf{e}_1 = [1, 0, \ldots, 0]^T$); then let $\mathbf{Q}_1$ be $\mathbf{Q}_{\mathbf{w}_1}$. As a consequence

$$\mathbf{Q}_1 \mathbf{a}_1 = \|\mathbf{a}_1\|\mathbf{e}_1$$

Thus the matrix $\mathbf{Q}_1 \mathbf{A}$ has its first column equal to the first column of **R** and may be expressed in partitioned form as

$$\mathbf{Q}_1 \mathbf{A} = \begin{bmatrix} \|\mathbf{a}_1\| & \mathbf{r}_1^T \\ \mathbf{0} & \mathbf{A}^{(1)} \end{bmatrix}$$

In this expression, $\mathbf{r}_1^T$ is a row vector (with $n-1$ rows) and $\mathbf{A}^{(1)}$ is an $(n-1)\times(n-1)$ matrix. Now the process is continued by operating on the submatrix $\mathbf{A}^{(1)}$ of $\mathbf{Q}_1 \mathbf{A}$ in exactly the same way: let $\mathbf{a}_2$ be the first column of $\mathbf{A}^{(1)}$ extended to be an n-vector by adding 0 as its first component; take $\mathbf{w}_2$ to be $\mathbf{a}_2 - \|\mathbf{a}_2\|\mathbf{e}_2$. Then let $\mathbf{Q}_2 = \mathbf{Q}_{\mathbf{w}_2}$. The matrix $\mathbf{A}^{(2)}$ is defined as the $(n-2)\times(n-2)$ lower right submatrix of $\mathbf{Q}_2 \mathbf{Q}_1 \mathbf{A}$ whose first two columns equal the first two columns of **R**. The process then repeats, and after n steps the entire **R** matrix has been obtained.

One property of the matrices in the QR factorization is immediate. Since an orthogonal matrix preserves the length of vectors that it multiplies, the lengths of the columns of **A** and of **R** are the same. As already mentioned, a key fact that is important in computing the QR factorization is that products of orthogonal matrices are still orthogonal, so the determination of a particular **Q** and **R** can be done in a sequence of "elementary" steps. Not only can this be beneficial for reducing computations, but it has

Sec. 1.2   Linear Equations and Matrix Factorizations   29

also been found that certain choices of these elementary steps are particularly amenable to parallel computing architectures. This approach has proven to be valuable in the design of special-purpose VLSI hardware for implementing the QR factorization for particular families of matrices.

To see how the QR factorization can be used in obtaining the best approximate solution of overdetermined linear equations, suppose that $\mathbf{A}$ is $m \times n$ with $m > n$, and suppose that $\mathbf{A}$ has linearly independent columns so that the Gram matrix $\mathbf{A}^T\mathbf{A}$ is invertible. The system of linear equations may be written as

$$\mathbf{Ax} = \mathbf{y} + \mathbf{z} \qquad (1.65)$$

where $\mathbf{z}$ represents the error $\mathbf{Ax} - \mathbf{y}$ whose length is to be minimized. The squared length of $\mathbf{z}$ may be expressed as follows:

$$\begin{aligned}\mathbf{z}^T\mathbf{z} &= (\mathbf{Ax}-\mathbf{y})^T(\mathbf{Ax}-\mathbf{y}) \\ &= (\mathbf{A}^T\mathbf{Ax}-\mathbf{A}^T\mathbf{y})^T(\mathbf{A}^T\mathbf{A})^{-1}(\mathbf{A}^T\mathbf{Ax}-\mathbf{A}^T\mathbf{y}) + \mathbf{y}^T\mathbf{y} - \mathbf{y}^T\mathbf{A}\mathbf{A}^T\mathbf{y}\end{aligned} \qquad (1.66)$$

Now $\mathbf{x}$ is to be chosen to minimize the squared length of $\mathbf{z}$, and we see that the only term on the right that can be affected by choice of $\mathbf{x}$ is the first one. We introduce the QR factorization of $\mathbf{A}$ to obtain

$$(\mathbf{A}^T\mathbf{A})^{-1} = (\mathbf{R}^T\mathbf{Q}^T\mathbf{Q}\mathbf{R})^{-1} = (\mathbf{R}^T\mathbf{R})^{-1} = (\mathbf{R}^{-1})(\mathbf{R}^T)^{-1} \qquad (1.67)$$

where we have used $\mathbf{Q}^T\mathbf{Q} = \mathbf{I}$ because $\mathbf{Q}$ has orthonormal columns. Thus the all-important term in the expression for the squared length of $\mathbf{z}$ is

$$\begin{aligned}(\mathbf{A}^T\mathbf{Ax}-\mathbf{A}^T\mathbf{y})^T & (\mathbf{A}^T\mathbf{A})^{-1}(\mathbf{A}^T\mathbf{Ax}-\mathbf{A}^T\mathbf{y}) \\ &= ((\mathbf{R}^T)^{-1}(\mathbf{A}^T\mathbf{Ax}-\mathbf{A}^T\mathbf{y}))^T((\mathbf{R}^T)^{-1}(\mathbf{A}^T\mathbf{Ax}-\mathbf{A}^T\mathbf{y}))\end{aligned} \qquad (1.68)$$

which is the squared length of the vector $((\mathbf{R}^T)^{-1}(\mathbf{A}^T\mathbf{Ax}-\mathbf{A}^T\mathbf{y}))$. Since a squared length is always a nonnegative quantity, this term is minimized by forcing it to be zero, that is, by setting

$$((\mathbf{R}^T)^{-1}(\mathbf{A}^T\mathbf{Ax}-\mathbf{A}^T\mathbf{y})) = \mathbf{0} \qquad (1.69)$$

This means that

$$\mathbf{A}^T\mathbf{Ax} - \mathbf{A}^T\mathbf{y} = \mathbf{0} \qquad (1.70)$$

Because the Gram matrix $\mathbf{A}^T\mathbf{A}$ is assumed to be invertible,

$$\mathbf{x} = (\mathbf{A}^T\mathbf{A})^{-1}\mathbf{A}^T\mathbf{y} \qquad (1.71)$$

which provides a derivation of the formula for the pseudo-inverse matrix for this case given earlier: $\mathbf{A}^\dagger = (\mathbf{A}^T\mathbf{A})^{-1}\mathbf{A}^T$. Going back one step and introducing the QR factorization of $\mathbf{A}$ we have

$$(\mathbf{QR})^T(\mathbf{QR})\mathbf{x} = (\mathbf{QR})^T\mathbf{y} \qquad (1.72)$$

Simplifying this equation by using the orthonormality of the columns of $\mathbf{Q}$ and the rule for transposing matrix products gives

$$Rx = Q^T y \qquad (1.73)$$

as the equations to be solved for **x**; the upper triangular form of **R** means that back-substitution may be used to easily find the solution. It turns out that this use of the QR factorization often provides much better numerical results than forming the Gram matrix and solving the associated linear equations as derived above:

$$A^T A x = A^T y \qquad (1.74)$$

This is because computing the Gram matrix $A^T A$ can result in substantial loss in numerical accuracy.

Another connection between QR factorization and Gram matrices is seen from the following equations:

$$A^T A = R^T Q^T Q R = R^T R \qquad (1.75)$$

The **R** matrix in the QR factorization of a matrix **A** determines a (symmetric!) LU factorization of the symmetric Gram matrix $A^T A$.

As an indication of how the QR factorization is used to compute eigenvalues, consider the following sequence of operations on a square matrix **A**.

1. Let $A_1 = A$ and take $i = 1$.
2. Compute the QR factorization $A_i = Q_i R_i$.
3. Compute the product $R_i Q_i = A_{i+1}$, increment i and go to step 2.

Here we encounter a first example of an iterative algorithm that generates a sequence of matrices. If we adopt the reasonable notion for the *limit* of a sequence of matrices, namely that it is the matrix of elementwise limits, assuming that all of the elemental sequences have limits, we then have a notion for convergence of sequences of matrices that may be applied to study the properties of iterative techniques like the one described above. Amazingly to a first glance, the sequence of matrices above does converge, and the diagonal elements of the limiting **R** factor approach the eigenvalues of **A** in the limit. (Notice that it's easy to see that the eigenvalues of each of the matrices in the $\{A_i\}$ sequence are the same; the matrices are all similar to each other.)

The actual implementation of the QR factorization for finding eigenvalues involves some variations on this theme. A preliminary transformation to "almost triangular" form is done to introduce many zero elements so as to speed up computation and convergence. Judicious use of a sequence of "shifting operations" where a scaled identity matrix is added to elements of the sequence A_i also speeds convergence. This takes advantage of the easily verified fact that the eigenvalues of the matrix $A + \delta I$ are the numbers $\lambda_i + \delta$, where the λ_i are the eigenvalues of **A**. Also **A** and $A + \delta I$ have the same set of eigenvectors. The QR algorithm may be supplemented with an additional iterative step for finding the associated eigenvectors if desired; more details are given later in this section.

Singular Value Decomposition (SVD)

A factorization that may be applied to any matrix, the Singular Value Decomposition amounts to the triple product factorization

Sec. 1.2 Linear Equations and Matrix Factorizations

$$\mathbf{A} = \mathbf{VDU}^T \tag{1.76}$$

where $\mathbf{U}$ and $\mathbf{V}$ are orthogonal matrices (to be described later) and $\mathbf{D}$ is a generalized diagonal matrix: the upper left corner of $\mathbf{D}$ is a diagonal matrix with nonnegative diagonal elements, and the remaining rows or columns of $\mathbf{D}$ (needed to make $\mathbf{D}$ rectangular with the same shape as $\mathbf{A}$) are all zeros.

The matrices in the Singular Value Decomposition are closely related to Gram matrices associated with the columns of $\mathbf{A}$ and $\mathbf{A}^T$. Notice that

$$\mathbf{AA}^T = \mathbf{VDD}^T\mathbf{V}^T \tag{1.77}$$

and since $\mathbf{DD}^T$ is a square diagonal matrix, we can recognize this as the eigenvector-eigenvalue decomposition of $\mathbf{AA}^T$. Thus $\mathbf{V}$ is an orthogonal matrix whose columns are eigenvectors of $\mathbf{AA}^T$, and the diagonal elements of $\mathbf{D}$ are the positive square roots of the eigenvalues of $\mathbf{AA}^T$. A similar analysis shows that the matrix $\mathbf{U}$ is an orthogonal matrix whose columns are the eigenvectors of $\mathbf{A}^T\mathbf{A}$, and

$$\mathbf{A}^T\mathbf{A} = \mathbf{UD}^T\mathbf{DU}^T \tag{1.78}$$

The diagonal elements of the matrix $\mathbf{D}$ in the Singular Value Decomposition of $\mathbf{A}$ are called the *singular values* of $\mathbf{A}$. As noted above, the singular values are the square roots of the nonzero eigenvalues of $\mathbf{AA}^T$ and of $\mathbf{A}^T\mathbf{A}$. When $\mathbf{A}$ is a square matrix, its largest singular value is its induced Euclidean norm (see (A.44)) and when $\mathbf{A}$ is invertible, the ratio of its largest and smallest singular values is its condition number, $\kappa(\mathbf{A})$ (see (A.48)).

The Singular Value Decomposition is useful in a number of applications. It is used as a representation for determining the rank of the matrix $\mathbf{A}$ (the number of linearly independent columns or rows). If $\mathbf{A}$ has k nonzero singular values, then the rank of $\mathbf{A}$ is k; in practice, where $\mathbf{D}$ must be determined by numerical means and is therefore subject to computational inaccuracies, judging the size of elements of $\mathbf{D}$ in comparison to the "machine accuracy" of the computer being used provides a sound basis for computing rank. The SVD can be used to generate "low rank" approximations to the matrix $\mathbf{A}$ in the following way. If the rank of $\mathbf{A}$ is k, then for $r < k$ the best rank r approximation (in a suitable least-squares sense) is formed by setting the smallest $k-r$ elements of $\mathbf{D}$ to zero and multiplying by $\mathbf{V}$ and $\mathbf{U}^T$ as in the SVD formula. Finally, the SVD also provides a way of computing the pseudo-inverse:

$$\mathbf{A}^\dagger = \mathbf{UD}^\dagger\mathbf{V}^T \tag{1.79}$$

where $\mathbf{D}^\dagger$, the pseudo-inverse of $\mathbf{D}$, is obtained from $\mathbf{D}^T$ by inverting its nonzero elements. (Again, in numerical applications, the test for a zero element is based on machine accuracy.)

Example

We give an example where the SVD is used for the computation of the pseudo-inverse of a matrix. This example also illustrates that one should not "blindly" accept the (approximate) solution to linear equations obtained with the help of the pseudo-inverse, since it may not provide the kind of approximation that is reasonable in the context of the problem under consideration. Consider the rbi and home run totals of six of baseball's great sluggers:

Aaron (2297,755)
Ruth (2192,714)
Mays (1903,660)
Robinson (1812,586)
Mantle (1509,536)
Foxx (1921,534)

From these data we cannot determine the percentages of solo, two-run, three-run, and grand slam home runs that might be typical for these players. Taking these percentages as the unknowns x_1, x_2, x_3, x_4 and taking x_5 as the ratio of "non-home-run rbi's" to home runs, we may write two equations for each of the players:

$$x_1 + x_2 + x_3 + x_4 = 1$$
$$x_1 + 2x_2 + 3x_3 + 4x_4 + x_5 = \frac{\text{rbi's}}{\text{home runs}}$$

Combining these equations for all six players and solving using the pseudo-inverse gives the answer $x_1 = 0.1041$, $x_2 = 0.2014$, $x_3 = 0.2986$, $x_4 = 0.3959$, and $x_5 = 0.0972$. (The matrix involved in the system of linear equations has only two linearly independent rows and columns, so the explicit formulas for the pseudo-inverse in terms of Gram matrices may not be applied to this problem.) This is obviously a very poor way to approximate the percentages of the various types of home runs! A little reflection indicates that there was no reason to believe that the least-squares, minimum length property of approximate solutions obtained by applying the pseudo-inverse has anything to do with the reality of home run hitting. Finding a better approximate solution is left as a problem for the reader. Alternatively, we might modify the experiment by replacing rbi's with the following list of rbi's resulting from home runs only: (1240, 1209, 1039, 931, 861, 944) for Aaron, Ruth, Mays, Robinson, Mantle, and Foxx, respectively. And we might also include the stats for H. Killebrew, whose 573 homers generated 957 rbi's, or the stats of other players.*

Iterative Solution of Linear Equations

In certain situations, it is essential to employ iterative methods for computing solutions to linear equations, relying on tests of convergence to provide a guide for the number of iterative steps that are performed before stopping the computation and accepting the result as a sufficiently good approximation to the "true" answer. For an extremely large system of equations, especially in the case of a sparse system whose PLU factorization and inverse are not sparse, tremendous computational efficiency may be realized by exploiting the ease of repeated matrix-vector multiplications in carrying out each step of

---

*For a list of home run statistics, including a breakdown by type, see "The Henry Aaron Home Run Analysis," by John C. Tattersall, in *Insider's Baseball*, L. Robert Davids, ed. (New York: Charles Scribner's Sons, 1983), pp. 252–256. The career rbi totals were taken from *Hall of Fame Fact Book*, 1982 edition, Paul MacFarlane, ed. (St. Louis: The Sporting News Press, 1982).

Sec. 1.2 Linear Equations and Matrix Factorizations

a suitable iteration. Even for large, unstructured systems the ability to get good approximate solutions can provide a necessary trade-off when it is computationally intractable to obtain an "exact" (in quotes because of the ever-present numerical errors) solution. Finally, for implementation on highly parallel computers, certain iterative methods offer significant performance improvements.

Splitting methods, in which the linear equations are decomposed additively, form the basis for many iterative solution methods. For the case of n linear equations in n unknowns,

$$\mathbf{A}\mathbf{x} = \mathbf{y} \tag{1.80}$$

the idea is to write $\mathbf{A} = \mathbf{A}_1 - \mathbf{A}_2$ so that the equations may be rewritten as

$$\mathbf{A}_1 \mathbf{x} = \mathbf{A}_2 \mathbf{x} + \mathbf{y} \tag{1.81}$$

We think of $\mathbf{A}_1$ as a good approximation to $\mathbf{A}$ with $\mathbf{A}_2$ being an error term that will hopefully have a small effect. (Notice that with $\mathbf{A}_1 = \mathbf{A}$ the error term vanishes.) Ignoring the first term on the right side of the equation gives the simplified equation

$$\mathbf{A}_1 \mathbf{x}_1 = \mathbf{y} \tag{1.82}$$

and we view $\mathbf{x}_1$ as an approximation to the true solution $\mathbf{x}$. As a second step, $\mathbf{x}_1$ may be used to make a (hopefully!) better approximation to the right side of the equation, namely $\mathbf{A}_2 \mathbf{x}_1 + \mathbf{y}$, and a second approximation to the solution can be obtained by solving

$$\mathbf{A}_1 \mathbf{x}_2 = \mathbf{A}_2 \mathbf{x}_1 + \mathbf{y} \tag{1.83}$$

Obviously, the process can be repeated over and over, leading to the iterative scheme: For $k > 1$, solve for $\mathbf{x}_{k+1}$ in the linear equations

$$\mathbf{A}_1 \mathbf{x}_{k+1} = \mathbf{A}_2 \mathbf{x}_k + \mathbf{y} \tag{1.84}$$

Notice that we may define $\mathbf{x}_0 = \mathbf{0}$ and obtain $\mathbf{x}_1$ from this formula also.

To analyze the sequence of approximate solutions obtained by this scheme, we examine the "error vector" $\mathbf{x} - \mathbf{x}_k$, noting that

$$\mathbf{A}_1 (\mathbf{x} - \mathbf{x}_{k+1}) = \mathbf{A}_2 (\mathbf{x} - \mathbf{x}_k) \tag{1.85}$$

Using $\mathbf{x}_0 = \mathbf{0}$, successive substitution leads to the following expression:

$$(\mathbf{x} - \mathbf{x}_k) = (\mathbf{A}_1^{-1} \mathbf{A}_2)^k \mathbf{x} \tag{1.86}$$

Thus it is the matrix $(\mathbf{A}_1^{-1} \mathbf{A}_2)$ that determines the properties of the iterative algorithm. Assuming that the eigenvectors of $(\mathbf{A}_1^{-1} \mathbf{A}_2)$, say $\{\mathbf{v}_1, \ldots, \mathbf{v}_n\}$, form a basis, the spectral decomposition of $(\mathbf{x} - \mathbf{x}_k)$ takes the form

$$(\mathbf{x} - \mathbf{x}_k) = \sum_{i=1}^{n} \alpha_i \lambda_i^k \mathbf{v}_i \tag{1.87}$$

where $\{\lambda_1, \ldots, \lambda_n\}$ are the corresponding eigenvalues of $(\mathbf{A}_1^{-1} \mathbf{A}_2)$. From this expression, we see that the convergence condition, $\mathbf{x}_k \to \mathbf{x}$, is satisfied if and only if the magnitude of every eigenvalue of $(\mathbf{A}_1^{-1}\mathbf{A}_2)$ is less than 1. Furthermore, the rate of convergence is determined by the eigenvalue of largest magnitude.

For the Jacobi method, where $\mathbf{A}_1$ is chosen to be the diagonal part of $\mathbf{A}$, a sufficient condition for convergence is *diagonal dominance*: each diagonal element of $\mathbf{A}$ is larger in magnitude than the sum of the magnitudes of all of the remaining elements in the same row.

Examples

As a trivial example, consider the solution of $3.5x = 7$. Writing $3.5 = 4 - 0.5$, and using the iteration gives

$$4 x_{k+1} = 0.5 x_k + 7$$

and the first few terms of the approximating sequence of solutions are (to four decimal places) { 1.75, 1.9688, 1.9961, 1.9995, ... } and indeed the error in the approximate solution decreases by a factor of $(0.5/4) = 1/8$ at each iteration. (So almost one additional decimal place of accuracy is gained from each iteration.)

For a bit more sophisticated example, let

$$\begin{bmatrix} 4 & 2 \\ 1 & 4 \end{bmatrix} \mathbf{x} = \begin{bmatrix} 6 \\ 5 \end{bmatrix}$$

We first examine the Jacobi iteration, choosing the diagonal part of the matrix as $\mathbf{A}_1$, which leads to the following approximating sequence:

$$\begin{bmatrix} 1.5000 \\ 1.2500 \end{bmatrix}, \begin{bmatrix} 0.8750 \\ 0.8750 \end{bmatrix}, \begin{bmatrix} 1.0625 \\ 1.0313 \end{bmatrix}, \begin{bmatrix} 0.9844 \\ 0.9844 \end{bmatrix}, \ldots$$

For this choice the eigenvalues of $(\mathbf{A}_1^{-1}\mathbf{A}_2)$ are $\pm\sqrt{2}/4 \approx \pm 0.3536$. Next we instead choose $\mathbf{A}_1$ as the lower triangular part of the matrix; the following approximating sequence is obtained:

$$\begin{bmatrix} 1.5000 \\ 0.8750 \end{bmatrix}, \begin{bmatrix} 1.0625 \\ 0.9844 \end{bmatrix}, \begin{bmatrix} 1.0078 \\ 0.9980 \end{bmatrix}, \begin{bmatrix} 1.0010 \\ 0.9998 \end{bmatrix}, \ldots$$

For this choice, the eigenvalues of $(\mathbf{A}_1^{-1}\mathbf{A}_2)$ are 0 and 1/8; thus the convergence rate is $(\sqrt{2}/4)/(1/8) \approx 3$ times faster than for the Jacobi method.

It may be unexpected, but the Gauss-Seidel method, the name given to the method employing the lower triangular part of $\mathbf{A}$, achieves improved performance with no additional computational cost. To see how this is accomplished, we rewrite the Jacobi method, replacing $\mathbf{A}_2$ by the sum of its upper and lower triangular parts, so that

$$\mathbf{A} = \mathbf{A}_1 - \mathbf{L} - \mathbf{U} \tag{1.88}$$

and the Jacobi iteration for the approximate solutions is

$$\mathbf{A}_1 \mathbf{x}_{k+1} = \mathbf{L}\mathbf{x}_k + \mathbf{U}\mathbf{x}_k + \mathbf{y} \tag{1.89}$$

Since $\mathbf{A}_1 - \mathbf{L}$ is the lower triangular part of $\mathbf{A}$, the Gauss-Seidel method takes the form

$$(\mathbf{A}_1 - \mathbf{L})\mathbf{x}_{k+1} = \mathbf{U}\mathbf{x}_k + \mathbf{y} \tag{1.90}$$

Moving the term involving $\mathbf{L}$ to the right side gives

$$\mathbf{A}_1 \mathbf{x}_{k+1} = \mathbf{L}\mathbf{x}_{k+1} + \mathbf{U}\mathbf{x}_k + \mathbf{y} \tag{1.91}$$

Sec. 1.2 Linear Equations and Matrix Factorizations 35

Now the difference between Jacobi and Gauss-Seidel is easily seen from a comparison of (1.89) and (1.91): When the components of the vector $\mathbf{x}_{k+1}$ are computed in natural order ("top to bottom"), Gauss-Seidel uses the partial results of the updating as they become available. (This follows from the strictly lower triangular form of $\mathbf{L}$; write out the equations for the components of $\mathbf{x}_{k+1}$ to check it.) The immediate use of partial new results leads to performance improvement over the Jacobi method, where "stale" values are used.

Another method, known as Successive Overrelaxation (SOR), results from the combination of a simple extrapolation method with the Gauss-Seidel method in an effort to speed up convergence. The method is described by the equation

$$\mathbf{A}_1 \mathbf{x}_{k+1} = (1-\gamma)\mathbf{A}_1 \mathbf{x}_k + \gamma(\mathbf{L}\mathbf{x}_{k+1} + \mathbf{U}\mathbf{x}_k + \mathbf{y}) \tag{1.92}$$

where the extrapolation parameter, γ, is suitably chosen. There is an elegant theory for determining the best value of γ, but we will not elaborate on it here. The range of possibilities is $0 < \gamma < 2$, with a choice near 2, say 1.9, often providing good performance.

Iterative Calculation of Eigenvectors

Iterative methods are also important in the numerical determination of eigenvectors. The power method involves the iteration

$$\mathbf{x}_{k+1} = \frac{\mathbf{A}\mathbf{x}_k}{\|\mathbf{A}\mathbf{x}_k\|} \tag{1.93}$$

which converges, for a sufficiently general choice of initial $\mathbf{x}_0$, to an eigenvector of $\mathbf{A}$ corresponding to the eigenvalue of largest magnitude.

The analysis of convergence of the power method is instructive. First notice that the vectors generated by the power method may be expressed in terms of the initial vector as

$$\mathbf{x}_{k+1} = \frac{\mathbf{A}^k \mathbf{x}_0}{\|\mathbf{A}^k \mathbf{x}_0\|} \tag{1.94}$$

Suppose that the eigenvectors of $\mathbf{A}$, $\{\mathbf{u}_1, \ldots, \mathbf{u}_n\}$, are linearly independent. Suppose also that the corresponding eigenvalues are $\{\lambda_1, \ldots, \lambda_n\}$, that $|\lambda_1| > |\lambda_i|$ for $i \neq 1$, and suppose that $\mathbf{x}_0$ is such that $\alpha_1 \neq 0$ in its spectral decomposition:

$$\mathbf{x}_0 = \sum_{i=1}^{n} \alpha_i \mathbf{u}_i \tag{1.95}$$

(This last condition is what is meant by $\mathbf{x}_0$ being sufficiently general.)

Using the spectral decomposition of $\mathbf{x}_0$, it is easy to see that the form of the spectral decomposition of $\mathbf{A}\mathbf{x}_k$ is

$$\mathbf{A}\mathbf{x}_k = c \sum_{i=1}^{n} \alpha_i \lambda_i^k \mathbf{u}_i \tag{1.96}$$

where $c = 1/\|\mathbf{A}^{k-1}\mathbf{x}_0\|$. Separating out the dominant term in this sum, we have

$$\mathbf{A}\mathbf{x}_k = c\,\alpha_1 \lambda_1^k \mathbf{u}_1 + c\,\lambda_1^k \left[\sum_{i=2}^{n} \alpha_i \left(\frac{\lambda_i}{\lambda_1} \right)^k \mathbf{u}_i \right] \tag{1.97}$$

As $k \to \infty$, the term in brackets goes to zero. Thus the $\mathbf{x}_k$ sequence tends to an eigenvector of $\mathbf{A}$ corresponding to λ_1, the eigenvalue of largest magnitude. The eigenvalue itself is obtained from the limiting relation

$$\lim_{k \to \infty} \mathbf{A}\mathbf{x}_k = \lambda_1 \mathbf{x}_k \tag{1.98}$$

choosing k large enough so that the power method iteration has converged (i.e., choose k so that $x_{k+1} \approx x_k$).

To find all of the eigenvalues and eigenvectors of $\mathbf{A}$, we could rely on a "deflation" procedure. From the spectral representation of $\mathbf{A}$,

$$\mathbf{A} = \sum_{i=1}^{n} \lambda_i \mathbf{u}_i \mathbf{v}_i \tag{1.99}$$

we notice that $\mathbf{A} - \lambda_1 \mathbf{u}_1 \mathbf{v}_1$ has the same eigenvectors and eigenvalues as $\mathbf{A}$ except that the eigenvalue λ_1 is replaced by 0. Thus we can "deflate" $\mathbf{A}$ by this operation and repeat the power method on the modified matrix to find the second largest (in magnitude) eigenvalue of $\mathbf{A}$. This is not carried to its logical conclusion in practice because the QR factorization, when applied to solving for the eigenvalues of $\mathbf{A}$ as mentioned earlier, essentially amounts to an organized application of the power method to get all of the eigenvalues at once.

In a similar way, an iteration known as the inverse power method involves the sequence of solutions to the equations

$$\mathbf{A}\mathbf{x}_{k+1} = \mathbf{x}_k \tag{1.100}$$

(Actually, the sequence of vectors is also normalized to unit length at each step as in the power method.) The inverse power method converges to an eigenvector corresponding to the eigenvalue of smallest magnitude. Notice that the PLU factorization of $\mathbf{A}$ can be used repeatedly to generate the iterates of the inverse power method. Also notice that when good estimates of an eigenvalue are available, say from the QR method, using shifting to subtract off this multiple of the identity matrix will leave a matrix with a very small eigenvalue (i.e., one that is ideally suited for using the inverse power method to find a "polished" value for its smallest eigenvalue and the corresponding eigenvector). Indeed, the inverse power method finds widespread use in numerical algorithms for finding eigenvectors and eigenvalues.

1.3 MATRIX CALCULUS

Another important topic concerning matrices is worth a special mention. This has to do with extension of the basic ideas of calculus to matrices. For the case of a single independent variable, all of this material is relatively easy to derive from elementary calculus of real- or complex-valued functions, since it is only necessary to remember

Sec. 1.3 Matrix Calculus

that the operations of differentiation and integration are carried out elementwise when dealing with matrices. Nevertheless, a quick survey of some of the important results will be helpful both for later use and for a review of some of the calculus material that might be a bit unfamiliar in the matrix setting.

To start, let's suppose we have a matrix $\mathbf{X}(t)$ of time functions; for now, we'll leave the dimensions of $\mathbf{X}$ unspecified. An equivalent viewpoint is to think of $\mathbf{X}$ as a matrix-valued time function. It makes no real difference to what follows which of the two ways of thinking about $\mathbf{X}(t)$ we adopt, but the first one is our choice for the moment. We can easily define the derivative and integral of $\mathbf{X}$ by letting the calculus operations work elementwise. Thus if the (i,j)th element of $\mathbf{X}(t)$ is $x_{ij}(t)$, then the (i,j)th element of the derivative matrix is

$$\left[\frac{d}{dt}\mathbf{X}(t)\right]_{ij} = \frac{dx_{ij}(t)}{dt} \tag{1.101}$$

and a similar elementwise formula is used to define the integral of the matrix $\mathbf{X}(t)$:

$$\left[\int \mathbf{X}(t)\,dt\right]_{ij} = \int x_{ij}(t)\,dt \tag{1.102}$$

An important fact to note is that the statement of the Fundamental Theorem of Calculus is preserved by these definitions:

$$\int_{t_0}^{t_1} \frac{d}{dt}\mathbf{X}(t)\,dt = \mathbf{X}(t_1) - \mathbf{X}(t_0) \tag{1.103}$$

The relation of differentiation with various algebraic operations is also of considerable practical importance and interest. To describe these facts, let $\mathbf{Y}(t)$ be a second matrix of time functions. Then we have the following two important results. Both follow by applying the definition of matrix multiplication (in terms of matrix elements), and they reflect the fact that matrix multiplication is not commutative.

The first result concerns the linearity of differentiation and integration.

$$\frac{d}{dt}\left[\mathbf{A}_1\mathbf{X}(t)\mathbf{A}_2 + \mathbf{A}_3\mathbf{Y}(t)\mathbf{A}_4\right] = \mathbf{A}_1\left[\frac{d}{dt}\mathbf{X}(t)\right]\mathbf{A}_2 + \mathbf{A}_3\left[\frac{d}{dt}\mathbf{Y}(t)\right]\mathbf{A}_4 \tag{1.104}$$

Here $\mathbf{A}_1$, $\mathbf{A}_2$, $\mathbf{A}_3$, and $\mathbf{A}_4$ are constant matrices dimensioned so that the sum and products in the expression make sense. A similar result holds for integration, which is also linear.

The second result is the matrix version of the product rule for differentiation.

$$\frac{d}{dt}\left[\mathbf{X}(t)\mathbf{Y}(t)\right] = \left[\frac{d}{dt}\mathbf{X}(t)\right]\mathbf{Y}(t) + \mathbf{X}(t)\left[\frac{d}{dt}\mathbf{Y}(t)\right] \tag{1.105}$$

Here it is required that $\mathbf{X}$ and $\mathbf{Y}$ are dimensioned so that the matrix product is meaningful.

A result of particular interest can be deduced from the product rule in the following important special case: Suppose that in the formula for the product rule, $\mathbf{Y}(t)$ is a square matrix and is the inverse of $\mathbf{X}(t)$. Then we find that

$$\frac{d}{dt}\left[\mathbf{X}^{-1}(t)\right] = -\mathbf{X}^{-1}(t)\left[\frac{d}{dt}\mathbf{X}(t)\right]\mathbf{X}^{-1}(t) \tag{1.106}$$

This expression should be compared to the familiar calculus formula for the derivative of the reciprocal of a function, to which it reduces as a special case:

$$\frac{d}{dt}\left[\frac{1}{f(t)}\right] = -\frac{1}{f^2(t)}\frac{df(t)}{dt} \tag{1.107}$$

where $f(t)$ is a function (i.e., a 1×1 matrix).

Derivatives play an important role in providing power series expressions for functions. A function $f(t)$ that is sufficiently "well-behaved" near $t=0$, and this includes being infinitely differentiable at $t=0$, can be expressed as a power series (known as the Taylor series)

$$f(t) = \sum_{i=0}^{\infty} f_i t^i \tag{1.108}$$

where the coefficients in the power series are determined from the derivatives of $f(t)$ at $t=0$ according to the formula

$$f_i = \frac{f^{(i)}(0)}{i!} \tag{1.109}$$

and the subscript represents ith derivative; $i!$ is read as "i factorial," the symbol used for the product of the i integers $\{1, \ldots, i\}$. The range of validity of a power series representation is determined by examination of each individual case, the region of convergence being some interval around 0, $\{t : |t| < T_0\}$.

Example

We recall the power series representation of one particular function that is well worth remembering. It follows from the expression for the sum of a sequence whose terms form a geometric progression:

$$\sum_{i=0}^{\infty} \rho^i = \frac{1}{1-\rho}$$

which is valid for $|\rho| < 1$. Taking $\rho = \alpha t$ gives a power series for the function $f(t) = (1 - \alpha t)^{-1}$, valid for $\{t : |t| < |1/\alpha|\}$. The coefficients give a formula for the derivatives of the function at $t=0$.

$$f^{(i)}(0) = i! \, \alpha^i$$

For a matrix of "well-behaved" functions, $\Psi(t)$, a simple extension of the Taylor series expansion may be written

$$\Psi(t) = \sum_{i=0}^{\infty} \Psi_i t^i \tag{1.110}$$

where

$$\Psi_i = \frac{\Psi^{(i)}(0)}{i!} \tag{1.111}$$

Sec. 1.3   Matrix Calculus

and we again use a superscript of (i) to denote the ith derivative. Just as in the case of functions such as $(1-\alpha t)^{-1}$, it is sometimes possible to recognize power series expressions for "known" matrices of functions. For example, the matrix version of the geometric series formula will arise in later work. Letting $\mathbf{P}$ be a square matrix, we have

$$\sum_{i=0}^{\infty} \mathbf{P}^i = (\mathbf{I}-\mathbf{P})^{-1} \tag{1.112}$$

which is valid for matrices $\mathbf{P}$ whose eigenvalues are all smaller than 1 in magnitude. To see this, recall that the infinite sum is defined as the limit of the sequence of partial sums:

$$\sum_{i=0}^{\infty} \mathbf{P}^i = \lim_{n \to \infty} \sum_{i=0}^{n} \mathbf{P}^i \tag{1.113}$$

We may evaluate the partial sum by noting that

$$(\mathbf{I}-\mathbf{P})\sum_{i=0}^{n} \mathbf{P}^i = \mathbf{I}-\mathbf{P}^{n+1} \tag{1.114}$$

(which follows by a direct calculation). If we assume that no eigenvalue of $\mathbf{P}$ equals 1 so that $(\mathbf{I}-\mathbf{P})$ is invertible, then we have

$$\sum_{i=0}^{n} \mathbf{P}^i = (\mathbf{I}-\mathbf{P})^{-1}(\mathbf{I}-\mathbf{P}^{n+1}) = (\mathbf{I}-\mathbf{P})^{-1} - (\mathbf{I}-\mathbf{P})^{-1}\mathbf{P}^{n+1} \tag{1.115}$$

We see that the second term on the right will tend to a limit, in particular to the zero matrix, provided that all eigenvalues of $\mathbf{P}$ have magnitude less than 1. When this is the case, the limit of the entire expression gives the claimed result.

Letting $\mathbf{P} = \mathbf{A}t$, we have the matrix power series

$$\Psi(t) = \sum_{i=0}^{\infty} (\mathbf{A}t)^i = (\mathbf{I}-\mathbf{A}t)^{-1} \tag{1.116}$$

which defines a meaningful matrix of functions of t for all $|t| < |1/\lambda|_{\max}$, where we have used suggestive notation to indicate that the largest value over all eigenvalues of $\mathbf{A}$ is to be used in the inequality. Just as above, we may identify the derivatives of $\Psi(t)$ from the power series coefficients, obtaining

$$\Psi^{(i)}(0) = i!\,\mathbf{A}^i \tag{1.117}$$

We will see some other important examples of matrices of functions defined by power series in following sections. For now, we will simply point out an important fact that follows from the Cayley-Hamilton theorem. Suppose that $\mathbf{M}(t)$ is a square matrix of functions defined in terms of a power series involving an $n \times n$ square matrix $\mathbf{A}$:

$$\mathbf{M}(t) = \sum_{i=0}^{\infty} \mu_i (\mathbf{A}t)^i \tag{1.118}$$

for some sequence of real coefficients $\{\mu_i\}$. The Cayley-Hamilton theorem states that $\mathbf{A}^n$ may be expressed as a linear combination of the lower powers of $\mathbf{A}$, and it follows by induction that all powers of $\mathbf{A}$ greater than n may also be expressed in terms of $\mathbf{A}^k$, $0 \leq k < n$, so that we may write

$$\mathbf{M}(t) = \sum_{i=0}^{n-1} m_i(t)\mathbf{A}^i \qquad (1.119)$$

for some suitably chosen functions $\{m_i(t)\}$.

A little analysis should convince you that closed-form expressions for these functions of t are not easily obtained. For one thing, some kind of analytical expression for the μ_i sequence would be required. Sometimes there is an approach to finding the $m_i(t)$ functions by using the derivative of $\mathbf{M}(t)$. Suppose that we can express the derivative in terms of $\mathbf{M}(t)$, $\mathbf{A}$, and t; then we can (in principle) obtain from the finite sum expression for $\mathbf{M}(t)$ a coupled set of differential equations for the functions $m_i(t)$. Assuming an appropriate boundary condition is known, these differential equations may be solved; if there is no analytical solution, a numerical solution is possible.

Example

Consider the example already introduced:

$$\Psi(t) = \sum_{i=0}^{\infty} (\mathbf{A}t)^i$$

which we desire to rewrite as a finite sum

$$\Psi(t) = \sum_{i=0}^{n-1} m_i(t)\mathbf{A}^i$$

For this case we have $\Psi(0) = \mathbf{I}$ and we obtain the following from the formula for differentiating an inverse matrix:

$$\frac{d}{dt}\Psi(t) = \Psi(t)\mathbf{A}\Psi(t)$$

Taking a concrete example, let

$$\mathbf{A} = \begin{bmatrix} 0 & 1 \\ -a_2 & -a_1 \end{bmatrix}$$

For this companion matrix, the Cayley-Hamilton theorem gives

$$\mathbf{A}^2 = -a_1 \mathbf{A} - a_2 \mathbf{I}$$

Letting

$$\Psi(t) = m_0(t)\mathbf{I} + m_1(t)\mathbf{A}$$

we take the derivative of the equation expressing $\Psi(t)$ in terms of Ψ and $\mathbf{A}$ and simplify by using the Cayley-Hamilton theorem to obtain an equation of the form

$$\dot{m}_0(t)\mathbf{I} + \dot{m}_1(t)\mathbf{A} = p_0(t)\mathbf{I} + p_1(t)\mathbf{A}$$

where a dot indicates derivative and where $p_0(t)$ and $p_1(t)$ are functions involving sums of products of $m_0(t)$, $m_1(t)$, a_1, and a_2. Equating coefficients of powers of $\mathbf{A}$ on both sides gives the desired differential equations for $m_0(t)$ and $m_1(t)$:

Sec. 1.3 Matrix Calculus

$$\dot{m}_0(t) = p_0(t); \quad m_0(0) = 1$$
$$\dot{m}_1(t) = p_1(t); \quad m_1(0) = 0$$

where the initial conditions are obtained from the condition $\Psi(0) = \mathbf{I}$.

This example is not intended to suggest that this approach is a good way of computing $\Psi(t)$!

Matrix differential equations provide an implicit description of matrix functions. Assuming suitable regularity conditions are satisfied by the matrix-valued function $\mathbf{F}$, for a differential equation of the form

$$\frac{d}{dt}\mathbf{X}(t) = \mathbf{F}(\mathbf{X}(t), t) \tag{1.120}$$

together with an initial condition, $\mathbf{X}(t_0) = \mathbf{X}_0$, there exists a unique solution (i.e., a matrix function $\mathbf{X}(t)$ with the specified initial condition and whose derivative satisfied the differential equation in some neighborhood of t_0). Differential equations, especially for the case when $\mathbf{X}(t)$ is a vector quantity, will be the main underlying mathematical model employed throughout this book when studying dynamic systems.

Chapter 1 Appendix

Some Basic Facts About Matrices and Linear Algebra

This appendix is meant to be read in conjunction with Chapter 1. It provides a summary of some of the important fundamentals concerning vectors and matrices. The appendix will serve as a reference for terminology and notation, but not as a real substitute for the kind of background that comes from a course in linear algebra.

Matrices

A *matrix* is a rectangular array of complex numbers. The numbers making up a matrix are called its *elements*. We denote a matrix with a boldfaced letter, say **A**, and if the array has m rows and n columns we label the entries of the array, the matrix elements, with a double subscript, the first indicating the row position and the second indicating the column position. In writing a matrix we group the array in brackets.

Examples

$$\mathbf{W} = \begin{bmatrix} 0 & 1 \\ 0 & 1 \\ 1 & 1 \end{bmatrix}$$

$$\mathbf{X} = \begin{bmatrix} 3 & 0.5 \\ 2 & \pi \end{bmatrix}$$

$$\mathbf{Y} = \begin{bmatrix} 1+j & 0.5 & 1+2j \\ 1-j & -1 & 4 \end{bmatrix}$$

$$\mathbf{Z} = \begin{bmatrix} 3 & \pi \\ 2 & 0.5 \end{bmatrix}$$

If all of its elements are real numbers, a matrix is called a real matrix; in the examples above, **W**, **X**, and **Z** are real matrices while **Y** is not a real matrix. The number $1+2j$ is the 1,3 element of **Y**, which is a matrix with two rows and three columns. When we want to refer to a matrix of a general form without giving its elements numerically, we use doubly subscripted variables to denote matrix elements and we write

$$\mathbf{A} = \begin{bmatrix} a_{11} & a_{12} & \cdots & a_{1n} \\ a_{21} & a_{22} & \cdots & a_{2n} \\ \vdots & \vdots & & \vdots \\ a_{m1} & a_{m2} & \cdots & a_{mn} \end{bmatrix} \tag{A.1}$$

A matrix with m rows and n columns is said to be an $m \times n$ matrix, read as "an m by n matrix." We also say that **A** has *dimensions* m and n, giving the number of rows first. To denote a particular matrix element, say the (i,j)th, we use the following notation:

$$(\mathbf{A})_{ij} = a_{ij} \tag{A.2}$$

We use equations involving matrices with the understanding that two matrices are equal if and only if they have the same dimensions and the same elements.

Three "shapes" of matrices arise often enough in practice to warrant special terminology. A $m \times 1$ matrix is called a *column m-vector*, an *m-vector*, or simply a *vector*. We will usually use a lowercase bold symbol for vectors and drop the unnecessary second subscript in denoting its elements. Hence

$$\mathbf{x} = \begin{bmatrix} x_1 \\ x_2 \\ \vdots \\ x_m \end{bmatrix} \tag{A.3}$$

is an m-vector. A $1 \times n$ matrix is called a *row vector*. Again, only one subscript is necessary to label its elements. Row vectors have obvious advantages for typography since they can be written in a line of text quite easily. For instance, $[(1+j) \ 2 \ -0.7]$ is a row vector. Finally, a matrix with equal numbers of rows and columns is called a *square matrix*, for obvious reasons. The example **X** above is a 2×2 square matrix, or more simply a square matrix of dimension 2. It is important to point out that a 1×1 matrix can be regarded as a vector, a row vector, a square matrix, or simply as an ordinary number; it is only a matter of interpretation.

Algebraic Operations Involving Matrices

Scalar multiplication. The product of a matrix **A** and a complex number z may always be formed. The result is a matrix with the same dimensions as **A** whose elements are the products of z with corresponding elements of **A**. In terms of a formula,

$$(z\mathbf{A})_{ij} = za_{ij} = (\mathbf{A}z)_{ij} \tag{A.4}$$

Addition of matrices. Two matrices, say **A** and **B**, may be added if and only if they have the same dimensions. In this case we have $\mathbf{A} + \mathbf{B} = \mathbf{C}$, where **C** is another matrix with the same dimensions whose (i,j)th element is the sum of the (i,j)th elements of **A** and **B**. Symbolically,

$$(\mathbf{C})_{ij} = c_{ij} = a_{ij} + b_{ij} \tag{A.5}$$

where

$$(\mathbf{A})_{ij} = a_{ij} \quad \text{and} \quad (\mathbf{B})_{ij} = b_{ij} \tag{A.6}$$

Multiplication of matrices. Two matrices, say **A** and **B**, may be multiplied with **A** as the left factor and **B** as the right factor if and only if their dimensions are compatible: If **A** is $m_\mathbf{A} \times n_\mathbf{A}$ and **B** is $m_\mathbf{B} \times n_\mathbf{B}$, then it is required that $n_\mathbf{A} = m_\mathbf{B}$. In words, the number of columns of the left factor must equal the number of rows of the right factor. When this is the case, the product matrix $\mathbf{C} = \mathbf{AB}$ is $m_\mathbf{A} \times n_\mathbf{B}$, or again in words, the product is a matrix with the same number of rows as the left factor and the same number of columns as the right factor. Choosing some simpler notation, if **A** is $m \times n$ and **B** is $n \times p$, then the elements of the product matrix **C** are given by the formula

$$c_{ij} = \sum_{k=1}^{n} a_{ik} b_{kj} \tag{A.7}$$

where $1 \leq i \leq m$ and $1 \leq j \leq p$. (The origin of this curious-looking formula will become clear later when the connection between matrices and linear functions is discussed. It results from properties of the composition of two linear functions.)

There are two interpretations of matrix multiplication in terms of rows and columns of the factors. Each of these has its own importance, and it is a simple matter to verify them both. These interpretations arise by considering what the elementwise formula for matrix multiplication means in terms of the rows and columns of the product matrix **C**.

1. The columns of **C** are obtained by multiplying the matrix **A** times the corresponding columns of **B**. In more detail, the jth column of **C**, say $\mathbf{c}_j$, is $\mathbf{Ab}_j$, where $\mathbf{b}_j$ is the jth column of **B**:

$$\mathbf{C} = \begin{bmatrix} \mathbf{c}_1 & \mathbf{c}_2 & \cdots & \mathbf{c}_p \end{bmatrix} = \mathbf{A} \begin{bmatrix} \mathbf{b}_1 & \mathbf{b}_2 & \cdots & \mathbf{b}_p \end{bmatrix} = \begin{bmatrix} \mathbf{Ab}_1 & \mathbf{Ab}_2 & \cdots & \mathbf{Ab}_p \end{bmatrix} \tag{A.8}$$

2. The rows of **C** are obtained by multiplying the corresponding rows of **A** times the matrix **B**. The ith row of **C**, which will be denoted by $\mathbf{C}_i$, equals $\mathbf{A}_i \mathbf{B}$, where $\mathbf{A}_i$ is the ith row of **A**:

$$\mathbf{C} = \begin{bmatrix} \mathbf{C}_1 \\ \mathbf{C}_2 \\ \vdots \\ \mathbf{C}_m \end{bmatrix} = \begin{bmatrix} \mathbf{A}_1 \\ \mathbf{A}_2 \\ \vdots \\ \mathbf{A}_m \end{bmatrix} \mathbf{B} = \begin{bmatrix} \mathbf{A}_1 \mathbf{B} \\ \mathbf{A}_2 \mathbf{B} \\ \vdots \\ \mathbf{A}_m \mathbf{B} \end{bmatrix} \tag{A.9}$$

These interpretations are simple and their applications are widespread. The combination of the two interpretations provides a restatement of the definition of matrix multiplication:

$$(\mathbf{C})_{ij} = c_{ij} = \mathbf{A}_i \mathbf{b}_j \qquad (A.10)$$

which uses the notation introduced above to express the (i,j)th element of $\mathbf{C}$ as the product of the ith row of $\mathbf{A}$ times the jth column of $\mathbf{B}$.

The elementwise formula for matrix multiplication also leads to another representation of the matrix product. Consider one term in the sum defining the elements of the product:

$$(\mathbf{C})_{ij} = c_{ij} = \sum_{k=1}^{n} a_{ik} b_{kj} \qquad (A.11)$$

namely $a_{ik} b_{kj}$. A little thought shows that this is the expression for the (i,j)th element of the matrix product of the kth column of $\mathbf{A}$ times the kth row of $\mathbf{B}$:

$$a_{ik} b_{kj} = (\mathbf{a}_k \mathbf{B}_k)_{ij} \qquad (A.12)$$

Thus the sum over k which gives the formula for the (i,j)th element of $\mathbf{C}$ provides an additive decomposition of the matrix product $\mathbf{AB}$ as a sum of products of columns of $\mathbf{A}$ times rows of $\mathbf{B}$:

$$\mathbf{C} = \sum_{k=1}^{n} \mathbf{a}_k \mathbf{B}_k \qquad (A.13)$$

Generally, matrix multiplication is not commutative. Even when it makes sense according to the definition to form both of the products $\mathbf{AB}$ and $\mathbf{BA}$, the two results are not necessarily equal.

Examples

For $\mathbf{X}$ and $\mathbf{Z}$ given above we find that

$$\mathbf{XZ} = \begin{bmatrix} 10 & 3\pi + 0.25 \\ 6 + 2\pi & 5\pi/2 \end{bmatrix}$$

$$\mathbf{ZX} = \begin{bmatrix} 9 + 2\pi & 3/2 + \pi^2 \\ 7 & 1 + \pi/2 \end{bmatrix}$$

Notice also that for the matrices $\mathbf{X}$ and $\mathbf{Y}$, the product $\mathbf{XY}$ is defined while it is not possible to form the product $\mathbf{YX}$. For the matrices $\mathbf{W}$ and $\mathbf{Y}$, both of the products $\mathbf{WY}$ and $\mathbf{YW}$ are defined, but they do not even have the same dimensions, being 3×3 and 2×2, respectively.

On account of the noncommutativity of matrix multiplication, it is necessary to pay careful attention to preserving the order of multiplications in complicated expressions such as $(\mathbf{A}+\mathbf{B})(\mathbf{A}+3\mathbf{B})(2\mathbf{A}+5\mathbf{B})$. Fortunately, the associativity and distributivity properties of matrix addition and multiplication are the same as for addition and multiplication of real and complex numbers!

The Zero Matrix and the Identity Matrix

Many special classes of matrices arise in applications; rather than try to catalog them all at once, most will be described when they arise in the text. However, there are two forms of particular importance relative to matrix algebra.

We will denote by **0** the *zero matrix*, any matrix whose elements are all zero:

$$(\mathbf{0})_{ij} = 0 \tag{A.14}$$

Usually, we will leave the dimensions of **0** unspecified, leaving them to be understood from context. We write $\mathbf{0}_{m \times n}$ to specifically denote the $m \times n$ zero matrix. Obviously, **0** is a zero element for matrix addition:

$$\mathbf{0} + \mathbf{A} = \mathbf{A} = \mathbf{A} + \mathbf{0} \tag{A.15}$$

where **A** and **0** have the same dimensions. For matrix multiplication, if **A** is $m \times n$,

$$\mathbf{0}_{m \times m}\mathbf{A} = \mathbf{0}_{m \times n} = \mathbf{A}\mathbf{0}_{n \times n} \tag{A.16}$$

We will denote by **I** any *identity matrix*, that is, a square matrix whose only nonzero elements are the 1's along its main diagonal:

$$(\mathbf{I})_{ij} = \begin{cases} 0 & \text{for } i \neq j \\ 1 & \text{for } i = j \end{cases} \tag{A.17}$$

Again, the dimensions of **I** are usually obtained from the context of its use; to specifically denote an $n \times n$ identity matrix, we write $\mathbf{I}_n$. The name *identity matrix* arises because of its role as an identity element for matrix multiplication:

$$\mathbf{I}_m \mathbf{A} = \mathbf{A} = \mathbf{A}\mathbf{I}_n \tag{A.18}$$

for any $m \times n$ matrix **A**.

Matrix Inverse

Closely related to matrix multiplication is the notion of *matrix inverse*. If **A** and **X** are square matrices of the same dimensions and they satisfy $\mathbf{AX} = \mathbf{I} = \mathbf{XA}$, then we say that **X** is the matrix inverse of **A**. We denote the inverse matrix with the suggestive notation $\mathbf{A}^{-1}$, so that the inverse matrix satisfies

$$\mathbf{AA}^{-1} = \mathbf{I} = \mathbf{A}^{-1}\mathbf{A} \tag{A.19}$$

If a square matrix **A** has a matrix inverse, it is unique. Indeed, if $\mathbf{AX} = \mathbf{I}$ and if $\mathbf{YA} = \mathbf{I}$, then multiplying the first of these equations by **Y** on the left and using the second equation gives $\mathbf{Y} = \mathbf{X}$.

If the square matrix **A** has a matrix inverse, we say that **A** is *invertible*; the terms *nonsingular* and *regular* are also used as synonyms for invertible. If **A** has no matrix inverse, we say that it is *noninvertible* or *singular*.

Examples

If **A** is a 1×1 matrix, then it is invertible if it is nonzero, and $\mathbf{A}^{-1} = 1/\mathbf{A}$.

If **A** is a 2×2 matrix, say

$$\mathbf{A} = \begin{bmatrix} a_{11} & a_{12} \\ a_{21} & a_{22} \end{bmatrix}$$

then **A** is invertible if $\Delta = a_{11}a_{22} - a_{21}a_{12}$ is nonzero. In this case

$$\mathbf{A}^{-1} = \begin{bmatrix} a_{22}/\Delta & -a_{12}/\Delta \\ -a_{21}/\Delta & a_{11}/\Delta \end{bmatrix}$$

This is a formula well worth remembering.

An identity matrix is always invertible: $\mathbf{I}^{-1} = \mathbf{I}$.

A *lower triangular matrix*, one of the form

$$\mathbf{L} = \begin{bmatrix} l_{11} & 0 & \cdots & 0 \\ l_{21} & l_{22} & \cdot & \cdot \\ \cdot & \cdot & \cdot & \cdot \\ \cdot & \cdot & \cdot & 0 \\ l_{n1} & l_{n2} & \cdots & l_{nn} \end{bmatrix}$$

is invertible if and only if all of the *diagonal elements*, $l_{11}, l_{22}, \ldots, l_{nn}$ are nonzero. If a lower triangular matrix is invertible, then its inverse is also a lower triangular matrix (and the diagonal elements of the inverse matrix are $1/l_{11}, 1/l_{22}, \ldots, 1/l_{nn}$). Analogous results hold for *upper triangular* matrices.

The Determinant

One criterion for determining whether a square matrix **A** is invertible involves the determinant. The *determinant* of a square matrix is a scalar function of the matrix elements defined by the formidable expression

$$\det \mathbf{A} = \sum_{\pi} (-1)^{\sigma(\pi)} \prod_{i=1}^{n} a_{i\pi(i)} \tag{A.20}$$

where **A** is an n-dimensional square matrix with elements $(\mathbf{A})_{ij} = a_{ij}$, the sum is taken over all permutation mappings π on the set $\{1, 2, \ldots, n\}$, and σ is a zero-one function giving the parity of the minimal representation of the permutation as a product of transpositions. Because this formula is of little value for computing $\det \mathbf{A}$, we will not need to be more explicit about these definitions. We will instead give the explicit formulas for the cases $n = 1$ and 2 and then describe Laplace's expansion for computing determinants of large matrices from determinants of smaller ones. For $n = 1$, $\det \mathbf{A} = a_{11} = \mathbf{A}$. For $n = 2$, $\det \mathbf{A} = a_{11}a_{22} - a_{12}a_{21}$. For any n, we may express $\det \mathbf{A}$ as follows:

$$\det \mathbf{A} = \sum_{k=1}^{n} (-1)^{i+k} a_{ik} \Delta_{ik} \tag{A.21}$$

or alternatively as

$$\det \mathbf{A} = \sum_{k=1}^{n} (-1)^{i+k} a_{ki} \Delta_{ki} \tag{A.22}$$

These are the Laplace expansions for the determinant corresponding to the ith row and ith column of $\mathbf{A}$, respectively. In these formulas, the quantity Δ_{ik} is the determinant of the $(n-1)$-dimensional square matrix obtained by deleting the ith row and kth column of $\mathbf{A}$, and similarly for Δ_{ki}.

Examples

The determinant of any identity matrix is easily evaluated by the Laplace expansion corresponding to any row or column: $\det \mathbf{I} = 1$.

The determinant of the lower triangular matrix $\mathbf{L}$ given above is also easily evaluated by the Laplace expansion corresponding to the first row (repeatedly) or last column (repeatedly): $\det \mathbf{L} = l_{11} l_{22} \cdots l_{nn}$.

Properties of the Determinant and the Matrix Inverse

Many familiar properties of determinants can be verified directly from the defining formula for the determinant and from the Laplace expansion formulas. For example, replacing any row of a matrix by its sum with another row does not change the value of the determinant; replacing a row of a matrix with a multiple of itself changes the determinant by the same factor. (And similarly for columns!)

Perhaps the most important theoretical application of determinants is the following fundamental fact:

The matrix $\mathbf{A}$ is invertible if and only if $\det \mathbf{A} \neq 0$.

In fact, more can be said: there is an explicit formula, known as *Cramer's rule*, expressing each element of $\mathbf{A}^{-1}$ in terms of a ratio of determinants:

$$(\mathbf{A}^{-1})_{ij} = \frac{(-1)^{i+j} \Delta_{ji}}{\det \mathbf{A}} \tag{A.23}$$

where the Δ_{ji} is the determinant of an $(n-1)$-dimensional submatrix introduced earlier in connection with the Laplace expansion. However, Cramer's rule is almost never used in practice because of its computational complexity and numerical sensitivity. When a matrix inverse needs to be computed (and this is rarely the case) certain matrix factorization methods described in Section 1.2 are employed. Indeed, factorizations provide the best methods for numerical computation of determinants as well.

There are some important properties of matrix inverses and determinants in connection with matrix and scalar multiplication.

1. If $\mathbf{A}$ is a square matrix of dimension n and z is a complex number, then $\det(z\mathbf{A}) = z^n \det \mathbf{A}$. If, in addition, $\mathbf{A}$ is invertible and $z \neq 0$, then $(z\mathbf{A})$ is invertible and $(z\mathbf{A})^{-1} = z^{-1} \mathbf{A}^{-1}$.

2. If $\mathbf{A}$ and $\mathbf{B}$ are matrices for which both products $\mathbf{AB}$ and $\mathbf{BA}$ are defined, then $\det(\mathbf{AB}) = \det(\mathbf{BA})$. If, in addition, both matrices are square, then

$$\det(\mathbf{AB}) = \det(\mathbf{BA}) = \det \mathbf{A} \det \mathbf{B} = \det \mathbf{B} \det \mathbf{A} \tag{A.24}$$

This is called the *product rule for determinants*.

Chap. 1 Appendix Some Basic Facts About Matrices and Linear Algebra

3. If $\mathbf{A}$ and $\mathbf{B}$ are square, invertible matrices of the same dimension, then their products are invertible also and

$$(\mathbf{AB})^{-1} = \mathbf{B}^{-1}\mathbf{A}^{-1} \tag{A.25}$$

$$(\mathbf{BA})^{-1} = \mathbf{A}^{-1}\mathbf{B}^{-1} \tag{A.26}$$

This may be extended to products of more than two terms in the obvious way, giving the *product rule for matrix inverses*: The inverse of a product of square matrices is the product of their inverses taken in reverse order, provided that the inverses of all the factors exist.

An easy application of these results shows how the determinant of a matrix and the determinant of its inverse are related. If $\mathbf{A}$ is invertible, then

$$\det(\mathbf{A}^{-1}) = \frac{1}{\det \mathbf{A}} \tag{A.27}$$

This follows from the product rule applied to $\mathbf{A}\mathbf{A}^{-1} = \mathbf{I}$ and recalling that $\det \mathbf{I} = 1$.

Matrix Transposition

Another operation on matrices will be very useful in a number of applications, *matrix transposition*. If $\mathbf{A}$ is an $m \times n$ matrix with

$$(\mathbf{A})_{ij} = a_{ij} \tag{A.28}$$

the *transpose* of $\mathbf{A}$, denoted $\mathbf{A}^T$, is the $n \times m$ matrix with

$$(\mathbf{A}^T)_{ij} = a_{ji} \tag{A.29}$$

In words, the transpose of a matrix is formed by interchanging its rows and columns. One important property is that transposition takes column vectors into row vectors, and vice versa. Also notice that if $\mathbf{A}$ is 1×1, then $\mathbf{A}^T = \mathbf{A}$. If a square matrix $\mathbf{A}$ satisfies $\mathbf{A}^T = \mathbf{A}$, it is called a *symmetric matrix*. Symmetry is an important feature of many matrices arising in applications, particularly real matrices.

For matrices whose elements may possibly be complex numbers, a generalization of transposition is often more appropriate. The *Hermitian transpose* of a matrix $\mathbf{A}$, denoted $\mathbf{A}^H$, is defined by

$$(\mathbf{A}^H)_{ij} = a_{ji}^* \tag{A.30}$$

In words, the rows and columns are interchanged and complex conjugation of the elements is performed. The matrix $\mathbf{A}$ is *Hermitian symmetric* if $\mathbf{A}^H = \mathbf{A}$.

With regard to the relationships between transposition and other operations we have already described, we list the following properties:

1. $(\mathbf{A}^T)^T = \mathbf{A}$.
2. $(z\mathbf{A})^T = z\mathbf{A}^T$.
3. $(\mathbf{A}+\mathbf{B})^T = \mathbf{A}^T + \mathbf{B}^T$.

4. $(\mathbf{AB})^T = \mathbf{B}^T\mathbf{A}^T$. Also notice that the products $\mathbf{AA}^T$ and $\mathbf{A}^T\mathbf{A}$ are always defined.
5. If $\mathbf{A}$ is a square matrix, $\det(\mathbf{A}^T) = \det \mathbf{A}$.
6. If $\mathbf{A}$ is an invertible matrix, $\mathbf{A}^T$ is also invertible and $(\mathbf{A}^T)^{-1} = (\mathbf{A}^{-1})^T$.

A similar list of properties holds for Hermitian transposition.

Even for 2×2 matrices, transposition appears to be a much simpler operation than inversion. Therefore, the class of matrices for which $\mathbf{A}^T = \mathbf{A}^{-1}$ is quite remarkable; a matrix whose transpose is also its inverse is known as an *orthogonal* matrix. The reason for this terminology will be explained after the concept of orthogonality of vectors is defined below. A matrix for which $\mathbf{A}^H = \mathbf{A}^{-1}$ is called a *unitary* matrix.

Vector Spaces

Now we turn our attention to vectors and linear algebra. A *vector space* consists of an ordered tuple $(\mathbf{V}, \mathbf{F}, +, \cdot)$ having the following list of attributes:

1. $\mathbf{V}$ is a set of elements called vectors, containing a distinguished vector $\mathbf{0}$, the zero vector.
2. $\mathbf{F}$ is a field of scalars; in this book $\mathbf{F} = \mathbb{R}$ or $\mathbb{C}$, the real or complex numbers. Thus when we refer to a *scalar*, we mean either a real or a complex number, depending on the context.
3. The $+$ operation is a vector addition operation defined on $\mathbf{V}$. For all $\mathbf{v}_1, \mathbf{v}_2, \mathbf{v}_3 \in \mathbf{V}$, the following properties must hold: (a) $\mathbf{v}_1 + \mathbf{v}_2 = \mathbf{v}_2 + \mathbf{v}_1$, (b) $\mathbf{v}_1 + \mathbf{0} = \mathbf{v}_1$, and (c) $(\mathbf{v}_1 + \mathbf{v}_2) + \mathbf{v}_3 = \mathbf{v}_1 + (\mathbf{v}_2 + \mathbf{v}_3)$.
4. The $\cdot$ operation is a scalar multiplication of vectors (and usually the $\cdot$ is not written explicitly). For all $\mathbf{v}_1, \mathbf{v}_2 \in \mathbf{V}$ and $\alpha_1, \alpha_2 \in \mathbf{F}$, the following properties must hold: (a) $0\mathbf{v}_1 = \mathbf{0}$, (b) $1\mathbf{v}_1 = \mathbf{v}_1$, (c) $\alpha_1(\mathbf{v}_1 + \mathbf{v}_2) = \alpha_1\mathbf{v}_1 + \alpha_1\mathbf{v}_2$, (d) $(\alpha_1 + \alpha_2)\mathbf{v}_1 = \alpha_1\mathbf{v}_1 + \alpha_2\mathbf{v}_1$, (e) $\alpha_1(\alpha_2\mathbf{v}_1) = (\alpha_1\alpha_2)\mathbf{v}_1$.

For brevity we might paraphrase these details and say that a vector space is a set of elements that is closed under the operation of taking linear combinations.

If $\mathbf{V}$ is a vector space and $\mathbf{W}$ is a subset of vectors from $\mathbf{V}$, we say that $\mathbf{W}$ is a *subspace* of $\mathbf{V}$ if $\mathbf{W}$ contains $\mathbf{0}$ and is closed under the operations of vector addition and scalar multiplication. Notice that this means that $\mathbf{W}$ is a vector space itself. The set $\mathbf{W} = \{\mathbf{0}\}$ is always a subspace, and we can regard $\mathbf{V}$ as a subspace of itself. If $\mathbf{v}$ is a nonzero vector in a vector space $\mathbf{V}$, then the set $\{\alpha\mathbf{v} \mid \alpha \text{ a scalar}\}$ is a subspace of $\mathbf{V}$. The geometric intuition of subspaces is that they consist of "lines" or "planes" (often called "hyperplanes" in spaces of high dimension) passing through the origin $\mathbf{0}$.

In this appendix and in Chapter 1, we deal mainly with the concrete notion of vectors already introduced, namely m-vectors or $m \times 1$ matrices, where the scalar field is either the set of real numbers or the set of complex numbers. The conventional notation for the set $\mathbf{V}$ consisting of m-vectors of complex numbers is $\mathbb{C}^m$; for m-vectors of real numbers, the notation $\mathbb{R}^m$ is used.

Chap. 1 Appendix   Some Basic Facts About Matrices and Linear Algebra

Examples

A few other examples of vector spaces will be given to suggest that the theory is widely applicable.

1. The set of $m \times n$ matrices, with the usual rules for scalar multiplication and matrix addition, forms a vector space.

2. The set of polynomial functions of a real variable is a vector space because addition of two polynomials produces another polynomial, as does multiplication of a polynomial by a scalar.

3. The set of real-valued continuous functions defined on the closed interval $0 \leq t \leq T$ is a vector space because the sum of two continuous functions is another continuous function and scalar multiplication also preserves continuity.

We next introduce a number of important concepts from *linear algebra* (which is nothing more than the theory of vector spaces).

Linear independence.   A set of m-vectors $\{\mathbf{v}_1, \mathbf{v}_2, \ldots, \mathbf{v}_k\}$ is called *linearly independent* when the equation

$$\sum_{i=1}^{k} \alpha_i \mathbf{v}_i = \mathbf{0} \tag{A.31}$$

is satisfied only by the trivial choice of the scalars: $\alpha_1 = \alpha_2 = \cdots = \alpha_k = 0$. In words, the only linear combination giving the zero vector is the trivial one formed by adding up zero times each of the vectors.

A set of vectors that is not linearly independent is called, naturally, *linearly dependent*. Notice that any set containing $\mathbf{0}$ is linearly dependent.

Spanning set.   If every vector can be written as a linear combination of the vectors from some set, that set is called a *spanning set*.

Basis.   A *basis* for a vector space is any spanning set of linearly independent vectors.

Dimension.   The *dimension* of a vector space is the number of vectors in any basis.

For the case of $\mathbb{C}^m$, the concepts above are considerably simplified. It is clear that no set of more than m m-vectors can be linearly independent, and the set of "unit vectors" is a spanning set of size m. Hence the dimension of $\mathbb{C}^m$ is m, and we might as well define a basis as any set of m linearly independent m-vectors.

Examples

In $\mathbb{C}^2$, consider the following sets of vectors:

$$S_1 = \left\{ \begin{bmatrix} 1 \\ 0 \end{bmatrix} \right\}$$

$$S_2 = \left\{ \begin{bmatrix} 0 \\ 1+j \end{bmatrix} \right\}$$

$$S_3 = \left\{ \begin{bmatrix} 1 \\ 1+j \end{bmatrix}, \begin{bmatrix} 1 \\ 2 \end{bmatrix} \right\}$$

$$S_4 = \left\{ \begin{bmatrix} 1 \\ 1+j \end{bmatrix}, \begin{bmatrix} 1 \\ 1 \end{bmatrix}, \begin{bmatrix} 1 \\ 0 \end{bmatrix} \right\}$$

$$S_5 = \left\{ \begin{bmatrix} 1 \\ 0 \end{bmatrix}, \begin{bmatrix} 0 \\ 1 \end{bmatrix} \right\}$$

$$S_6 = \left\{ \begin{bmatrix} 1 \\ 0 \end{bmatrix}, \begin{bmatrix} 1+j \\ 0 \end{bmatrix} \right\}$$

All of these except S_4 and S_6 are linearly independent sets. Sets S_3, S_4, and S_5 are spanning sets. Sets S_3 and S_5 are bases for $\mathbb{C}^2$ (S_5 is also a basis for $\mathbb{R}^2$).

In the preceding example, S_5 is the *standard basis* of $\mathbb{C}^2$. An analogous set of m "unit vectors" or "principal axis vectors" make up the standard basis for $\mathbb{C}^m$. A commonly used notation for the standard basis is $\{\mathbf{e}_1, \ldots, \mathbf{e}_m\}$, where the ith member of the set, $\mathbf{e}_i$, is the m-vector having ith component 1 and all other components zero.

The notions of linear independence, spanning sets, and basis arise in the study of solutions of systems of linear equations; see Section 1.2. We will now describe one property of particular importance. If S is a linearly independent set of vectors, and if a vector $\mathbf{v}$ can be expressed as a linear combination of the vectors in S, then there is *only one* way of expressing $\mathbf{v}$ as a linear combination of the vectors in S. To show this is an instructive exercise. Suppose that $\mathbf{v}$ can be expressed in more than one way, say

$$\mathbf{v} = \sum_{i=1}^{k} \alpha_i \mathbf{v}_i = \sum_{i=1}^{k} a_i \mathbf{v}_i \tag{A.32}$$

where we assume that S has k elements, $\mathbf{v}_1, \ldots, \mathbf{v}_k$. From these equations we have

$$\mathbf{0} = \sum_{i=1}^{k} (a_i - \alpha_i) \mathbf{v}_i \tag{A.33}$$

Because S is a linearly independent set, we must have all of the coefficients in this linear combination equal to zero, so that $a_i = \alpha_i$ for all i, and the two expressions are exactly the same, as was to be shown. Since a basis is a spanning set of linearly independent vectors, we conclude from this development that every vector in a vector space has a unique representation as a linear combination of the vectors in any basis.

Finding the linear combination of basis vectors to express a given vector requires the solution of a set of linear equations. Indeed, suppose that $\{\mathbf{v}_1, \ldots, \mathbf{v}_m\}$ is a basis and $\mathbf{v}$ is any m-vector. Writing the desired linear combination with coefficients denoted by x_i gives the equation

$$\mathbf{v} = \sum_{i=1}^{m} x_i \mathbf{v}_i = \sum_{i=1}^{m} \mathbf{v}_i x_i \tag{A.34}$$

Chap. 1 Appendix   Some Basic Facts About Matrices and Linear Algebra   53

The purpose of reordering the terms in the sum is to recognize this as a matrix-vector product:

$$\mathbf{v} = \begin{bmatrix} \mathbf{v}_1 & \cdots & \mathbf{v}_m \end{bmatrix} \mathbf{x} \tag{A.35}$$

where $\mathbf{x}$ is the m-vector whose components are the x_i coefficients. In obvious notation, $\mathbf{v} = \mathbf{V}\mathbf{x}$, which is the matrix form of m linear equations in the m unknown components of $\mathbf{x}$. Because the columns of $\mathbf{V}$ form a basis, and hence span $\mathbb{C}^m$, a solution may always be found. By our previous argument, the solution is unique because the columns of $\mathbf{V}$ are also linearly independent. Indeed, we have essentially proved the important fact that a square matrix $\mathbf{A}$ of dimension m has a matrix inverse if and only if its columns form a basis for $\mathbb{C}^m$.

Inner Product and Norm

Vectors in $\mathbb{R}^2$ and $\mathbb{R}^3$ may be viewed geometrically as points in two- and three-dimensional Euclidean space, respectively. Geometric intuition may be extended to higher-dimensional vector spaces $\mathbb{R}^m$, $m > 3$ (and to vector spaces over the complex numbers such as $\mathbb{C}^m$). Points may in turn be associated with directed line segments emanating from the origin; the origin is the point associated with $\mathbf{0}$. This leads to two familiar notions: vector length and the angle between two vectors. Both of these are conveniently expressed for the two- and three-dimensional cases with the dot product for vectors. Extending this idea to any dimension, we define the *inner product* of two m-vectors $\mathbf{x}$ and $\mathbf{y}$ in $\mathbb{R}^m$ as $\mathbf{x}^T\mathbf{y}$. Writing this out in terms of the components of the vectors:

$$\mathbf{x} = \begin{bmatrix} x_1 \\ x_2 \\ \vdots \\ x_m \end{bmatrix}, \quad \mathbf{y} = \begin{bmatrix} y_1 \\ y_2 \\ \vdots \\ y_m \end{bmatrix} \tag{A.36}$$

$$\mathbf{x}^T\mathbf{y} = \sum_{i=1}^{m} x_i y_i \tag{A.37}$$

from which we see that $\mathbf{x}^T\mathbf{y} = \mathbf{y}^T\mathbf{x}$. (This can also be seen simply by noting that $\mathbf{x}^T\mathbf{y}$ is a 1×1 matrix and hence equal to its own transpose, which is $\mathbf{y}^T\mathbf{x}$.)

To generalize the geometric properties holding in the two- and three-dimensional cases, we define the *Euclidean norm* or *length* of an m-vector to be the square root of the sum of squares of its components. Using the symbol $\|\mathbf{x}\|$ to denote the length of $\mathbf{x}$, we have

$$\|\mathbf{x}\| = (\mathbf{x}^T\mathbf{x})^{1/2} \tag{A.38}$$

the square root of the inner product of $\mathbf{x}$ with itself. The Euclidean norm satisfies three important properties:

54 Some Basic Facts About Matrices and Linear Algebra   Chap. 1 Appendix

N1. $\|\mathbf{x}\| \geq 0$ with equality holding only for $\mathbf{x} = \mathbf{0}$.
N2. $\|\alpha \mathbf{x}\| = |\alpha| \|\mathbf{x}\|$, for any scalar α.
N3. *(Triangle inequality)* $\|\mathbf{x} + \mathbf{y}\| \leq \|\mathbf{x}\| + \|\mathbf{y}\|$.

Other functions may be used to define different norms, although they generally lack some of the nice geometric interpretations that go along with the inner product function (e.g., the Schwarz inequality given below). To provide a valid definition of a norm, a function must satisfy the three properties listed above. One alternative norm for m-vectors is the *uniform norm*, which will be denoted by $\|\mathbf{x}\|_\infty$ to distinguish it from the Euclidean (or any other) norm:

$$\|\mathbf{x}\|_\infty = \max_{1 \leq i \leq m} |x_i| \tag{A.39}$$

In words, $\|\mathbf{x}\|_\infty$ is the largest (in magnitude) of the elements of the vector $\mathbf{x}$. An entire family of norms for m-vectors, the p-norms, is defined for real numbers $1 \leq p < \infty$ by

$$\|\mathbf{x}\|_p = \left(\sum_{1}^{p} |x_i|^p \right)^{1/p} \tag{A.40}$$

Notice that the Euclidean norm is the p-norm for $p = 2$.

Example

It is interesting to compare graphical representations of various norms. Figure 1.6 depicts the "unit circle" in dimension 2 for the Euclidean and two non-Euclidean norms, $\|\mathbf{x}\|$ (which is also $\|\mathbf{x}\|_2$), $\|\mathbf{x}\|_1$, and $\|\mathbf{x}\|_\infty$; the figure shows the loci of points defined by $\|\mathbf{x}\| = 1$, $\|\mathbf{x}\|_1 = 1$, and $\|\mathbf{x}\|_\infty = 1$.

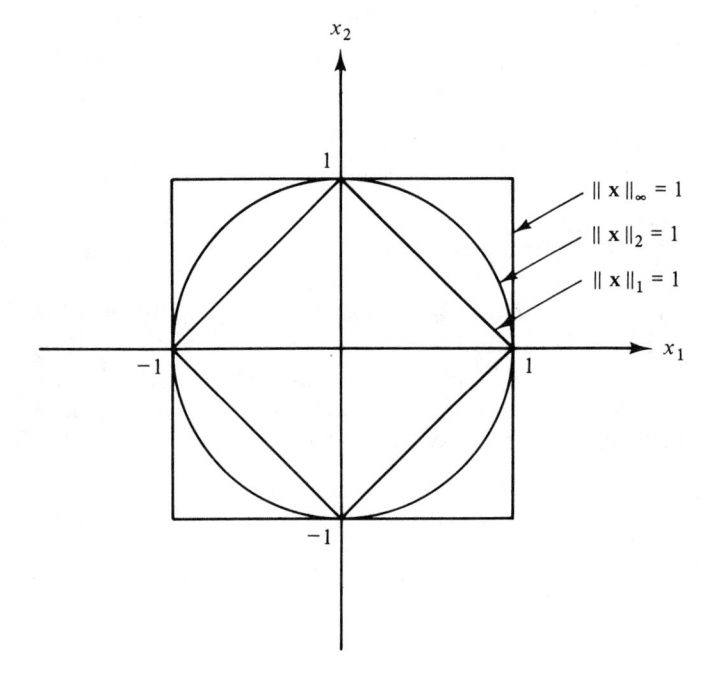

Figure 1.6 Loci of points having unit norm.

Various norms turn out to be appropriate for applications involving limits of sequences of vectors in other vector spaces; this topic arises in a few places in this book, including Sections 1.3 and 4.1. A vector space having an associated norm is called a *normed vector space*. As an example, a suitable norm for the space of real-valued continuous functions on the interval $0 \leq t \leq T$ is the uniform norm:

$$\|g(t)\|_\infty = \max_{0 \leq t \leq T} |g(t)| \tag{A.41}$$

The name and notation are the same as used for the uniform norm for m-vectors since the analogy is apparent. A notion of p-norm for vector spaces of functions can also be established. The 2-norm is the natural generalization of Euclidean norm:

$$\|g(t)\|_2 = \left(\int_0^T |g(t)|^2 \, dt\right)^{1/2} \tag{A.42}$$

For matrices, especially for square matrices, the notion of a *matrix norm* has important applications. The *Frobenius norm* of an $m \times n$ matrix, denoted $\|\mathbf{A}\|_F$, is simply the Euclidean norm of the nm-vector consisting of all elements of $\mathbf{A}$:

$$\|\mathbf{A}\|_F = \left(\sum_{i=1}^m \sum_{j=1}^n a_{ij}^2\right)^{1/2} \tag{A.43}$$

The *induced Euclidean norm* of $\mathbf{A}$ is quite useful for characterizing how "big" the effects of matrix multiplication by $\mathbf{A}$ are. It is defined in terms of the Euclidean norms of two vectors, $\mathbf{x}$ and $\mathbf{A}\mathbf{x}$:

$$\|\mathbf{A}\| = \max_{\|\mathbf{x}\|=1} \|\mathbf{A}\mathbf{x}\| \tag{A.44}$$

(This form may be used with any vector norm to define a corresponding induced matrix norm.) A consequence of this definition is the inequality

$$\|\mathbf{A}\mathbf{x}\| \leq \|\mathbf{A}\| \, \|\mathbf{x}\| \tag{A.45}$$

which holds for all vectors $\mathbf{x}$. This inequality also implies the following inequality for the induced norm of a matrix product:

$$\|\mathbf{A}\mathbf{B}\| \leq \|\mathbf{A}\| \, \|\mathbf{B}\| \tag{A.46}$$

Example

Explicit expressions for three of the most important induced matrix norms can be determined. Suppose that $\mathbf{A}$ is an $n \times n$ matrix with elements a_{ij}. For the induced Euclidean norm

$$\|\mathbf{A}\| = (\lambda_{\max}(\mathbf{A}^T \mathbf{A}))^{1/2}$$

In words, the induced Euclidean norm of $\mathbf{A}$ is the square root of the largest eigenvalue of the matrix $\mathbf{A}^T \mathbf{A}$. For the induced uniform norm

$$\|\mathbf{A}\|_\infty = \max_{1 \leq i \leq n} \sum_{j=1}^n |a_{ij}|$$

which is the largest of the absolute row-sums of $\mathbf{A}$. For the induced 1-norm

$$\|\mathbf{A}\|_1 = \max_{1 \le j \le n} \sum_{i=1}^{n} |a_{ij}|$$

which is the largest of the absolute column-sums of $\mathbf{A}$.

An important use of the induced Euclidean norm of a matrix arises in sensitivity analysis for matrix products. For a square, invertible matrix $\mathbf{A}$, consider the linear equations $\mathbf{A}\mathbf{x} = \mathbf{y}$, and suppose that because of numerical inaccuracies such as roundoff error, or possibly because of measurement errors, the vector $\mathbf{y}$ is perturbed to become $\mathbf{y} + \Delta\mathbf{y}$. Then $\mathbf{A}(\mathbf{x} + \Delta\mathbf{x}) = \mathbf{y} + \Delta\mathbf{y}$, where $\Delta\mathbf{x} = \mathbf{A}^{-1}\Delta\mathbf{y}$. Using norms to quantify the relative error in $\mathbf{x}$ arising from the relative error in $\mathbf{y}$, we have

$$\frac{\|\Delta\mathbf{x}\|}{\|\mathbf{x}\|} = \frac{\|\mathbf{A}^{-1}\Delta\mathbf{y}\|}{\|\mathbf{A}^{-1}\mathbf{y}\|} \le \frac{\|\mathbf{A}^{-1}\|\,\|\Delta\mathbf{y}\|}{\|\mathbf{A}^{-1}\mathbf{y}\|}$$

$$= \frac{\|\mathbf{A}^{-1}\|\,\|\Delta\mathbf{y}\|}{\|\mathbf{A}^{-1}\mathbf{y}\|}\frac{\|\mathbf{A}\|}{\|\mathbf{A}\|}$$

$$\le \|\mathbf{A}\|\frac{\|\mathbf{A}^{-1}\|\,\|\Delta\mathbf{y}\|}{\|\mathbf{A}\mathbf{A}^{-1}\mathbf{y}\|}$$

$$= \kappa(\mathbf{A})\frac{\|\Delta\mathbf{y}\|}{\|\mathbf{y}\|} \tag{A.47}$$

where $\kappa(\mathbf{A})$ denotes the *condition number* of $\mathbf{A}$, defined as

$$\kappa(\mathbf{A}) = \|\mathbf{A}\|\,\|\mathbf{A}^{-1}\| \tag{A.48}$$

Thus the condition number provides an indication of how relative errors in $\mathbf{y}$ scale to relative errors in $\mathbf{x}$. Since

$$\kappa(\mathbf{A}) = \|\mathbf{A}\|\,\|\mathbf{A}^{-1}\| \ge \|\mathbf{A}\mathbf{A}^{-1}\| = 1 \tag{A.49}$$

we can say that when $\kappa(\mathbf{A}) \approx 1$, the matrix $\mathbf{A}$ is well-conditioned, but when $\kappa(\mathbf{A}) \gg 1$, the matrix $\mathbf{A}$ is ill-conditioned. A similar argument shows that the condition number of $\mathbf{A}$ also serves as the multiplier scaling relative errors in $\mathbf{A}$ to relative errors in $\mathbf{x}$.

Angles and Orthogonality

Generalizing the relation holding between angles and dot products in two- and three-dimensional space, the quantity

$$\cos\theta_{\mathbf{xy}} = \frac{\mathbf{x}^T\mathbf{y}}{\|\mathbf{x}\|\,\|\mathbf{y}\|} = \frac{\mathbf{x}^T\mathbf{y}}{((\mathbf{x}^T\mathbf{x})(\mathbf{y}^T\mathbf{y}))^{1/2}} \tag{A.50}$$

represents the cosine of the angle between (the nonzero vectors) $\mathbf{x}$ and $\mathbf{y}$ (Fig. 1.7). The fact that the right side of the equation always takes a value between -1 and 1 is the result of the *Schwarz inequality*:

$$|\mathbf{x}^T\mathbf{y}| \le \|\mathbf{x}\|\,\|\mathbf{y}\| \tag{A.51}$$

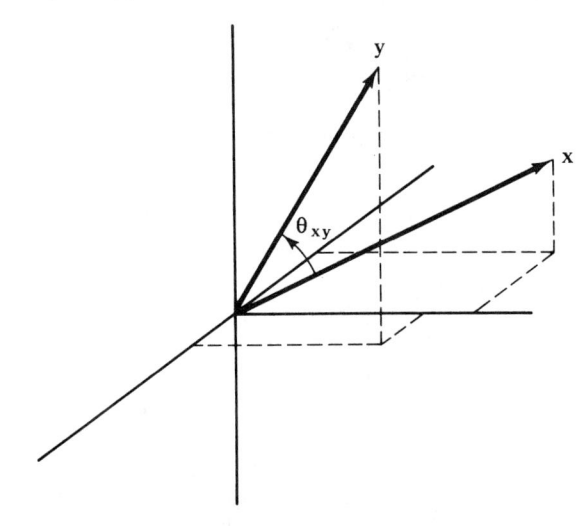

Figure 1.7 Angle between vectors **x** and **y**.

The most important special cases of the angle formula are given by the conditions for the cosine to be −1, 0, and 1. As expected, if $\mathbf{x} = \alpha \mathbf{y}$ (collinear vectors), then the angle between the two vectors is 0 or π radians, depending on whether the vectors point in the same direction ($\alpha > 0$) or in opposite directions ($\alpha < 0$). The condition for the cosine to be 0 is that the angle between the vectors be $\pm \pi/2$, a right angle. This occurs when $\mathbf{x}^T \mathbf{y} = 0$. Thus we say that two vectors are *orthogonal* (or perpendicular) if their inner product is zero.

A set of mutually orthogonal vectors is known as an orthogonal set, and a basis consisting of mutually orthogonal vectors is known as an orthogonal basis. An orthogonal basis consisting of vectors whose norms are all 1 (i.e., consisting of vectors having unit length) is called an orthonormal basis. There is an important constructive procedure for obtaining an orthonormal basis starting from an arbitrary basis, the *Gram-Schmidt procedure*. Starting with a basis $\{\mathbf{v}_1, \mathbf{v}_2, \ldots, \mathbf{v}_k\}$, the orthonormal basis $\{\mathbf{w}_1, \mathbf{w}_2, \ldots, \mathbf{w}_k\}$ is constructed sequentially according to the following steps:

1. $\mathbf{w}_1 = \mathbf{v}_1 / \|\mathbf{v}_1\|$.
2. For $2 \leq i \leq k$, $\mathbf{w}_i = \mathbf{z}_i / \|\mathbf{z}_i\|$, where $\mathbf{z}_i = \mathbf{v}_i - \sum_{j=1}^{i-1} (\mathbf{v}_i^T \mathbf{w}_j) \mathbf{w}_j$.

Example

We give an example of the Gram-Schmidt procedure to illustrate the computations involved. Let $\{\mathbf{v}_1, \mathbf{v}_2, \mathbf{v}_3\}$ be the original basis:

$$\{\mathbf{v}_1, \mathbf{v}_2, \mathbf{v}_3\} = \left\{ \begin{bmatrix} 1 \\ 0 \\ 1 \end{bmatrix}, \begin{bmatrix} 0 \\ 1 \\ -1 \end{bmatrix}, \begin{bmatrix} 2 \\ 1 \\ 0 \end{bmatrix} \right\}$$

In the first step, $\mathbf{w}_1 = \mathbf{v}_1 / \|\mathbf{v}_1\|$, and $\|\mathbf{v}_1\| = \sqrt{2}$, giving

$$\mathbf{w}_1 = \begin{bmatrix} \sqrt{2}/2 \\ 0 \\ \sqrt{2}/2 \end{bmatrix}$$

Next we compute the inner product $\mathbf{v}_2^T\mathbf{w}_1 = -\sqrt{2}/2$. Thus we find

$$\mathbf{z}_2 = \mathbf{v}_2 + \frac{\sqrt{2}}{2}\mathbf{w}_1 = \begin{bmatrix} 1/2 \\ 1 \\ -1/2 \end{bmatrix}$$

Since $\|\mathbf{z}_2\| = \sqrt{6}/2$ we obtain

$$\mathbf{w}_2 = \frac{\mathbf{z}_2}{\|\mathbf{z}_2\|} = \begin{bmatrix} \sqrt{6}/6 \\ \sqrt{6}/3 \\ -\sqrt{6}/6 \end{bmatrix}$$

Finally, we compute the inner products $\mathbf{v}_3^T\mathbf{w}_1 = \sqrt{2}$ and $\mathbf{v}_3^T\mathbf{w}_2 = 2\sqrt{6}/3$, so that

$$\mathbf{z}_3 = \mathbf{v}_3 - \sqrt{2}\mathbf{w}_1 - \frac{2\sqrt{6}}{3}\mathbf{w}_2 = \begin{bmatrix} 1/3 \\ -1/3 \\ -1/3 \end{bmatrix}$$

Since $\|\mathbf{z}_3\| = \sqrt{3}/3$ we obtain

$$\mathbf{w}_3 = \frac{\mathbf{z}_3}{\|\mathbf{z}_3\|} = \begin{bmatrix} \sqrt{3}/3 \\ -\sqrt{3}/3 \\ -\sqrt{3}/3 \end{bmatrix}$$

Why is an orthonormal basis of interest? One important property is that expressing a vector as a linear combination of orthonormal basis vectors does not require the solution of coupled linear equations for the coefficients. To see this, suppose that $\{\mathbf{w}_1, \ldots, \mathbf{w}_k\}$ is an orthonormal basis, and that the vector $\mathbf{v}$ is given by

$$\mathbf{v} = \sum_{i=1}^{k} \alpha_i \mathbf{w}_i \tag{A.52}$$

To determine the coefficients in this linear combination, premultiply the equation for $\mathbf{v}$ by each of the $\mathbf{w}_i^T$ and use the orthogonality and unit norm properties of these basis vectors to obtain

$$\mathbf{w}_i^T \mathbf{v} = \alpha_i \mathbf{w}_i^T \mathbf{w}_i = \alpha_i \tag{A.53}$$

Thus the coefficients are the inner products of the orthonormal basis vectors with the $\mathbf{v}$ vector.

The columns of an orthogonal matrix, a square matrix $\mathbf{O}$ satisfying $\mathbf{O}^{-1} = \mathbf{O}^T$, comprise an orthonormal basis. This follows easily by writing the expression for the (i,j)th entry of the product $\mathbf{O}^T\mathbf{O} = \mathbf{I}$:

$$\mathbf{o}_i^T \mathbf{o}_j = \begin{cases} 1 & \text{for } i = j \\ 0 & \text{for } i \neq j \end{cases} \tag{A.54}$$

where the lowercase subscripted symbols denote the columns of $\mathbf{O}$. Another important fact about orthogonal matrices is that multiplying a vector by an orthogonal matrix preserves the vector's length. This follows from the calculation

$$\|\mathbf{Ox}\|^2 = (\mathbf{Ox})^T(\mathbf{Ox}) = \mathbf{x}^T\mathbf{O}^T\mathbf{Ox}$$

Then, since $\mathbf{O}^T = \mathbf{O}^{-1}$,

$$\|\mathbf{Ox}\|^2 = \mathbf{x}^T\mathbf{O}^{-1}\mathbf{Ox} = \mathbf{x}^T\mathbf{x} = \|\mathbf{x}\|^2 \tag{A.55}$$

Linear Functions

A great deal of the theory of linear algebra concerns the relationships between vector spaces and linear functions. Some of the results of particular importance for this book will be introduced here.

First we define a *linear function* (sometimes called a linear transformation, linear operator, or linear mapping), in the concrete context of column vectors. A function f mapping the vector space $\mathbb{C}^n$ into the vector space $\mathbb{C}^m$ is simply a rule that assigns to each n-vector (each element of the domain of the function) an m-vector (i.e., the image of the function is contained in $\mathbb{C}^m$). Such a function is *linear* if

$$f(\alpha_1\mathbf{v}_1 + \alpha_2\mathbf{v}_2) = \alpha_1 f(\mathbf{v}_1) + \alpha_2 f(\mathbf{v}_2) \tag{A.56}$$

for every choice of $\mathbf{v}_1$ and $\mathbf{v}_2$ in $\mathbb{C}^n$ and for every choice of complex numbers α_1 and α_2. (The analogous property for the real case is obvious.)

There is a one-to-one correspondence between linear functions from $\mathbb{C}^n$ to $\mathbb{C}^m$ and $m \times n$ matrices. For the linear function f, let $\mathbf{A}_f$ be an $m \times n$ matrix whose ith column is $f(\mathbf{e}_i)$, the value of the function for the ith unit vector, for $1 \leq i \leq n$. Then the value of the function for any vector in $\mathbb{C}^n$ can be obtained by matrix multiplication:

$$f(\mathbf{v}) = \mathbf{A}_f\mathbf{v} \tag{A.57}$$

Thus we have an even stronger statement:

> **Every linear function from $\mathbb{C}^n \to \mathbb{C}^m$ is equivalent to multiplication by a (uniquely determined) $m \times n$ matrix. As a special case, every linear function from $\mathbb{R}^n \to \mathbb{R}$ is equivalent to taking the inner product with a uniquely determined real m-vector.**

This statement provides an interpretation of a matrix as the representation of a linear function with respect to the standard bases for the domain and image spaces. If we consider the linear function h mapping $\mathbb{C}^n \to \mathbb{C}^p$ defined by the $f(\mathbf{v}) = g(f(\mathbf{v}))$, the composition of two other linear functions, f mapping $\mathbb{C}^n \to \mathbb{C}^m$ and g mapping $\mathbb{C}^m \to \mathbb{C}^p$, it is a straightforward exercise to show that

$$h(\mathbf{v}) = \mathbf{A}_h\mathbf{v} = \mathbf{A}_g\mathbf{A}_f\mathbf{v} \tag{A.58}$$

In words, matrix multiplication corresponds to composition of the associated linear functions. Indeed, the formula for matrix multiplication was chosen precisely to obtain this correspondence with composition of linear functions.

There is an invertible linear function that relates any two choices of basis of a vector space. If $\{x_1, \ldots, x_m\}$ and $\{z_1, \ldots, z_m\}$ are two bases for $\mathbb{C}^m$, let $\mathbf{T_x}$ be the $m \times m$ matrix whose columns are the $\mathbf{x}_i$ basis vectors (in natural order) and similarly for $\mathbf{T_z}$. Both $\mathbf{T_x}$ and $\mathbf{T_z}$ are invertible because each has linearly independent columns. Now since

$$\begin{bmatrix} z_1 & \cdots & z_m \end{bmatrix} = \mathbf{T_z} = (\mathbf{T_z T_x^{-1}}) \mathbf{T_x} = (\mathbf{T_z T_x^{-1}}) \begin{bmatrix} x_1 & \cdots & x_m \end{bmatrix} \quad (A.59)$$

we see that each of the $\mathbf{z}_i$ basis vectors is obtained by multiplying the matrix $(\mathbf{T_z T_x^{-1}})$ times the corresponding $\mathbf{x}_i$ basis vector. Furthermore, the matrix is invertible, with inverse $(\mathbf{T_x T_z^{-1}})$. By the correspondence between matrix multiplication and linear functions, we see that, as claimed, an invertible linear function relates any two choices of basis.

Associated with every linear function f mapping $\mathbb{C}^n \to \mathbb{C}^m$ are two subspaces of particular importance. The *nullspace* or *kernel* of f is the subspace $\{\mathbf{x}|f(\mathbf{x}) = \mathbf{0}\}$, which is a subspace of $\mathbb{C}^n$. In words, the nullspace of f is the set of vectors in $\mathbb{C}^n$ that get mapped to the zero vector (the zero vector in $\mathbb{C}^m$). The *range* or *image* of f is the subspace $\{\mathbf{y}|\mathbf{y} = f(\mathbf{x}) \text{ for some } \mathbf{x}\}$, which is a subspace of $\mathbb{C}^m$. In words, the range of f is the set of all vectors in $\mathbb{C}^m$ that are obtained by applying f to all of the vectors $\mathbf{x}$ in $\mathbb{C}^n$. We often refer directly to the kernel, nullspace, range, or image of a matrix $\mathbf{A}$, in which case we are relying on the natural correspondence between a matrix and the linear function $f(\mathbf{x}) = \mathbf{Ax}$. The term *column space* of a matrix is sometimes used in place of range or image for obvious reasons: The range of a matrix $\mathbf{A}$ is the subspace of all linear combinations of the columns of $\mathbf{A}$.

1.4 NOTES AND REFERENCES

Matrices and linear algebra are covered in a wide variety of books, so we will mention only a few particular favorites. Strang's book [7] offers a concrete approach to all of the basic material. The book by Golub and Van Loan [3] is a good source for information about computational aspects. For a thorough coverage of parallel and distributed computational algorithms, see Bertsekas and Tsitsiklis [1]; special-purpose computer architectures for many matrix computations are described in Kung [6]. The book by Hirsch and Smale [4] includes material having particular relevance to differential equations. Many books on linear systems include coverage of some of the topics found in this chapter; noteworthy are the first chapter of Delchamps' book [2] and the comprehensive appendix on matrices in Kailath's book [5].

BIBLIOGRAPHY

[1] D.P. Bertsekas and J.N. Tsitsiklis, *Parallel and Distributed Computation: Numerical Methods*, Prentice-Hall, Englewood Cliffs, NJ, 1989.

[2] D.F. Delchamps, *State Space and Input-Output Linear Systems*, Springer-Verlag, New York, 1988.

[3] G.H. Golub and C.F. Van Loan, *Matrix Computations*, 2nd ed., The Johns Hopkins University Press, Baltimore, MD, 1989.
[4] M.W. Hirsch and S. Smale, *Differential Equations, Dynamical Systems, and Linear Algebra*, Academic Press, New York, 1974.
[5] T. Kailath, *Linear Systems*, Prentice-Hall, Englewood Cliffs, NJ, 1980.
[6] S.Y. Kung, *VLSI Array Processors*, Prentice-Hall, Englewood Cliffs, NJ, 1988.
[7] G. Strang, *Linear Algebra and Its Applications*, Academic Press, New York, 1976.

PROBLEMS

1. Show that the matrix $\mathbf{A}$ in equation (1.7) does not have two linearly independent eigenvectors.
2. Suppose that $\mathbf{L}$ is an $n \times n$ lower triangular matrix. (See the examples following equation (A.19).) Find the eigenvalues of $\mathbf{L}$ and describe the form of its eigenvectors.
3. Verify the formula in equation (1.25).
4. Find the eigenvalues of the matrix
$$\mathbf{A} = \begin{bmatrix} 13 & 5 \\ 5 & 2 \end{bmatrix}$$
5. Find a matrix with real entries whose eigenvalues are $\lambda_1 = 3$, $\lambda_2 = 2+j$, $\lambda_3 = 2-j$.
6. Show that all eigenvalues of an orthogonal matrix have unit magnitude.
7. Show that if λ is a zero of the characteristic polynomial of $\mathbf{A}$ having multiplicity 1, then $\mathbf{A}\mathbf{u}_1 = \lambda \mathbf{u}_1$ and $\mathbf{A}\mathbf{u}_2 = \lambda \mathbf{u}_2$, for nonzero $\mathbf{u}_1$ and $\mathbf{u}_2$, implies that $\mathbf{u}_2 = c\mathbf{u}_1$.
8. Use the Gram matrix of the vectors
$$\begin{bmatrix} 1 \\ 0 \\ 1 \end{bmatrix}, \begin{bmatrix} 1 \\ 1 \\ 0 \end{bmatrix}, \begin{bmatrix} 2 \\ 1 \\ 1 \end{bmatrix}$$
to determine if the vectors are linearly independent or not.
9. Suppose that the $n \times n$ matrices $\mathbf{X}$, $\mathbf{Y}$, and $\mathbf{X}+\mathbf{Y}$ are all invertible. Show that
$$(\mathbf{X}^{-1}+\mathbf{Y}^{-1})^{-1} = \mathbf{X}(\mathbf{X}+\mathbf{Y})^{-1}\mathbf{Y}$$
(Notice that the roles of $\mathbf{X}$ and $\mathbf{Y}$ are interchangeable.)
10. Verify the following matrix identity, assuming that the matrices have suitable dimensions and that the indicated inverses all exist:
$$(\mathbf{A}+\mathbf{BPC})^{-1} = \mathbf{A}^{-1} - \mathbf{A}^{-1}\mathbf{B}(\mathbf{P}^{-1}+\mathbf{CA}^{-1}\mathbf{B})^{-1}\mathbf{CA}^{-1}$$
Simplify for the special case when $\mathbf{P} = 1$, $\mathbf{B}$ is a column vector, and $\mathbf{C}$ is a row vector, and show how the solutions of $\mathbf{A}\mathbf{x} = \mathbf{y}$ and $(\mathbf{A}+\mathbf{BPC})\mathbf{x} = \mathbf{y}$ are related in this special case.
11. Show that the following system of n linear equations in n unknowns over the complex numbers,
$$(\mathbf{A}_1 + j\mathbf{A}_2)(\mathbf{x} + j\mathbf{y}) = \mathbf{u} + j\mathbf{v}$$

can be written in an equivalent form as a system of $2n$ linear equations in $2n$ unknowns over the real numbers. (*Hint:* Separate the real and imaginary parts of the "knowns" and "unknowns.")

12. Suppose that $\mathbf{A}$ is an $n \times n$ matrix, not necessarily invertible, which can be diagonalized by a similarity transformation. Suppose that $\mathbf{y}$ is selected so that the linear equations $\mathbf{Ax} = \mathbf{y}$ are consistent. Show that

$$\lim_{\varepsilon \to 0}(\mathbf{A}+\varepsilon\mathbf{I})^{-1}\mathbf{y}$$

exists and that the limit is a solution to $\mathbf{Ax} = \mathbf{y}$.

13. Formulate the problem of fitting a quadratic function to a set of ordered pairs $\{(x_1,y_1), \ldots, (x_N,y_N)\}$ in terms of solving an inconsistent system of linear equations.

14. Express the function $2x_1^2 + 10x_1x_2 + 13x_2^2$ as a sum of squares.

15. Find the QR factorization of the matrix $\mathbf{A}$ in Problem 4.

16. Verify the four properties of the Householder transformation $\mathbf{Q}_w$ mentioned in the example following equation (1.64).

17. Let $\mathbf{A} = \mathbf{QR}$ be the QR factorization of a square matrix $\mathbf{A}$. Show that the matrix $\mathbf{RQ}$ is similar to $\mathbf{A}$.

18. Investigate the changes suggested at the end of the baseball problem given in the example following equation (1.79).

19. Consider the Jacobi iteration method for solving linear equations when $\mathbf{A}$ is the 2×2 matrix

$$\mathbf{A} = \begin{bmatrix} a_{11} & a_{12} \\ a_{21} & a_{22} \end{bmatrix}$$

Find the eigenvalues of $\mathbf{A}_1^{-1}\mathbf{A}_2$ and find the conditions for convergence of this iterative solution method. Does the method always converge if $\mathbf{A}$ is symmetric and has two real positive eigenvalues?

20. Let $p(x)$ be a polynomial function and let $\mathbf{X}(t)$ be a square matrix of differentiable time functions. Then $p(\mathbf{X}(t))$ is a well-defined matrix of time functions. Find an expression for the time derivative of $p(\mathbf{X}(t))$. Simplify the expression for the case when $\mathbf{X}(t) = \mathbf{A}t$, where $\mathbf{A}$ is a constant matrix.

21. Let $\mathbf{x}(t)$ be an n-vector of differentiable time functions and let $\mathbf{Q}$ be an $n \times n$ symmetric matrix. Show that

$$\frac{d}{dt}\mathbf{x}^T(t)\mathbf{Q}\mathbf{x}(t) = 2\mathbf{x}^T(t)\mathbf{Q}\dot{\mathbf{x}}(t)$$

where the dot denotes time derivative.

22. Let $\mathbf{P}_3$ denote the vector space of polynomials in the variable λ having degree not larger than 3, with real coefficients. Thus a typical vector in $\mathbf{P}_3$ takes the form

$$\mathbf{v} = a_0\lambda^3 + a_1\lambda^2 + a_2\lambda + a_3$$

Show that the vectors $\mathbf{v}_1 = 1$, $\mathbf{v}_2 = \lambda$, $\mathbf{v}_3 = \lambda^2$, and $\mathbf{v}_4 = \lambda^3$ form a basis.

23. Let $\mathbf{Q}$ be an $n \times n$ real symmetric matrix. Show that any n-vector $\mathbf{x}$ can be written uniquely as $\mathbf{x} = \mathbf{y}+\mathbf{z}$, where $\mathbf{y}$ is in the nullspace of $\mathbf{Q}$ and $\mathbf{z}$ is in the range space of $\mathbf{Q}$.

CHAPTER 2

Linear Systems

Differential equations provide an important class of mathematical models for studies of the dynamic behavior of systems evolving in continuous time. The special case of linear differential equations deserves particular attention because of its tremendous importance for applications, both in the many situations where it provides an appropriate model for real system behavior, and in cases where it serves as an essential tool used in approximating genuinely nonlinear systems for purposes such as quantitative and qualitative analysis of behavior. Thanks to the analytical and computational tractability of many aspects of linear differential equations, the study of linear differential equation models, the main focus of this chapter on *linear systems*, is readily accessible to undergraduate engineers; not surprisingly, linear algebra and matrices will play important roles throughout the chapter.

For discrete-time systems, difference equation models provide the mathematical framework for modeling dynamic behavior. Since discretization is a principal concern for a variety of computer-related areas of system analysis and design, such as simulation, signal processing, and digital control, there is considerable practical motivation for studying many aspects of discrete-time systems. A quick overview of the topic of linear discrete-time dynamic systems is presented in the last section of this chapter.

The common mathematical framework that is adopted for modeling of continuous-time linear systems involves the following set of differential equations with initial conditions:

$$\begin{bmatrix} \dot{x}_1(t) \\ \dot{x}_2(t) \\ \cdot \\ \cdot \\ \cdot \\ \dot{x}_n(t) \end{bmatrix} = \dot{\mathbf{x}}(t) = \mathbf{A}\mathbf{x}(t) + \mathbf{B}\mathbf{u}(t); \qquad \mathbf{x}(0) = \mathbf{x}_0$$

We usually think of t, the independent variable, as representing time. We call $\mathbf{x}(t)$ the *state vector* of the system; it is an n-vector of time functions that "solve" this differential equation and satisfy the initial conditions. $\mathbf{u}(t)$ is an m-vector of (known) forcing terms (i.e., *inputs*), and the matrices $\mathbf{A}$ and $\mathbf{B}$ are $n \times n$ and $n \times m$, respectively.

Strictly speaking, what we have given is the description of a *finite-dimensional, time-invariant linear system*, where "finite-dimensional" refers to the fact that the state vector $\mathbf{x}(t)$ has $n < \infty$ components, and "time-invariant" refers to the fact that $\mathbf{A}$ and $\mathbf{B}$ are matrices of constants rather than more general functions of t. Both kinds of generalization can be made, and a few examples will be found later in the book. In this chapter, and in most places in this book, we will deal with finite-dimensional, time-invariant linear systems. We will usually omit the qualifying terms for the sake of convenience, relying on the notation to prevent any confusion.

Examples

To provide some explicit motivation for the linear system model just introduced, we will provide a range of examples of physical systems that are commonly described in exactly this form.

Electric circuits composed of interconnected resistors, capacitors, inductors, and sources are one important area of application of linear systems. They also provide an important illustration of another concept of fundamental importance for mathematical modeling of complex systems: The differential equations describing a large network are obtained by a systematic application of a basic set of interconnection relations that govern how the variables associated with the elementary circuit elements are related. Thus the simple differential equations that describe the circuit elements may be combined in a straightforward way to obtain the overall circuit equations. Breaking down a complex system into a collection of simpler, interacting subsystems is a key idea for mathematical modeling of dynamic systems that we will put to use in a wide variety of examples throughout this book.

Before providing some "real" examples, we review the equations that govern the behavior of the elements that are used in electric circuits. For *two-terminal devices* like those to be described here, the physical variables used to describe the elements are the *voltage* across the device—choosing one terminal arbitrarily as a reference—and the *current* through the device—taking positive current to denote current flowing toward the reference terminal. In all discussions of electric circuits we will use the symbols v and i to denote voltage and current variables, respectively; these variables are functions of time whose dynamic behavior is governed by differential equations that form the class of mathematical models for circuit behavior.

Now we list the current-voltage relationships that characterize five important circuit elements (Fig. 2.1).

Resistor. The voltage across a resistor is proportional to the current through it:

$$v = R\,i$$

For v in volts and i in amperes, R is the resistance in ohms. The equation $v = R\,i$ is commonly known as *Ohm's law*.

Capacitor. The current through a capacitor is proportional to the time derivative of voltage across it:

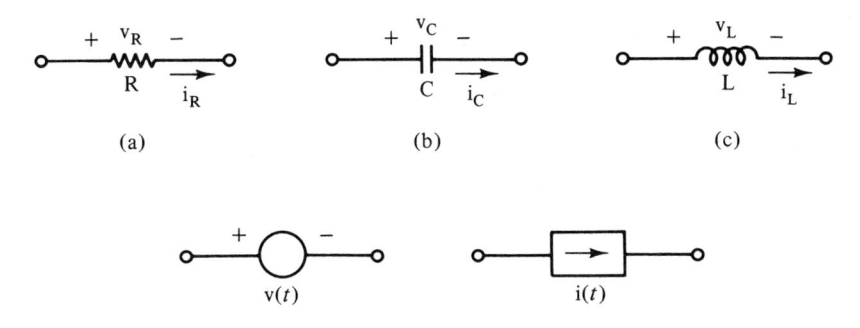

Figure 2.1 Circuit elements: (a) resistor; (b) capacitor; (c) inductor; (d) voltage source; (e) current source.

$$i = C\frac{dv}{dt}$$

The capacitance, C, is commonly given in farads, with 1 farad being 1 ampere-second per volt.

Inductor. The voltage across an inductor is proportional to the time derivative of the current through it:

$$v = L\frac{di}{dt}$$

The inductance, L, is commonly given in henries, with 1 henry being 1 volt-second per ampere.

Voltage source. The voltage across a voltage source is given by a specified function of time; there is no constraint on the resulting current through the voltage source. A *battery* is a constant-voltage source (i.e., the source voltage is constant in time). Time-varying voltage sources correspond to the waveform generators found in the laboratory. A *dependent voltage source* is one for which the voltage is determined by (i.e., depends on) some other variable arising in the overall circuit. Dependent sources are used in "small-signal" models of electronic devices such as transistors, and more complicated collections of devices such as operational amplifiers.

Current source. The current through a voltage source is given by a specified function of time; there is no constraint on the resulting voltage across the current source. Current sources may be dependent just as was described in the case of voltage sources.

Kirchhoff's laws provide the interconnection rules that determine how the elemental variables are related in a network. The Kirchhoff current law, traditionally abbreviated KCL, states that the algebraic sum of the currents at any node of a circuit must equal zero (Fig. 2.2a). The Kirchhoff voltage law, KVL, states that the algebraic sum of the voltages across the elements in any loop of a circuit must equal zero (Fig. 2.2b). Given the description of a circuit as a graph consisting of branches (two-terminal devices) connected at nodes, the equations that can result from combining the simple descriptions of each branch according to the constraints imposed by KCL and KVL are highly structured; this has been exploited in programs for computer-aided circuit analysis and design to automate the

$$0 = i_1 + i_2 + i_3 + i_4$$

(a)

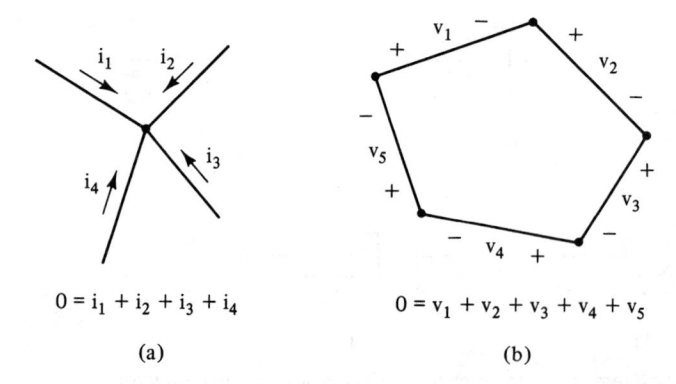

$$0 = v_1 + v_2 + v_3 + v_4 + v_5$$

(b)

Figure 2.2 (a) Kirchhoff's current law; (b) Kirchhoff's voltage law.

process of taking a circuit description as a graph and generating the corresponding circuit equations.

Now we turn to some specific examples of linear systems in the form of electric circuits. As a first example, consider a single-loop, series RLC circuit with an independent voltage source whose voltage is $u(t)$ (Fig. 2.3). KCL implies that the same current, say i, flows through all of the branches of the single-loop circuit. KVL gives the equation $u = v_R + v_L + v_C$. The equations for the three elements are available from considering the devices involved: $v_R = R\,i$, $v_L = L\,(\mathbf{D}i)$, and $i = C(\mathbf{D}v_C)$. Here we have used the **D** to symbolize differentiation with respect to time t. Combining these equations, and choosing $x_1 = i$ and $x_2 = v_C$, we obtain the following linear system describing the circuit:

$$\begin{bmatrix} \dot{x}_1(t) \\ \dot{x}_2(t) \end{bmatrix} = \begin{bmatrix} -\dfrac{R}{L} & -\dfrac{1}{L} \\ \dfrac{1}{C} & 0 \end{bmatrix} \begin{bmatrix} x_1(t) \\ x_2(t) \end{bmatrix} + \begin{bmatrix} \dfrac{1}{L} \\ 0 \end{bmatrix} u(t)$$

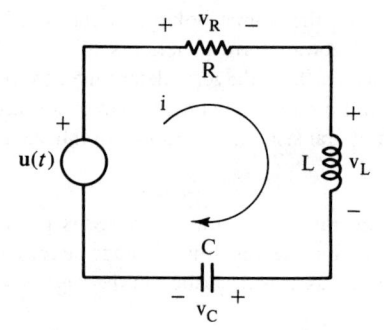

Figure 2.3 Series RLC circuit with voltage source input.

Next we consider an n-p-n bipolar junction transistor circuit (Fig. 2.4), whose hybrid-π model is shown in Fig. 2.5. This is a small-signal (or incremental) model, applicable for small changes in input voltages around a nominal operating point (determined by biasing). It is customary to use lowercase symbols to represent the resistors for small-signal transistor models such as this; the values of these resistors, as well as the value of the *transconductance*, the gain on the controlled current source, depend on the choice of operating point. The parameters appearing in the model are as follows: r_π, the base-emitter incremental resistance, C_π, the base-emitter capacitance, r_{bb}, r_{ee}, and r_{cc}, the incremental

Linear Systems   Chap. 2

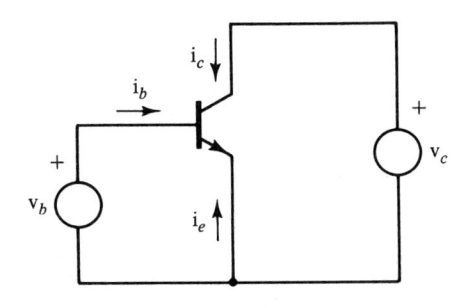

Figure 2.4 n-p-n bipolar junction transistor circuit.

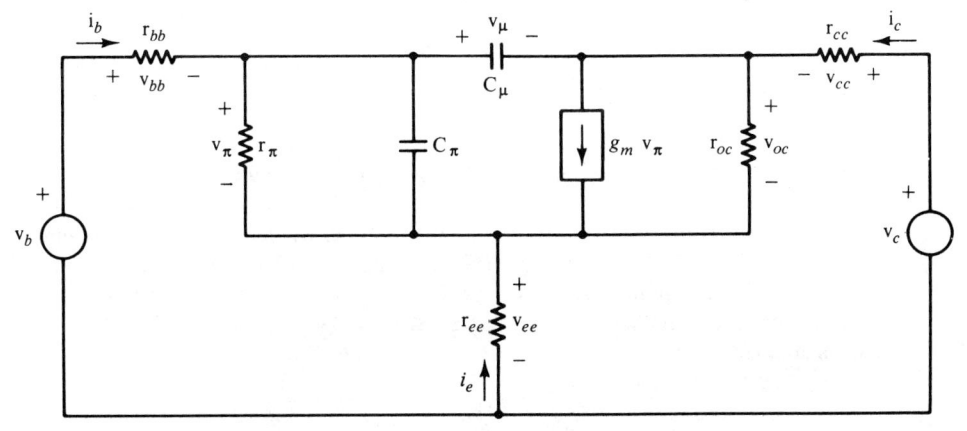

Figure 2.5 Hybrid-π model for n-p-n bipolar junction transistor circuit. Circuit parameters depend on the operating point.

ohmic bulk resistances for the base, emitter, and collector regions, respectively, C_μ, the base-collector capacitance, r_{oc}, the collector output resistance (to model base-width-modulation effects), and g_m, the transconductance. The incremental current gain, β_o, also denoted h_{fe}, equals the product $r_\pi g_m$.

We may use KCL and KVL to write four equations involving eight voltage and current variables, v_b, v_c, i_b, i_c, i_e, v_π, v_μ, and v_{oc}:

$$v_b = i_b r_{bb} + v_\pi - i_e r_{ee}$$

$$i_c = g_m v_\pi + \frac{v_{oc}}{r_{oc}} - C_\mu \dot{v}_\mu$$

$$v_c = i_c r_{cc} + v_{oc} - r_{ee} i_e$$

$$i_b = C_\mu \dot{v}_\mu + C_\pi \dot{v}_\pi + \frac{v_\pi}{r_\pi}$$

Writing a set of state space equations for the hybrid-π model amounts to eliminating i_b, i_c, i_e, and v_{oc}, and obtaining two differential equations for the capacitor voltages v_π and v_μ with the base and emitter voltages v_b and v_e playing the role of input variables. This is left as an exercise for the reader.

An LC "lumped" transmission line is a third circuit example. Each section of the line is a simple LC voltage divider; Fig. 2.6 shows the nth section, from which the following two equations are found by application of KVL and KCL:

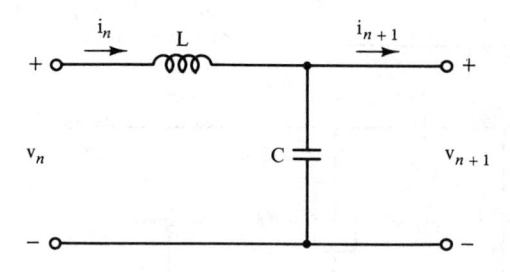

$$v_n = L\frac{di_n}{dt} + v_{n+1}$$

$$i_n = C\frac{dv_{n+1}}{dt} + i_{n+1}$$

Figure 2.6 One stage of a transmission line circuit.

These equations hold for $0 \leq n \leq N-1$, for an N-stage line, where v_0 is the voltage input to the first stage and $i_N = 0$ at the last stage, assuming that the line is terminated with an open circuit (Fig. 2.7). The components of the $2N$-dimensional state vector are the collection of current and voltage variables: $x_{2k+1} = i_k$, $0 \leq k \leq N-1$ and $x_{2k} = v_k$, $1 \leq k \leq N$. The state equations are then given by

$$\dot{x}_1 = -\frac{1}{L}(x_2 - v_0)$$

$$\dot{x}_{2k} = \frac{1}{C}(x_{2k-1} - x_{2k+1}), \qquad 1 < k < N$$

$$\dot{x}_{2k+1} = \frac{1}{L}(x_{2k} - x_{2k+2}), \qquad 1 < k < N-1$$

$$\dot{x}_{2N} = \frac{1}{C}x_{2N-1}$$

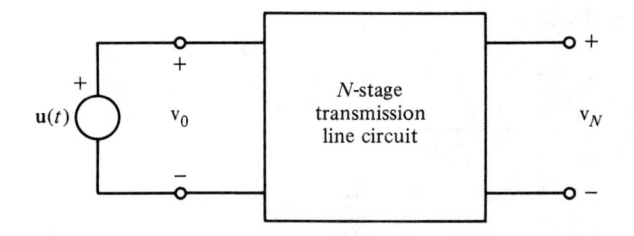

Figure 2.7 N-stage transmission line circuit. The input is the voltage source, the ouput is the capacitor voltage of the Nth stage (open-circuit termination).

Simple mechanical systems provide other examples of linear systems. Consider masses of m_1 and m_2 kilograms connected by a spring with force constant k newtons/meter, moving on a horizontal surface, each subject to a frictional force proportional to its velocity and acted on by an external force (see Fig. 2.8). Choose the components of the state vector as follows: x_1 is the position of the first mass, x_2 is the velocity of the first mass, x_3 is the position of the second mass, and x_4 is the velocity of the second mass. Then the state equations are given by

Linear Systems   Chap. 2

$$\begin{bmatrix} \dot{x}_1 \\ \dot{x}_2 \\ \dot{x}_3 \\ \dot{x}_4 \end{bmatrix} = \begin{bmatrix} 0 & 1 & 0 & 0 \\ -k/m_1 & -f/m_1 & k/m_2 & 0 \\ 0 & 0 & 0 & 1 \\ k/m_1 & 0 & -k/m_2 & -f/m_2 \end{bmatrix} \begin{bmatrix} x_1 \\ x_2 \\ x_3 \\ x_4 \end{bmatrix} + \begin{bmatrix} 0 & 0 \\ 1/m_1 & 0 \\ 0 & 0 \\ 0 & 1/m_2 \end{bmatrix} \begin{bmatrix} F_1 \\ F_2 \end{bmatrix}$$

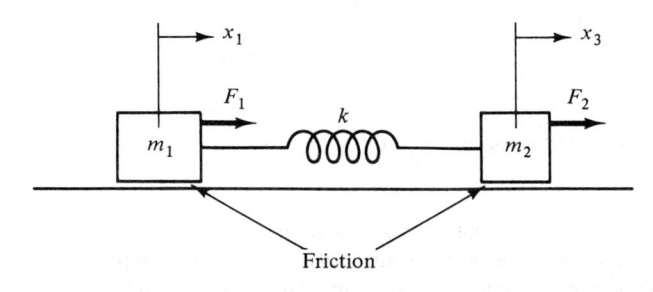

Figure 2.8 Two coupled masses. The inputs are the external forces applied to masses.

Electric circuit equations are combined with Newtonian equations of motion for physical objects to provide a complete description of electromechanical systems such as an electric motor. Consider the example of an "armature-controlled" electric motor (i.e., a motor with a constant current source driving its field circuit or with a permanent magnet providing a constant field). The armature circuit (Fig. 2.9a) is a series RL circuit with a voltage source to provide the controlling input signal and a dependent source to model the "back emf" voltage due to the motor, v_b, which is proportional to the angular velocity of the motor shaft. The mechanical equations of the motor shaft are obtained from Newton's laws, assuming that the motor shaft and load have an angular inertia of J and that there is a frictional torque on the shaft that is proportional to the angular velocity (Fig. 2.9b). The driving torque applied to the shaft by the motor is proportional to the current in the armature circuit. The motor equations thus take the following form:

$$Ri + L\frac{di}{dt} + v_b = \mathbf{u}(t)$$

$$v_b = k_b \frac{d\theta}{dt}$$

$$J\frac{d^2\theta}{dt^2} + f\frac{d\theta}{dt} = T(t)$$

$$T(t) = k_t i$$

These equations may be combined into a linear system equation by choosing $x_1 = i$, $x_2 = \theta$, and $x_3 = d\theta/dt$. The results are

$$\dot{x}_1(t) = -\frac{R}{L}x_1 - \frac{k_b}{L}x_3 + \mathbf{u}$$

$$\dot{x}_2(t) = x_3$$

$$\dot{x}_3(t) = \frac{k_t}{J}x_1 - \frac{f}{J}x_3$$

(The reader can easily put this in matrix form if desired.)

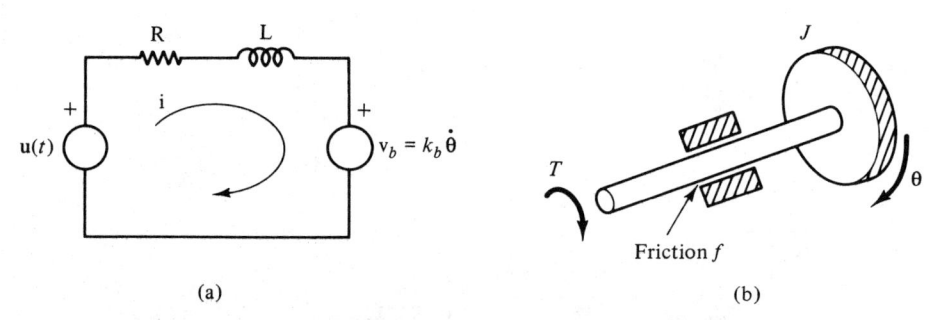

Figure 2.9 (a) Armature circuit for an electric motor; (b) rotational dynamics of a motor shaft.

The next example is a simple, but representative model from aerodynamics. It describes the pitch response of an airplane in steady (constant velocity) flight (Fig. 2.10). The state variables are q and α, the pitch rate (the angular velocity about the center of gravity) and angle of attack, respectively. Two control inputs are included, the elevator and flap deflections, e and f. The state equations take the form

$$\begin{bmatrix} \dot{q}(t) \\ \dot{\alpha}(t) \end{bmatrix} = \begin{bmatrix} m_q & m_\alpha \\ 1 & k_\alpha \end{bmatrix} \begin{bmatrix} q(t) \\ \alpha(t) \end{bmatrix} + \begin{bmatrix} m_e & m_f \\ k_e & k_f \end{bmatrix} \begin{bmatrix} e(t) \\ f(t) \end{bmatrix}$$

Figure 2.10 Airplane pitch dynamics. The inputs are elevator and flap deflections, and the states are angle of attack and pitch rate. V is the nominal horizontal velocity.

The m coefficients are the pitch acceleration proportionality constants for small changes in the corresponding subscripted variables, and the k coefficients are the lift acceleration proportionality constants divided by the negative of the nominal velocity V. (These coefficients depend on the nominal velocity and altitude of the airplane, just as the coefficients in the small-signal model for the transistor depend on operating point currents and voltages.)

2.1 DIFFERENTIAL EQUATIONS AND LINEAR STATE SPACE SYSTEMS

The examples already introduced, and many others that occur in a wide variety of applications, provide some basic motivation for the study of models taking the form

$$\dot{\mathbf{x}}(t) = \mathbf{A}\mathbf{x}(t) + \mathbf{B}\mathbf{u}(t); \quad \mathbf{x}(0) = \mathbf{x}_0 \tag{2.1}$$

Sec. 2.1 Differential Equations and Linear State Space Systems

Without further ado, we begin our study of the quantitative and qualitative characteristics of linear systems.

The first, most basic, facts pertain to differential equation models of all kinds. A *solution* of this set of differential equations is a vector $\mathbf{x}(t)$ of time functions satisfying the differential equations and the given initial conditions.

The general theory of differential equations tells us that solutions for linear differential equations of this sort *exist* and are *unique*. Indeed, we will be quite explicit in describing the form that a solution must take, and this serves as a constructive proof of the existence of solutions. Uniqueness is a property that will be very useful in some of our later analysis, and we simply accept it without a detailed technical justification. Not that a proof is "hard," but it adds little to the development of other ideas in this book.

Example

It is easy to reduce the uniqueness question to its simplest form. Suppose that $\mathbf{x}_1(t)$ and $\mathbf{x}_2(t)$ are two solutions to the equations describing a linear system, with $\mathbf{x}_1(0) = \mathbf{x}_2(0)$. Then $\mathbf{z}(t) = \mathbf{x}_1(0) - \mathbf{x}_2(0)$ is a solution of the differential equation

$$\dot{\mathbf{z}}(t) = \mathbf{A}\mathbf{z}(t); \qquad \mathbf{z}(0) = \mathbf{0}$$

Only if the zero function is the unique solution to this equation will there be a unique solution to the linear system so that $\mathbf{x}_1(t) = \mathbf{x}_2(t)$. Rather than directly proving that $\mathbf{z}(t)$ is identically zero, it turns out to be easier to show that its squared norm, $\mathbf{q}(t) = \|\mathbf{z}(t)\|^2 = \mathbf{z}^T(t)\mathbf{z}(t)$, is zero. This approach is left for the interested reader to pursue in detail.

A first thing to notice about the solution of a linear system has to do with *time invariance*. If $\mathbf{x}(t)$ is the solution to the linear system above, $\mathbf{x}(t - t_0)$ is the solution to the closely related linear system whose input is the "delayed" input $\mathbf{u}(t - t_0)$ and whose initial condition is $\mathbf{x}(t - t_0) = \mathbf{x}_0$ at time $t = t_0$. This means that there was no loss of generality in assuming that the initial condition is specified at time $t = 0$.

The next thing to notice is that there is an additive representation that applies here. The solution can be written as a sum of two pieces, say $\mathbf{x}(t) = \mathbf{x}_h(t) + \mathbf{x}_p(t)$, where the subscripts denote "homogeneous" and "particular." The *homogeneous solution* is the solution to the unforced equation having the correct initial condition, that is,

$$\dot{\mathbf{x}}_h(t) = \mathbf{A}\mathbf{x}_h(t); \qquad \mathbf{x}_h(0) = \mathbf{x}_0 \qquad (2.2)$$

The *particular solution*,* on the other hand, satisfies the forced equation subject to zero initial conditions:

$$\dot{\mathbf{x}}_p(t) = \mathbf{A}\mathbf{x}_p(t) + \mathbf{B}\mathbf{u}(t); \qquad \mathbf{x}_p(0) = \mathbf{0} \qquad (2.3)$$

The verification of this additive representation is an easy consequence of uniqueness of solutions for the differential equation. Since this line of reasoning will be used in many places in this book, we want to emphasize the importance of understanding the (simple and elegant) approach by giving the details. Suppose that $\mathbf{x}(t)$ is a solution to

---

*The reader should be aware that the term *particular solution* is used in a somewhat different way in some textbooks on differential equations.

the differential equation with $\mathbf{x}(0) = \mathbf{x}_0$. No assumption about the form of $\mathbf{x}(t)$ is made. It is easily checked by differentiating $\mathbf{x}_h(t) + \mathbf{x}_p(t)$ and using the expressions for the derivatives of the two terms that this sum satisfies the differential equation with initial condition $\mathbf{x}_0$ also. Since the differential equation has a unique solution, it follows that $\mathbf{x}(t) = \mathbf{x}_h(t) + \mathbf{x}_p(t)$. This kind of argument is summarized in the following general terms: If two expressions for the solution of a differential equation both satisfy the differential equation and both have the same value at some instant t_0, uniqueness of solutions for the differential equation implies that the two expressions are equal for all instants t.

The Homogeneous Solution

Knowing that the solution can be written as a sum is helpful because it allows us to focus attention on the two pieces individually. Turning first to the homogeneous solution, we need to determine the solution to equation (2.2):

$$\dot{\mathbf{x}}_h(t) = \mathbf{A}\mathbf{x}_h(t); \qquad \mathbf{x}_h(0) = \mathbf{x}_0$$

The first thing to notice about this equation is that the solutions for all possible values of the initial condition $\mathbf{x}_0$ form a vector space of dimension n. To see this, suppose that $\mathbf{x}_1(t)$ and $\mathbf{x}_2(t)$ are two homogeneous solutions with different initial conditions, say $\mathbf{x}_1(0) = \mathbf{x}_1$ and $\mathbf{x}_2(0) = \mathbf{x}_2$. Then a direct calculation shows that $\alpha \mathbf{x}_1(t) + \beta \mathbf{x}_2(t)$ is also a homogeneous solution with initial condition $\alpha \mathbf{x}_1 + \beta \mathbf{x}_2$.

Since the set of all homogeneous solutions is a vector space, it is quite natural to ask if there is convenient basis for this space. The answer is yes, and it is easy to see why: Each homogeneous solution is determined by its initial condition, an n-vector, so if we know the homogeneous solutions corresponding to the initial conditions $\mathbf{e}_1, \mathbf{e}_2, \ldots, \mathbf{e}_n$, where $\mathbf{e}_i$ is the ith standard basis vector, then we can find the solution for any $\mathbf{x}_0$ by taking an appropriate linear combination of these "basic" homogeneous solutions.

The important outcome of the development above is that we can say that the general homogeneous solution takes the form

$$\mathbf{x}_h(t) = \Phi(t)\mathbf{x}_0 \qquad (2.4)$$

where the $n \times n$ matrix $\Phi(t)$ has as its columns the basic homogeneous solutions. In particular, notice that $\Phi(t)$ satisfies two important properties: (1) $\Phi(0) = \mathbf{I}$, and (2) $\Phi(t_1 + t_2) = \Phi(t_1)\Phi(t_2) = \Phi(t_2)\Phi(t_1)$. The first property follows directly from equation (2.4) by setting $t = 0$ and noting that the equation must hold for all possible choices of $\mathbf{x}_0$. The second property follows by noting that $\mathbf{x}_h(t_1 + t_2)$ is the homogeneous solution at time $t_1 + t_2$ to the differential equation with the initial condition $\mathbf{x}_0$ at time 0 (Fig. 2.11). This solution takes the value $\mathbf{x}_h(t_1)$ at time t_1. Therefore, if the initial condition at time 0 is $\mathbf{x}_h(t_1)$, the homogeneous solution takes the value $\mathbf{x}_h(t_1 + t_2)$ at time t_2. (This is the time-invariance property mentioned above.) This gives $\Phi(t_1 + t_2)\mathbf{x}_0 = \Phi(t_2)\mathbf{x}_h(t_1) = \Phi(t_2)\Phi(t_1)\mathbf{x}_0$, and this must hold for all possible choices of $\mathbf{x}_0$. The same argument holds with the roles of t_1 and t_2 reversed. Since the one-

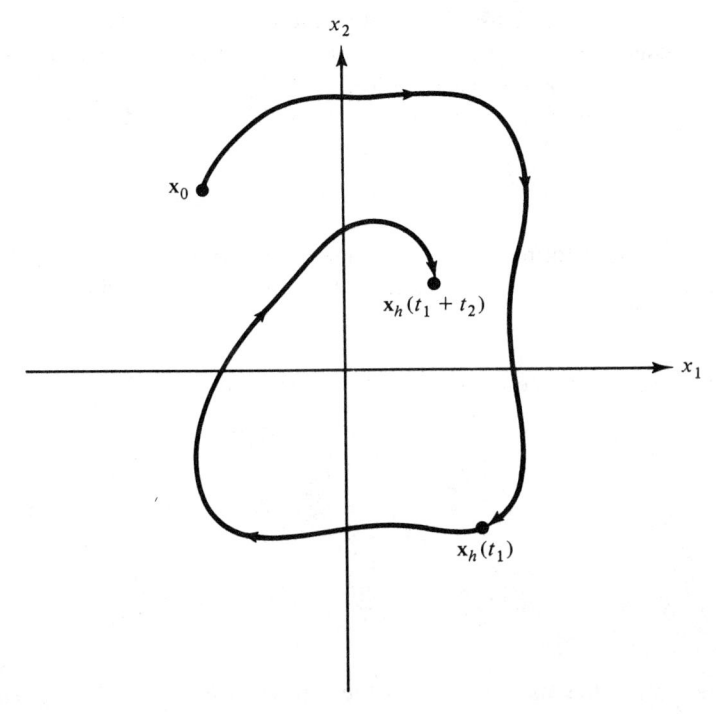

Figure 2.11 Trajectory in state space representing the solution as a function of time, $0 < t < t_1 + t_2$, where $t_1 > 0$ and $t_2 > 0$.

dimensional version of these conditions, namely $\phi(0) = 1$ and $\phi(t_1 + t_2) = \phi(t_1)\phi(t_2)$, is satisfied by the exponential function (and indeed the conditions uniquely determine the exponential function), it appears that what we are looking for is an appropriate generalization of the exponential function.

Matrix Exponential Function

We will be led to replace the vague notation involving $\Phi(t)$ with the more suggestive $e^{\mathbf{A}t}$, and we will provide some ways of defining and computing this quantity that do not involve solving differential equations at all, much as the ordinary exponential function can be defined without regard to its (useful) properties with respect to differentiation. Our first approach to this task makes use of the eigenvalues and eigenvectors of the matrix $\mathbf{A}$. As usual, we assume that $\mathbf{A}$ has a set of n linearly independent eigenvectors $\mathbf{u}_1, \mathbf{u}_2, \ldots, \mathbf{u}_n$ corresponding to eigenvalues $\lambda_1, \lambda_2, \ldots, \lambda_n$. Then by a transformation of coordinates, the homogeneous linear equations can be brought to diagonal form: If $\mathbf{x}_h(t) = \mathbf{T}\mathbf{z}(t)$, where $\mathbf{T} = [\mathbf{u}_1 \ \mathbf{u}_2 \ \cdots \ \mathbf{u}_n]$, then $\dot{\mathbf{x}}_h(t) = \mathbf{T}\dot{\mathbf{z}}(t)$ and the equations for $\mathbf{z}(t)$ are

$$\dot{\mathbf{z}}(t) = \Lambda \mathbf{z}(t) \tag{2.5}$$

where, as before, we use Λ to denote the diagonal matrix whose diagonal entries are the eigenvalues of $\mathbf{A}$, and $\Lambda = \mathbf{T}^{-1}\mathbf{A}\mathbf{T}$. The appropriate initial condition for $\mathbf{z}(t)$ is $\mathbf{z}(0) = \mathbf{T}^{-1}\mathbf{x}_0$, which we will denote as $\mathbf{z}_0$.

The differential equation for $\mathbf{z}(t)$ is easily solved since, for $1 \leq i \leq n$, the ith component satisfies

$$\dot{z}_i(t) = \lambda_i z_i(t), \qquad z_i(0) = z_{0i} \tag{2.6}$$

Thus the transformation of coordinates has produced a completely decoupled set of scalar (i.e., one-variable) equations that may be solved using basic calculus, giving

$$z_i(t) = e^{\lambda_i t} z_{0i}, \qquad 1 \leq i \leq n \tag{2.7}$$

Rewriting these n equations in matrix-vector form, we have

$$\mathbf{z}(t) = \begin{bmatrix} e^{\lambda_1 t} & 0 & \cdots & 0 \\ 0 & e^{\lambda_2 t} & & 0 \\ \vdots & & \ddots & \vdots \\ 0 & 0 & \cdots & e^{\lambda_n t} \end{bmatrix} \mathbf{z}_0 \tag{2.8}$$

The form of this matrix suggests the following notational "shorthand":

$$\mathbf{z}(t) = e^{\Lambda t} \mathbf{z}_0 \tag{2.9}$$

Thus for diagonal matrices we have a quite natural definition for the matrix exponential function, which also turns out to be diagonal in this case.

We can return to the original coordinate system, since $\mathbf{x}_h(t) = \mathbf{T}\mathbf{z}(t) = \mathbf{T}e^{\Lambda t}\mathbf{z}_0 = [\mathbf{T}e^{\Lambda t}\mathbf{T}^{-1}]\mathbf{x}_0$, which leads to the definition

$$e^{\mathbf{A}t} = \mathbf{T}e^{\Lambda t}\mathbf{T}^{-1} \tag{2.10}$$

which holds when $\Lambda = \mathbf{T}^{-1}\mathbf{A}\mathbf{T}$, so that $\mathbf{A} = \mathbf{T}\Lambda\mathbf{T}^{-1}$. Notice that this definition gives $e^{\Lambda t} = \mathbf{T}^{-1}e^{\mathbf{A}t}\mathbf{T}$, or in words, when Λ is obtained by similarity transformation from $\mathbf{A}$, $e^{\Lambda t}$ is obtained by the same similarity transformation from $e^{\mathbf{A}t}$.

We don't quite have a general definition for the matrix exponential function because we haven't covered cases when the matrix $\mathbf{A}$ cannot be diagonalized by a similarity transformation. To get a general definition, we recall the (Taylor) series expansion for the scalar exponential function:

$$e^{at} = 1 + at + \frac{(at)^2}{2} + \frac{(at)^3}{6} + \cdots = \sum_{i=0}^{\infty} \frac{(at)^i}{i!} \tag{2.11}$$

This may be used to express each diagonal term, $e^{\lambda_i t}$, of $e^{\Lambda t}$, and the result may be written as

$$e^{\Lambda t} = \mathbf{I} + \Lambda t + \frac{(\Lambda t)^2}{2} + \frac{(\Lambda t)^3}{6} + \cdots = \sum_{i=0}^{\infty} \frac{(\Lambda t)^i}{i!} \tag{2.12}$$

Sec. 2.1 Differential Equations and Linear State Space Systems

This works because all powers of the matrix Λt remain diagonal; in particular the diagonal elements of $(\Lambda t)^i$ are $(\lambda_1 t)^i, (\lambda_2 t)^i, \ldots, (\lambda_n t)^i$. Now we can get a power series expansion for $e^{\mathbf{A}t}$ when $\mathbf{A}$ is diagonalizable (i.e., when $\Lambda = \mathbf{T}^{-1}\mathbf{A}\mathbf{T}$):

$$e^{\mathbf{A}t} = \mathbf{T} e^{\Lambda t} \mathbf{T}^{-1} = \mathbf{T} \left[\sum_{i=0}^{\infty} \frac{(\Lambda t)^i}{i!} \right] \mathbf{T}^{-1} \tag{2.13}$$

Using $\Lambda = \mathbf{T}^{-1}\mathbf{A}\mathbf{T}$, successive products give $\Lambda^2 = \mathbf{T}^{-1}\mathbf{A}^2\mathbf{T}$, $\Lambda^3 = \mathbf{T}^{-1}\mathbf{A}^3\mathbf{T}$, and so on, so $\Lambda^i = \mathbf{T}^{-1}\mathbf{A}^i\mathbf{T}$ for all $i \geq 0$. (The $i = 0$ term involves $\mathbf{A}^0 = \Lambda^0 = \mathbf{I}$.) Thus we may absorb the $\mathbf{T}$ and $\mathbf{T}^{-1}$ factors into each term of the sum and obtain

$$e^{\mathbf{A}t} = \mathbf{I} + \mathbf{A}t + \frac{(\mathbf{A}t)^2}{2} + \frac{(\mathbf{A}t)^3}{6} + \cdots = \sum_{i=0}^{\infty} \frac{(\mathbf{A}t)^i}{i!} \tag{2.14}$$

Since this formula depends only on $\mathbf{A}$, even though our derivation involved Λ, we will adopt this as our definition of $e^{\mathbf{A}t}$ for all (square) matrices $\mathbf{A}$.

There are some important properties of $e^{\mathbf{A}t}$ that are easily verified by calculations using the defining expression. First, we note that $e^{\mathbf{A}0} = \mathbf{I}$, and $e^{(-\mathbf{A})t} = e^{\mathbf{A}(-t)}$. The latter property means that we can use the notation $e^{-\mathbf{A}t}$ without ambiguity. The next property involves the elementwise derivative of the matrix exponential function:

$$\frac{d}{dt} e^{\mathbf{A}t} = \mathbf{A} e^{\mathbf{A}t} = e^{\mathbf{A}t} \mathbf{A} \tag{2.15}$$

To see how this comes about, consider

$$\frac{d}{dt} e^{\mathbf{A}t} = \frac{d}{dt} \sum_{i=0}^{\infty} \frac{(\mathbf{A}t)^i}{i!} = \sum_{i=0}^{\infty} \frac{d}{dt} \left[\frac{(\mathbf{A}t)^i}{i!} \right] \tag{2.16}$$

which follows from linearity of the derivative operation with respect to sums and the uniform convergence of the infinite sum. Continuing,

$$\frac{d}{dt} e^{\mathbf{A}t} = \sum_{i=1}^{\infty} \frac{\mathbf{A}^i t^{i-1}}{(i-1)!} = \mathbf{A} \sum_{i=0}^{\infty} \frac{(\mathbf{A}t)^i}{i!} = \mathbf{A} e^{\mathbf{A}t} \tag{2.17}$$

or alternatively,

$$= \sum_{i=0}^{\infty} \frac{(\mathbf{A}t)^i}{i!} \mathbf{A} = e^{\mathbf{A}t} \mathbf{A} \tag{2.18}$$

Another property involves inversion of $e^{\mathbf{A}t}$, namely

$$\left[e^{\mathbf{A}t} \right]^{-1} = e^{-\mathbf{A}t} \tag{2.19}$$

The easy way to verify this result is to differentiate the product $e^{\mathbf{A}t} e^{-\mathbf{A}t}$ using the chain rule:

$$\frac{d}{dt} \left[e^{\mathbf{A}t} e^{-\mathbf{A}t} \right] = \frac{d}{dt} \left[e^{\mathbf{A}t} \right] e^{-\mathbf{A}t} + e^{\mathbf{A}t} \frac{d}{dt} \left[e^{-\mathbf{A}t} \right]$$

$$= \mathbf{A} e^{\mathbf{A}t} e^{-\mathbf{A}t} + e^{\mathbf{A}t} (-\mathbf{A}) e^{-\mathbf{A}t}$$

$$= \mathbf{A} e^{\mathbf{A}t} e^{-\mathbf{A}t} - \mathbf{A} e^{\mathbf{A}t} e^{-\mathbf{A}t} = 0 \tag{2.20}$$

Thus the product matrix $e^{\mathbf{A}t}e^{-\mathbf{A}t}$ is a constant matrix (as a function of t). Since the value of this matrix at $t=0$ is $\mathbf{I}$, we have $e^{\mathbf{A}t}e^{-\mathbf{A}t} = \mathbf{I}$, which means that $e^{-\mathbf{A}t}$ is the inverse of $e^{\mathbf{A}t}$ (and vice versa).

The final property to be mentioned corresponds to the time-invariance property of the underlying linear system described earlier: The matrix exponential function satisfies $e^{\mathbf{A}(t+\tau)} = e^{\mathbf{A}t}e^{\mathbf{A}\tau}$. This holds for all values of t and τ, both positive and negative. It can be verified in a way similar to the previous calculation, by showing that the product $e^{\mathbf{A}(t+\tau)}e^{-\mathbf{A}\tau}e^{-\mathbf{A}t} = \mathbf{I}$. The details are left as an exercise.

Returning to our original problem of solving for the homogeneous solution to the linear system, these properties make it easy to verify that $\mathbf{x}_h(t) = e^{\mathbf{A}t}\mathbf{x}_0$ provides the desired solution. To check this, notice that $\mathbf{x}_h(0) = \mathbf{I}\mathbf{x}_0 = \mathbf{x}_0$, as desired. And

$$\dot{\mathbf{x}}_h(t) = \frac{d}{dt}\left[e^{\mathbf{A}t}\mathbf{x}_0\right] = \frac{d}{dt}\left[e^{\mathbf{A}t}\right]\mathbf{x}_0 = \mathbf{A}e^{\mathbf{A}t}\mathbf{x}_0 = \mathbf{A}\mathbf{x}_h(t) \qquad (2.21)$$

as desired.

The Particular Solution

Turning to the determination of the particular solution, which must satisfy equation (2.3),

$$\dot{\mathbf{x}}_p(t) = \mathbf{A}\mathbf{x}_p(t) + \mathbf{B}\mathbf{u}(t); \qquad \mathbf{x}_p(0) = \mathbf{0}$$

we will employ another transformation, a *time-dependent* change of coordinates, to find the desired result. We let $\mathbf{w}(t) = e^{-\mathbf{A}t}\mathbf{x}_p(t)$. Then by the chain rule

$$\dot{\mathbf{w}}(t) = \frac{d}{dt}\left[e^{-\mathbf{A}t}\right]\mathbf{x}_p(t) + e^{-\mathbf{A}t}\dot{\mathbf{x}}_p(t)$$

$$= -\mathbf{A}e^{-\mathbf{A}t}\mathbf{x}_p(t) + e^{-\mathbf{A}t}\left[\mathbf{A}\mathbf{x}_p(t) + \mathbf{B}\mathbf{u}(t)\right] = e^{-\mathbf{A}t}\mathbf{B}\mathbf{u}(t) \qquad (2.22)$$

Since $\mathbf{w}(t)$ doesn't appear on the right-hand side of this equation, we can easily integrate both sides to obtain

$$\mathbf{w}(t) - \mathbf{w}(0) = \int_0^t e^{-\mathbf{A}\tau}\mathbf{B}\mathbf{u}(\tau)\,d\tau \qquad (2.23)$$

Now $\mathbf{w}(0) = e^{-\mathbf{A}0}\mathbf{x}_p(0) = \mathbf{0}$, so

$$\mathbf{w}(t) = \int_0^t e^{-\mathbf{A}\tau}\mathbf{B}\mathbf{u}(\tau)\,d\tau \qquad (2.24)$$

Returning to the original coordinate system,

$$\mathbf{x}_p(t) = e^{\mathbf{A}t}\int_0^t e^{-\mathbf{A}\tau}\mathbf{B}\mathbf{u}(\tau)\,d\tau = \int_0^t e^{\mathbf{A}(t-\tau)}\mathbf{B}\mathbf{u}(\tau)\,d\tau \qquad (2.25)$$

The Total Solution

Combining the results obtained above, we now have a formula for the total solution of the linear system described by equation (2.1):

$$\dot{\mathbf{x}}(t) = \mathbf{A}\mathbf{x}(t) + \mathbf{B}\mathbf{u}(t); \qquad \mathbf{x}(0) = \mathbf{x}_0$$

Sec. 2.2   Inputs, Outputs, and States   77

The solution is

$$\mathbf{x}(t) = e^{\mathbf{A}t}\mathbf{x}_0 + \int_0^t e^{\mathbf{A}(t-\tau)} \mathbf{B}\mathbf{u}(\tau)\,d\tau \qquad (2.26)$$

Example

Consider the series RLC circuit introduced as the first example at the beginning of this chapter. We will use some "artificial" component values for convenience of exposition; let $R = 1\,\Omega$, $L = 0.5$ H, and $C = 1$ F. The state equations are

$$\begin{bmatrix} \dot{x}_1(t) \\ \dot{x}_2(t) \end{bmatrix} = \begin{bmatrix} -2 & -2 \\ 1 & 0 \end{bmatrix} \begin{bmatrix} x_1(t) \\ x_2(t) \end{bmatrix} + \begin{bmatrix} 2 \\ 0 \end{bmatrix} \mathbf{u}(t)$$

We will assume that $x_1(0) = i_0$ and $x_2(0) = v_0$ are the initial current and voltage values. The input $\mathbf{u}(t)$ is a constant voltage v_u for $t \geq 0$.

The first step in determining the solution is to find $e^{\mathbf{A}t}$, and to do this analytically we will use eigenvalues and eigenvectors. We find that

$$\det(\lambda \mathbf{I} - \mathbf{A}) = \lambda^2 + 2\lambda + 2$$

and the eigenvalues of $\mathbf{A}$ are the zeros of this polynomial, namely $\lambda_1 = -1 + j$ and $\lambda_2 = -1 - j$. The corresponding eigenvectors of $\mathbf{A}$ are the nonzero solutions of the homogeneous equations

$$(\lambda_i \mathbf{I} - \mathbf{A})\mathbf{v}_i = \mathbf{0}, \qquad i = 1, 2$$

The eigenvectors obtained are used as the columns of a coordinate transformation matrix $\mathbf{T}$:

$$\mathbf{T} = \begin{bmatrix} \mathbf{v}_1 & \mathbf{v}_2 \end{bmatrix} = \begin{bmatrix} -1+j & -1-j \\ 1 & 1 \end{bmatrix}$$

Then

$$e^{\mathbf{A}t} = \mathbf{T} \begin{bmatrix} e^{(-1+j)t} & 0 \\ 0 & e^{(-1-j)t} \end{bmatrix} \mathbf{T}^{-1}$$

After some algebraic simplifications, this gives

$$e^{\mathbf{A}t} = \begin{bmatrix} e^{-t}(\cos t - \sin t) & -2e^{-t}\sin t \\ e^{-t}\sin t & e^{-t}(\cos t + \sin t) \end{bmatrix}$$

Thus we obtain the following expression for the solution:

$$\mathbf{x}(t) = e^{\mathbf{A}t}\mathbf{x}_0 + 2v_u \begin{bmatrix} \int_0^t e^{-t+\tau}(\cos(t-\tau) - \sin(t-\tau))\,d\tau \\ \int_0^t e^{-t+\tau}\sin(t-\tau)\,d\tau \end{bmatrix}$$

This may be simplified a bit more by evaluating the integrals, but we leave that for the reader.

2.2 INPUTS, OUTPUTS, AND STATES

Just as we have allowed our system models to have inputs, in the form of forcing functions, we often need to include certain *outputs* as part of an overall system model. Indeed, many systems come equipped with easily identifiable inputs and outputs (think of an audio amplifier or an automatic transmission). In such cases, the system is often designed to ensure that a certain relationship between the input and output signals is achieved. In other cases, the output signals may represent the only measurable information that is available about the state vector; often it is necessary to control a system described by a complicated set of state equations based on a limited set of measured quantities due to the expense of instrumenting the system with many sensors. Such problems are common in industrial process control applications.

For the moment, we adopt a simple model for outputs to include as part of a linear system model. We denote the output vector as $\mathbf{y}(t)$, a p-dimensional vector that is linearly related to the state vector $\mathbf{x}(t)$ according to the equation

$$\mathbf{y}(t) = \mathbf{C}\mathbf{x}(t) \tag{2.27}$$

where $\mathbf{C}$ is a $p \times n$ matrix. From the development in the preceding section, in particular equation (2.26), we can immediately write down the functional form for the output signals:

$$\mathbf{y}(t) = \mathbf{C}e^{\mathbf{A}t}\mathbf{x}_0 + \int_0^t \mathbf{C}e^{\mathbf{A}(t-\tau)}\mathbf{B}\mathbf{u}(\tau)\,d\tau \tag{2.28}$$

An Important Property of the State Vector

The equations (2.1) and (2.27), repeated here for convenience,

$$\dot{\mathbf{x}}(t) = \mathbf{A}\mathbf{x}(t) + \mathbf{B}\mathbf{u}(t); \qquad \mathbf{x}(0) = \mathbf{x}_0$$

$$\mathbf{y}(t) = \mathbf{C}\mathbf{x}(t)$$

comprise what is called an *internal description* of the associated system. This terminology makes reference to the use of the state vector $\mathbf{x}(t)$ to express the "history" or "memory" of the system about its operation before the initial time instant $t = 0$. Notice that given $\mathbf{x}(0) = \mathbf{x}_0$, the knowledge of inputs and states for times before $t = 0$ is not necessary to express the output for $t \geq 0$ as a function of the input for $t \geq 0$. So no matter how the initial state $\mathbf{x}_0$ was arrived at, whether by some past history of inputs or by some physical mechanisms associated with the state equations, only the value of $\mathbf{x}_0$ plays any role in determining $\mathbf{x}(t)$ and $\mathbf{y}(t)$ for $t \geq 0$. Also notice that we could make the same argument no matter what instant we chose to call the initial instant, since for any t_0

$$\mathbf{x}(t) = e^{\mathbf{A}(t-t_0)}\mathbf{x}(t_0) + \int_{t_0}^t e^{\mathbf{A}(t-\tau)}\mathbf{B}\mathbf{u}(\tau)\,d\tau \tag{2.29}$$

provided that this expression also gives $\mathbf{x}(0) = \mathbf{x}_0$. Of course, this condition is satisfied when

$$\mathbf{x}(t_0) = e^{\mathbf{A}t_0}\mathbf{x}_0 + \int_0^{t_0} e^{\mathbf{A}(t_0-\tau)}\mathbf{B}\mathbf{u}(\tau)\,d\tau \tag{2.30}$$

Thus the state vector at any time instant t summarizes all of the past information about the system that is needed for computing future values of the state and outputs from future values of the inputs.

Sec. 2.2 Inputs, Outputs, and States

Zero-State Output Response

For systems whose initial conditions are **0** (i.e., $\mathbf{x}_0$ is the zero vector) we point out one important feature. We note that the output in this case can be written as

$$\mathbf{y}(t) = \int_0^t h(t-\tau)\mathbf{u}(\tau)\,d\tau \tag{2.31}$$

where the *weighting function* $h(t)$ is given by

$$h(t) = \mathbf{C}e^{\mathbf{A}t}\mathbf{B}\mathbf{1}(t) \tag{2.32}$$

(The notation $\mathbf{1}(t)$ is used to denote the unit step function, the function that takes the value 0 for $t<0$ and the value 1 for $t\geq 0$. By abuse of notation, for systems with more than one input we will use $\mathbf{1}(t)$ to denote an m-vector whose elements are all equal to the unit step function.)

The integral in equation (2.31) is called a *convolution integral*. From this expression it is easy to see that linear systems obey a *superposition principle*:

The zero-state output response of a linear system is a linear mapping between inputs and outputs. If input $\mathbf{u}_i(t)$ produces zero-state output response $\mathbf{y}_i(t)$ for $1\leq i\leq k$, then for any choice of scalars $\{\alpha_i\}$ the input $\mathbf{u}(t) = \sum_{i=1}^{k}\alpha_i\mathbf{u}_i(t)$ produces zero-state output response $\mathbf{y}(t) = \sum_{i=1}^{k}\alpha_i\mathbf{y}_i(t)$.

This same conclusion can be reached by studying the state equations directly.

Example

For the series RLC circuit example given at the end of the preceding section, suppose that capacitor voltage, $x_2(t)$, is the system output, so that

$$\mathbf{y}(t) = [\,0\ \ 1\,]\mathbf{x}(t)$$

Using the calculations already carried out for this example, the form of the zero-state output response is

$$\mathbf{y}(t) = \int_0^t 2e^{-t+\tau}\sin(t-\tau)\mathbf{u}(\tau)\,d\tau$$

since the weighting function is

$$h(t) = 2e^{-t}\sin t\,\mathbf{1}(t)$$

Input-Output Models

In many applications, a system model may not arise in the form of a set of state equations. A typical example concerns the motion of a 1-kg mass connected to rigid walls with an ideal spring of total strength k newtons/meter and acted upon by an external force of strength $\mathbf{u}(t)$ newtons (Fig. 2.12). The equation for the displacement of the unit mass from equilibrium, $\mathbf{y}(t)$, in meters, is given by (Newton's law)

$$\ddot{\mathbf{y}}(t) = -k\,\mathbf{y}(t) + \mathbf{u}(t) \tag{2.33}$$

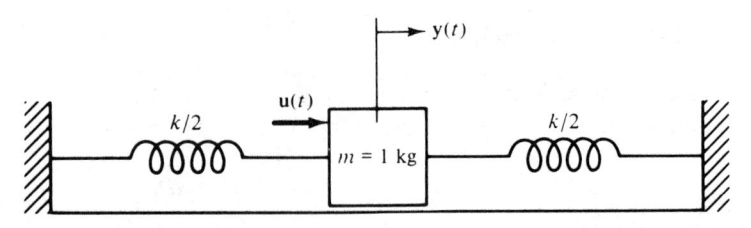

Figure 2.12 Mass-spring system. The input is the external force applied to mass; the output is the displacement of mass from its equilibrium position.

which is a second-order differential equation. If we view the displacement of the mass as the system output and the applied force as the input, we have a perfectly respectable description of a system, but not one in the form of state equations. This example is a simple *input-output model* that provides an *external description* of a system, that is, a description in terms of the system inputs and outputs only.

We can obtain a model in state equation form by defining the state vector

$$\mathbf{x}(t) = \begin{bmatrix} \mathbf{y}(t) \\ \dot{\mathbf{y}}(t) \end{bmatrix} \qquad (2.34)$$

the vector whose elements give the position and velocity of the mass. Then the state equations can be written in the usual form:

$$\dot{\mathbf{x}}(t) = \begin{bmatrix} 0 & 1 \\ -k & 0 \end{bmatrix} \mathbf{x}(t) + \begin{bmatrix} 0 \\ 1 \end{bmatrix} \mathbf{u}(t) \qquad (2.35)$$

and

$$\mathbf{y}(t) = [\,1 \;\; 0\,]\,\mathbf{x}(t) \qquad (2.36)$$

The initial conditions of this system are the initial position and velocity of the mass, namely $\mathbf{y}(0)$ and $\dot{\mathbf{y}}(0)$.

A quite general form for an input-output model for single-input, single-output systems is the following differential equation:

$$\mathbf{y}^{(n)}(t) + a_1 \mathbf{y}^{(n-1)}(t) + \cdots + a_{n-1} \mathbf{y}^{(1)}(t) + a_n \mathbf{y}(t)$$
$$= b_1 \mathbf{u}^{(n-1)}(t) + b_2 \mathbf{u}^{(n-2)}(t) + \cdots + b_{n-1} \mathbf{u}^{(1)}(t) + b_n \mathbf{u}(t) \qquad (2.37)$$

where $\mathbf{y}^{(i)}(t)$ denotes the *i*th derivative of $\mathbf{y}(t)$, with similar notation for derivatives of $\mathbf{u}(t)$. For this differential equation, the initial conditions are the values of $\mathbf{y}$ and its first $n-1$ derivatives at $t=0$ together with the values of $\mathbf{u}$ and its first $n-2$ derivatives at $t=0$. (Remember that $\mathbf{y}$ is the unknown output function and $\mathbf{u}$ is the known input or forcing function. Thus there are n initial condition terms associated with $\mathbf{y}$ when it is described by the *n*th-order equation above.) Again, we can show that the solution function $\mathbf{y}(t)$ can be written as the sum of a homogeneous solution and a particular solution. We will not go into details here since our aim is to give a state equation description of this input-output model; using the state equation description we may obtain the solution for $\mathbf{y}(t)$ using the methods developed earlier in this chapter.

Sec. 2.2 Inputs, Outputs, and States

For notational convenience it is convenient to introduce the symbol $\mathbf{D}$ to denote the differentiation operator,

$$\mathbf{D} = \frac{d}{dt} \tag{2.38}$$

so that $\mathbf{D}\mathbf{y}(t) = \dot{\mathbf{y}}(t)$ and $\mathbf{D}\mathbf{u}(t) = \dot{\mathbf{u}}(t)$. Higher-order derivatives are naturally expressed as powers of $\mathbf{D}$ since, for example,

$$\frac{d^2}{dt^2} = \frac{d}{dt}\left(\frac{d}{dt}\right) = \mathbf{D}(\mathbf{D}) = \mathbf{D}^2 \tag{2.39}$$

Finally, because differentiation is a linear operation on functions, we may write the input-output differential equation, (2.37), in the more compact form

$$(\mathbf{D}^n + a_1 \mathbf{D}^{n-1} + \cdots + a_{n-1}\mathbf{D} + a_n)\mathbf{y}(t)$$
$$= (b_1 \mathbf{D}^{n-1} + \cdots + b_{n-1}\mathbf{D} + b_n)\mathbf{u}(t) \tag{2.40}$$

Since we have seen that polynomial expressions involving $\mathbf{D}$ are meaningfully interpreted using derivatives, it is tempting to go one step further and "solve" equation (2.40) for $\mathbf{y}(t)$ in terms of $\mathbf{u}(t)$, expressing the result symbolically as

$$\mathbf{y}(t) = H(\mathbf{D})\mathbf{u}(t) = \frac{b_1 \mathbf{D}^{n-1} + \cdots + b_{n-1}\mathbf{D} + b_n}{\mathbf{D}^n + a_1 \mathbf{D}^{n-1} + \cdots + a_{n-1}\mathbf{D} + a_n} \mathbf{u}(t) \tag{2.41}$$

Keep in mind that an equation of the form $\mathbf{y}(t) = H(\mathbf{D})\mathbf{u}(t)$, where $H(\mathbf{D})$ is a rational function of the "symbol" $\mathbf{D}$ (i.e., a ratio of polynomial functions of $\mathbf{D}$) is simply a shorthand way of denoting the solution to the corresponding input-output differential equation. We will find that this symbolic representation is useful notation when we discuss interconnections of systems later in this section.

Example

The symbolic notation provides a slick way of finding the input-output representation of a state space model. Writing the state space equations as

$$\mathbf{D}\mathbf{x}(t) = \mathbf{A}\mathbf{x}(t) + \mathbf{B}\mathbf{u}(t)$$
$$\mathbf{y}(t) = \mathbf{C}\mathbf{x}(t)$$

we collect the terms in $\mathbf{x}(t)$ in the state equation to give

$$\mathbf{D}\mathbf{x}(t) - \mathbf{A}\mathbf{x}(t) = \mathbf{B}\mathbf{u}(t)$$

which may be rewritten as

$$(\mathbf{D}\mathbf{I} - \mathbf{A})\mathbf{x}(t) = \mathbf{B}\mathbf{u}(t)$$

"Solving" for $\mathbf{x}(t)$, we obtain

$$\mathbf{x}(t) = (\mathbf{D}\mathbf{I} - \mathbf{A})^{-1}\mathbf{B}\mathbf{u}(t)$$

Using this expression for $\mathbf{x}(t)$ in the system output equation gives

$$\mathbf{y}(t) = \mathbf{C}(\mathbf{D}\mathbf{I} - \mathbf{A})^{-1}\mathbf{B}\mathbf{u}(t)$$

This gives a formula for the rational function $H(\mathbf{D})$ appearing in equation (2.41) in terms of the matrices appearing in the state space model.

Finding State Space Equations from Input-Output Models

Given an input-output system description in the form above, the construction of a corresponding set of state equations is a bit more complicated in general than might be supposed from the simple mass-spring system examined above. It is *not* possible to use the output function and its derivatives as the components of a state vector when derivatives of the input function appear in the input-output differential equation. (Try it and see!) The state vector must depend on both the output and the input, so we are faced with the question of how to choose appropriate combinations of the various quantities appearing in the input-output equation in order to obtain a suitable state vector.

Rather than derive the general result from first principles, we will simply provide the answer first. Since it is a simple matter to carry out the calculation to check that the state equations do correspond to the input-output differential equation above, and because the result simply amounts to a recipe for writing a set of state equations "by inspection," we prefer this approach. The corresponding state equations are obtained by taking the following form:

$$\dot{\mathbf{x}}(t) = \mathbf{A}_o \mathbf{x}(t) + \mathbf{B}_o \mathbf{u}(t) \tag{2.42}$$

$$\mathbf{y}(t) = \mathbf{C}_o \mathbf{x}(t) \tag{2.43}$$

The matrices involved in these equations are the following ones, which are easily obtained from the input-output differential equation:

$$\mathbf{A}_o = \begin{bmatrix} -a_1 & 1 & 0 & \cdots & 0 \\ -a_2 & 0 & 1 & & \cdot \\ \cdot & \cdot & \cdot & & \cdot \\ \cdot & \cdot & & \cdot & 0 \\ \cdot & \cdot & & & 1 \\ -a_n & 0 & \cdot & \cdots & 0 \end{bmatrix}, \quad \mathbf{B}_o = \begin{bmatrix} b_1 \\ b_2 \\ \cdot \\ \cdot \\ \cdot \\ b_n \end{bmatrix}, \quad \mathbf{C}_o = \begin{bmatrix} 1 & 0 & \cdots & 0 \end{bmatrix} \tag{2.44}$$

Example

We will illustrate this form with the example of a series RLC circuit driven by a voltage source (Fig. 2.13). Let $\mathbf{y}(t)$ be the current in the circuit and let $\mathbf{u}(t)$ be the source voltage function. Then we have

$$\mathbf{u}(t) = R\,\mathbf{y}(t) + L\,\dot{\mathbf{y}}(t) + \frac{1}{C}\int^t \mathbf{y}(\tau)\,d\tau$$

which we differentiate to obtain a pure differential equation:

$$\dot{\mathbf{u}}(t) = R\,\dot{\mathbf{y}}(t) + L\,\ddot{\mathbf{y}}(t) + \frac{1}{C}\,\mathbf{y}(t)$$

Dividing this equation by L brings it to the form studied above, leading to the state equations:

Sec. 2.2 Inputs, Outputs, and States

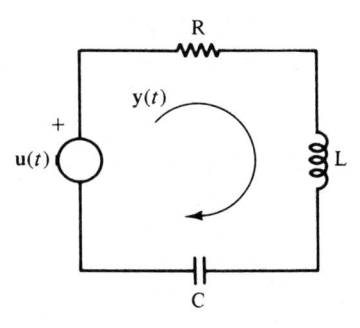

Figure 2.13 RLC circuit. The input is the voltage source, and the output is the loop current.

$$\dot{\mathbf{x}}(t) = \begin{bmatrix} -\dfrac{R}{L} & 1 \\ -\dfrac{1}{LC} & 0 \end{bmatrix} \mathbf{x}(t) + \begin{bmatrix} \dfrac{1}{L} \\ 0 \end{bmatrix} \mathbf{u}(t)$$

and

$$\mathbf{y}(t) = [\,1\ \ 0\,]\mathbf{x}(t)$$

Calling the matrices in these state equations **A**, **B**, and **C** as usual, we can also verify that the state equations correspond to the input-output model above by applying the general formula developed in the preceding example. There we showed that

$$\mathbf{y}(t) = \mathbf{C}(D\mathbf{I} - \mathbf{A})^{-1}\mathbf{B}\mathbf{u}(t)$$

In this case, the formula for the inverse of the 2×2 matrix $(D\mathbf{I} - \mathbf{A})$ can be applied, and the result is

$$\mathbf{y}(t) = \frac{CD}{LCD^2 + RCD + 1}\mathbf{u}(t)$$

which is the symbolic representation of the differential equation

$$(LCD^2 + RCD + 1)\mathbf{y}(t) = CD\mathbf{u}(t)$$

in agreement with the original input-output differential equation.

As this example suggests, it is just a matter of some routine algebraic manipulation to verify that these "canonical" state equations involving the matrices $\mathbf{A}_o$, $\mathbf{B}_o$, and $\mathbf{C}_o$, (2.44), produce the same input-output differential equation relating **y** and **u** as the nth-order differential equation. We will simply sketch the general argument. First, notice that $\mathbf{A}_o$ is a *companion matrix*, so we may obtain its characteristic polynomial "by inspection":

$$\det(\lambda\mathbf{I} - \mathbf{A}) = \lambda^n + a_1\lambda^{n-1} + \cdots + a_{n-1}\lambda + a_n \tag{2.45}$$

Next, we combine the equation for $\mathbf{y}(t)$ with the state equations to obtain equations for the derivatives $\mathbf{y}^{(1)}(t), \ldots, \mathbf{y}^{(n)}(t)$:

$$\mathbf{y}(t) = \mathbf{C}_o\mathbf{x}(t) \tag{2.46}$$

$$\mathbf{y}^{(1)}(t) = \mathbf{C}_o\dot{\mathbf{x}}(t) = \mathbf{C}_o\mathbf{A}_o\mathbf{x}(t) + \mathbf{C}_o\mathbf{B}_o\mathbf{u}(t) \tag{2.47}$$

$$\mathbf{y}^{(2)}(t) = \mathbf{C}_o\ddot{\mathbf{x}}(t) = \mathbf{C}_o\mathbf{A}_o^2\mathbf{x}(t) + \mathbf{C}_o\mathbf{A}_o\mathbf{B}_o\mathbf{u}(t) + \mathbf{C}_o\mathbf{B}_o\dot{\mathbf{u}}(t) \tag{2.48}$$

and so forth; the pattern should be clear. These expressions are substituted in the sum $\mathbf{y}^{(n)}(t) + a_1 \mathbf{y}^{(n-1)}(t) + \cdots + a_n \mathbf{y}(t)$, and it is easy to see that the terms multiplying $\mathbf{x}(t)$ can be grouped as $\mathbf{C}_o[\mathbf{A}_o^n + a_1 \mathbf{A}_o^{n-1} + \cdots + a_n \mathbf{I}]\mathbf{x}(t)$. The term in the brackets is zero because of the Cayley-Hamilton theorem applied to $\mathbf{A}_o$ (since the characteristic polynomial of $\mathbf{A}_o$ is $p(\lambda) = \lambda^n + a_1 \lambda^{n-1} + \cdots + a_n$). Finally, by multiplying out the products $\mathbf{C}_o \mathbf{A}_o^i \mathbf{B}_o$, the remaining terms are found to produce the desired result.

A slight variation on this development will enable the form of the state vector $\mathbf{x}(t)$ as a function of the system input and output variables to be determined. Besides its intrinsic importance, this is necessary to allow us to complete the description of the state equations, since we need to determine the appropriate initial conditions for $\mathbf{x}(0)$. To do this, we use the equations for the first $n-1$ derivatives of $\mathbf{y}(0)$ in terms of the state equation, leading to n linear equations for the unknown $\mathbf{x}(0)$. We obtain

$$\mathbf{y}(0) = \mathbf{C}_o \mathbf{x}(0), \quad \mathbf{y}^{(1)}(0) = \mathbf{C}_o \mathbf{A}_o \mathbf{x}(0) + \mathbf{C}_o \mathbf{B}_o \mathbf{u}(0), \ldots \quad (2.49)$$

and so on for $\mathbf{y}^{(2)}(0), \ldots, \mathbf{y}^{(n-1)}(0)$. Collecting these n equations into one vector equation to solve for the initial state $\mathbf{x}(0)$, we have

$$\begin{bmatrix} \mathbf{y}(0) \\ \mathbf{y}^{(1)}(0) \\ \vdots \\ \mathbf{y}^{(n-1)}(0) \end{bmatrix} = \mathbf{W}(\mathbf{A}_o, \mathbf{C}_o) \mathbf{x}(0) + \mathbf{T}(\mathbf{A}_o, \mathbf{B}_o, \mathbf{C}_o) \begin{bmatrix} \mathbf{u}(0) \\ \mathbf{u}^{(1)}(0) \\ \vdots \\ \mathbf{u}^{(n-2)}(0) \end{bmatrix} \quad (2.50)$$

where $\mathbf{T}(\mathbf{A}_o, \mathbf{B}_o, \mathbf{C}_o)$ is an $n \times (n-1)$ matrix whose form is of no particular interest at the moment and $\mathbf{W}(\mathbf{A}_o, \mathbf{C}_o)$ is the $n \times n$ matrix

$$\mathbf{W}(\mathbf{A}_o, \mathbf{C}_o) = \begin{bmatrix} \mathbf{C}_o \\ \mathbf{C}_o \mathbf{A}_o \\ \vdots \\ \mathbf{C}_o \mathbf{A}_o^{n-1} \end{bmatrix} \quad (2.51)$$

A calculation using the forms of $\mathbf{A}_o$ and $\mathbf{C}_o$ shows that this matrix is invertible (it's lower triangular with 1's on its diagonal and so has a determinant equal to 1), and thus the equations for $\mathbf{x}(0)$ always have a (unique) solution. (It should be pointed out that this procedure for determining $\mathbf{x}(0)$ can be carried out not only for the state equations involving the matrices $\mathbf{A}_o$, $\mathbf{B}_o$, and $\mathbf{C}_o$, but also when the state equations are described by any matrices for which $\mathbf{W}(\mathbf{A}, \mathbf{C})$ is invertible.)

Example

Completing the RLC circuit example started earlier, the equations for the initial conditions are

$$\begin{bmatrix} y(0) \\ \dot{y}(0) \end{bmatrix} = \begin{bmatrix} 1 & 0 \\ -\dfrac{R}{L} & 1 \end{bmatrix} \mathbf{x}(0) + \begin{bmatrix} 0 \\ \dfrac{1}{L} \end{bmatrix} u(0)$$

Sec. 2.2   Inputs, Outputs, and States

The solution for $\mathbf{x}(0)$ is

$$\mathbf{x}(0) = \begin{bmatrix} \mathbf{y}(0) \\ \dot{\mathbf{y}}(0) + \dfrac{R}{L}\mathbf{y}(0) - \dfrac{1}{L}\mathbf{u}(0) \end{bmatrix}$$

Notice that we could have used the same equations for a general time instant t rather than for $t=0$ to show what quantities make up the components of the state vector, namely

$$\mathbf{x}(t) = \begin{bmatrix} \mathbf{y}(t) \\ \dot{\mathbf{y}}(t) + \dfrac{R}{L}\mathbf{y}(t) - \dfrac{1}{L}\mathbf{u}(t) \end{bmatrix}$$

As indicated, the components of the state vector generally depend on both the input and output functions and their derivatives.

Interconnected Systems and Block Diagrams

Input-output models for systems provide a convenient framework for describing how systems can be interconnected: the inputs to a system can be taken to be the outputs of other systems. Viewing a system as an interconnection of subsystems is often helpful in analyzing, computing, and understanding system behavior, and in decomposing design problems for complex systems into sets of coupled, simpler problems.

For single-input, single-output linear systems, *block diagrams*, which provide a graphical representation of interconnections of input-output models, are convenient for many purposes. For use in block diagrams, an input-output model is represented as a block with an incoming line and an outgoing line for its input and output signals, respectively. It is convenient to label a block with the input-output differential equation description of the system it represents; for this purpose the label $H(\mathbf{D})$ is used to denote the system expressed in symbolic form as $\mathbf{y}(t) = H(\mathbf{D})\mathbf{u}(t)$ (Fig. 2.14a). Another kind of block is also used to represent the operation of multiplication of an input signal by a

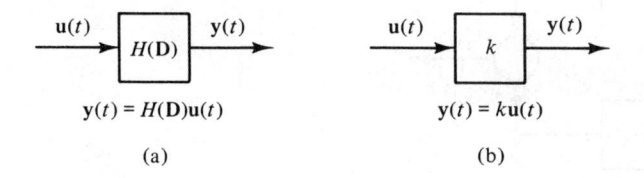

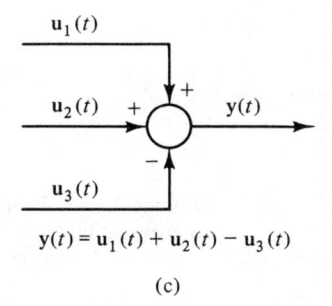

Figure 2.14 Block diagram components: (a) input-output system description; (b) scalar multiplication; (c) summing junction.

scalar, in which case the block is labeled with the value of the scalar multiplier (Fig. 2.14b). Finally, a summing junction is used to represent the operation of summing a number of input signals to obtain an output signal (Fig. 2.14c). With these basic elements available, interconnections of linear systems whose inputs are linear combinations of external input signals and outputs of subsystems can be drawn in graphical form as "block diagrams."

The three basic interconnection structures appearing in block diagrams are the *series combination*, *parallel combination*, and *feedback loop* structures (Fig. 2.15). For the series (or cascade) combination of two systems, the output of the first is used as the input of the second. Hence the equations are

$$\mathbf{y}_1(t) = H_1(\mathbf{D})\mathbf{u}_1(t) \qquad (2.52)$$

$$\mathbf{y}_2(t) = H_2(\mathbf{D})\mathbf{u}_2(t) \qquad (2.53)$$

$$\mathbf{u}_2(t) = \mathbf{y}_1(t) \qquad (2.54)$$

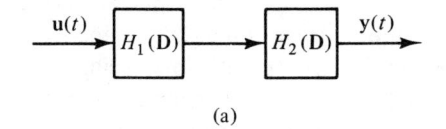

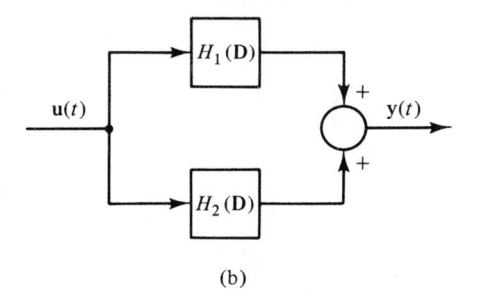

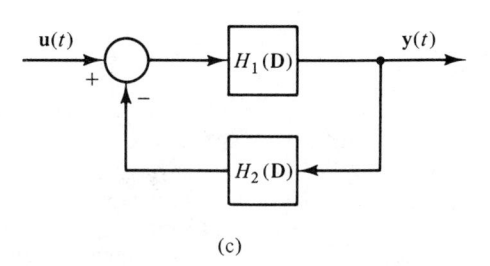

Figure 2.15 Basic block diagram configurations: (a) series connection; (b) parallel connection; (c) feedback loop.

The result of the series combination is another system whose input-output description may be obtained by eliminating the $\mathbf{y}_1(t)$ and $\mathbf{u}_2(t)$ variables, giving

$$\mathbf{y}_2(t) = H_2(\mathbf{D}) H_1(\mathbf{D}) \mathbf{u}_1(t) \tag{2.55}$$

From the expressions for H_1 and H_2 as fractions, an expression for the product as a fraction can easily be obtained; this provides the input-output differential equation for the series combination.

A similar development can be carried out for the parallel combination of two systems that receive a common input. The equations describing the parallel interconnection of two systems are

$$\mathbf{u}(t) = \mathbf{u}_1(t) = \mathbf{u}_2(t), \quad \mathbf{y}(t) = \mathbf{y}_1(t) + \mathbf{y}_2(t) \tag{2.56}$$

and the parallel combination is thus described by the input-output description

$$\mathbf{y}(t) = (H_1(\mathbf{D}) + H_2(\mathbf{D})) \mathbf{u}(t) \tag{2.57}$$

Finally, the feedback loop, with $H_1(\mathbf{D})$ in the "forward branch" and $H_2(\mathbf{D})$ in the "feedback branch," is described by the equations

$$\mathbf{u}_1(t) = \mathbf{u}(t) - \mathbf{y}_2(t) \tag{2.58}$$

$$\mathbf{u}_2(t) = \mathbf{y}_1(t) \tag{2.59}$$

and its input-output description is

$$\mathbf{y}_1(t) = \frac{H_1(\mathbf{D})}{1 + H_1(\mathbf{D}) H_2(\mathbf{D})} \mathbf{u}(t) \tag{2.60}$$

Example

As a simple example, suppose that

$$H_1(\mathbf{D}) = \frac{\mathbf{D} + 3}{\mathbf{D}^2 + 2\mathbf{D} + 2}$$

$$H_2(\mathbf{D}) = \frac{1}{\mathbf{D} + 4}$$

The parallel combination of $H_1(\mathbf{D})$ and $H_2(\mathbf{D})$ leads to the input-output description

$$\mathbf{y}(t) = \left(\frac{\mathbf{D} + 3}{\mathbf{D}^2 + 2\mathbf{D} + 2} + \frac{1}{\mathbf{D} + 4} \right) \mathbf{u}(t)$$

$$= \frac{2\mathbf{D}^2 + 9\mathbf{D} + 14}{\mathbf{D}^3 + 6\mathbf{D}^2 + 10\mathbf{D} + 8} \mathbf{u}(t)$$

For many complex block diagrams, a series of applications of the formulas derived above for the three basic interconnection structures will produce the simplified input-output description. The algebraic manipulations used to determine the input-output description of a complex interconnected system can be described in a highly systematic form that depends on the topological structure of its block diagram. The simplification process can easily be automated by using a symbolic computation program such as MACSYMA, REDUCE, or MATHEMATICA.

For systems described in terms of state space equations, the role of interconnections and subsystems is sometimes not appreciated, but it is equally important. For example, in writing down state equations for electric circuits, we employ the state space equations for the elemental subsystems, inductors and capacitors, along with the Kirchhoff network laws that describe how the variables involved in the elemental state space equations are related. Thus, a circuit is viewed quite naturally from the "bottom-up" perspective as a collection of interconnected subsystems. In writing down state equations for a mechanical system of masses and springs, the elemental subsystems are the second-order differential equations representing Newton's law for the positions of the masses, and the forces that represent the nature of the system interconnections are described in terms of positions. The velocity variables are added to obtain state equations, for both the subsystems and the overall system.

Both of these examples illustrate another point: The interconnected subsystems view of a linear state space system is directly related to the choice of state variables (i.e., by the choice of basis for the state space). On the other hand, there is an easy procedure for obtaining a block diagram that corresponds to any state space system of the usual form

$$\dot{\mathbf{x}}(t) = \mathbf{A}\mathbf{x}(t) + \mathbf{B}\mathbf{u}(t)$$

$$\mathbf{y}(t) = \mathbf{C}\mathbf{x}(t)$$

Write down n identical elementary input-output systems whose symbolic representation is $H(\mathbf{D}) = 1/\mathbf{D}$. These systems are simple "integrators" and if we choose to label the input to the ith block as $\dot{x}_i(t)$, then the output is $x_i(t)$. The ith row of the state equations describes how the input to the ith block is composed of a linear combination of outputs of the integrators and inputs, and hence how the integrators must be interconnected in the block diagram. The output equation describes how the system outputs are formed from linear combinations of the integrator outputs. Hence the state equations are simply an algebraic description of a particular block diagram structure portraying the system as an interconnection of simple dynamic subsystems.

State space basis changes can be viewed as transformations of the associated block diagrams. For example, consider a linear system whose **A** matrix can be brought to diagonal form by similarity transformation. The corresponding transformation of the state space basis provides a set of state variables that are completely decoupled from each other, so the transformed system turns out to be a particularly simple interconnection of subsystems; the block diagram corresponding to the diagonal form is a set of parallel branches (Fig. 2.16). Similarly, a system in the form described by the matrices $\mathbf{A}_o$, $\mathbf{B}_o$, and $\mathbf{C}_o$ may be drawn as a block diagram consisting of series and feedback interconnections (Fig. 2.17).

2.3 STABILITY

Before discussing some quantitative aspects of system responses, we mention the most important qualitative one, stability. While there are many mathematical formulations of the intuitive notion of a stable system, in the case of linear systems we really don't need to make many distinctions, since all of the formulations will turn out to coincide under

Sec. 2.3 Stability

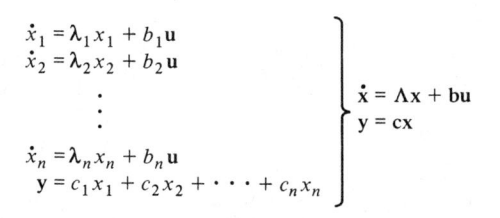

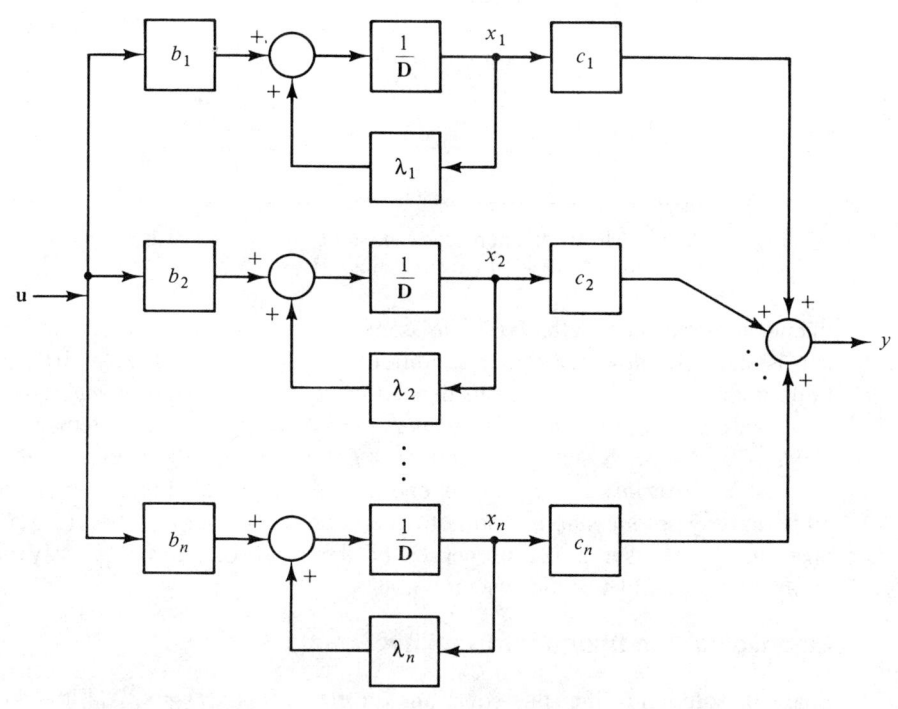

$$\left.\begin{array}{l}\dot{x}_1 = \lambda_1 x_1 + b_1 \mathbf{u} \\ \dot{x}_2 = \lambda_2 x_2 + b_2 \mathbf{u} \\ \quad \vdots \\ \dot{x}_n = \lambda_n x_n + b_n \mathbf{u} \\ y = c_1 x_1 + c_2 x_2 + \cdots + c_n x_n\end{array}\right\} \begin{array}{l}\dot{\mathbf{x}} = \Lambda \mathbf{x} + \mathbf{bu} \\ y = \mathbf{cx}\end{array}$$

Figure 2.16 Block diagram for a diagonal system showing parallel paths, each with a feedback loop.

reasonable assumptions. So without further elaboration, we proceed directly to the definitions.

Stability Definitions

We say that the linear system

$$\dot{\mathbf{x}}(t) = \mathbf{A}\mathbf{x}(t) + \mathbf{B}\mathbf{u}(t); \quad \mathbf{x}(0) = \mathbf{x}_0 \tag{2.61}$$

$$\mathbf{y}(t) = \mathbf{C}\mathbf{x}(t) \tag{2.62}$$

is *stable* if all unforced solutions of the state equations, (2.61), are bounded for all positive time (Fig. 2.18a). This condition is expressed in mathematical terms by using the

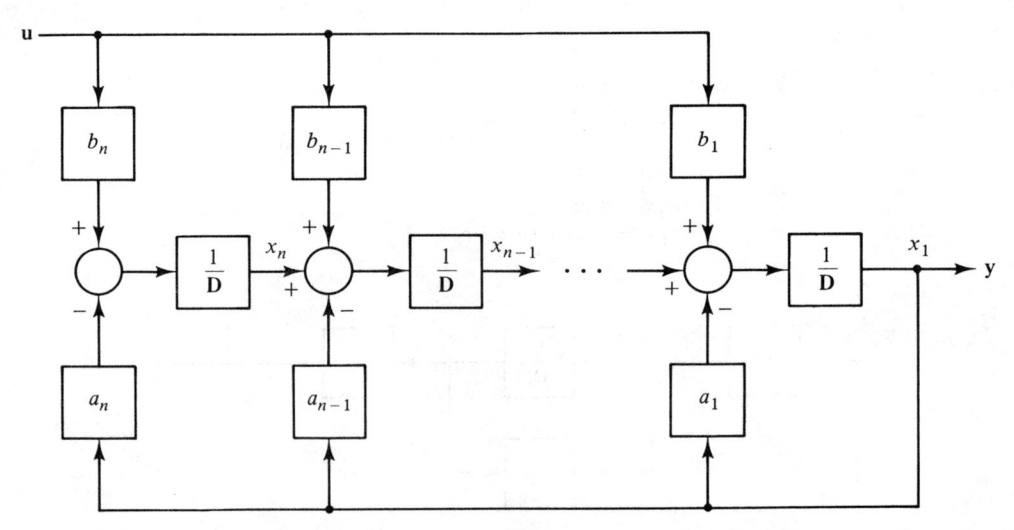

Figure 2.17 Block diagram corresponding to $\dot{\mathbf{x}} = \mathbf{A}_o\mathbf{x} + \mathbf{B}_o\mathbf{u}$, $y = \mathbf{C}_o\mathbf{x}$, showing series, parallel, and feedback loop structures.

Euclidean norm, or length, $\|\mathbf{x}(t)\|$, as a measure of the "size" of $\mathbf{x}(t)$ at time t; the system is stable if for every initial condition $\mathbf{x}_0$, when $\mathbf{u}(t) \equiv \mathbf{0}$ then $\|\mathbf{x}(t)\| < B$ for some finite number B. (B is allowed to depend on $\mathbf{x}_0$.) If a system is stable and in addition every unforced solution tends to $\mathbf{0}$ as $t \to \infty$, we say that the system is *asymptotically stable* (Fig. 2.18b). A stable system that is not asymptotically stable is said to be *marginally stable*. An *unstable system* is one that is not stable (Fig. 2.18c). Notice that stability and asymptotic stability do not depend on the output equation (2.62) at all; on the other hand, the asymptotic properties of the state vector $\mathbf{x}(t)$ clearly determine the asymptotic properties of the system output $\mathbf{y}(t)$.

Eigenvalue Conditions for Stability

Since the solution of the state equations for $\mathbf{u}(t) \equiv \mathbf{0}$ is $\mathbf{x}(t) = e^{\mathbf{A}t}\mathbf{x}_0$, it is the form of $e^{\mathbf{A}t}$ that determines stability and asymptotic stability. In other words, stability and asymptotic stability depend only on the matrix $\mathbf{A}$. A consideration of the diagonal case goes a long way toward showing exactly how stability depends on $\mathbf{A}$, and the answer turns out to involve eigenvalues (again!). Recall that for a diagonal matrix Λ, $e^{\Lambda t}$ is the diagonal matrix whose diagonal entries are $\{e^{\lambda_i t}\}$, where the $\{\lambda_i\}$ are the diagonal entries of Λ. For each exponential function (e.g., $e^{\lambda_i t}$) the limiting behavior for large positive t is determined by the real part of the corresponding eigenvalue (of $\mathbf{A}$), λ_i, since

$$e^{\lambda t} = e^{\lambda_R t} e^{j\lambda_I t} \tag{2.63}$$

where the subscripts R and I denote the real and imaginary parts, respectively, and j is the imaginary unit. Since the second term in the product is a complex number with unit magnitude for all t, the limiting behavior of the product is determined by the limiting behavior of the first term. To assure that this term tends to zero for large t, it is necessary and sufficient that $\lambda_R < 0$.

Sec. 2.3　Stability

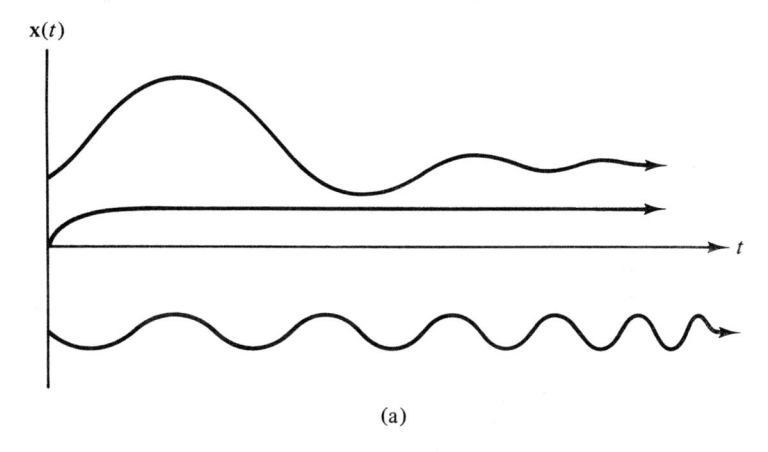

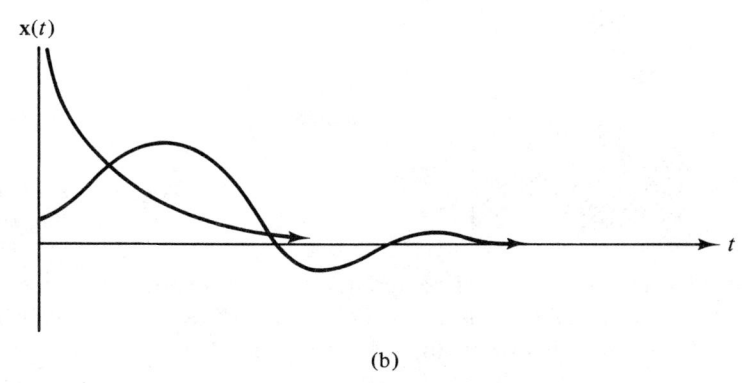

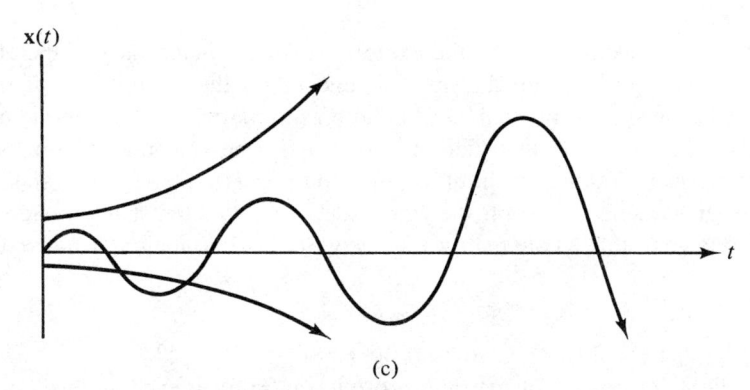

Figure 2.18　Qualitative properties of solutions: (a) stable; (b) asymptotically stable; (c) unstable.

This in fact provides the necessary and sufficient condition for asymptotic stability that holds in complete generality, regardless of whether **A** can be diagonalized by a similarity transformation or not: The linear system given above is asymptotically stable if and only if all of the eigenvalues of **A** lie in the open left half of the complex plane (Fig. 2.19), or in other words, $\text{Re}(\lambda_i) < 0$ for every eigenvalue, λ_i, of **A**. (Here Re is used to denote the real part of a complex quantity.) To verify this, let **A** be $n \times n$, and consider the equation

$$e^{\mathbf{A}t} = e^{\mathbf{A}(\tau+k)} = e^{\mathbf{A}\tau}e^{\mathbf{A}k} \tag{2.64}$$

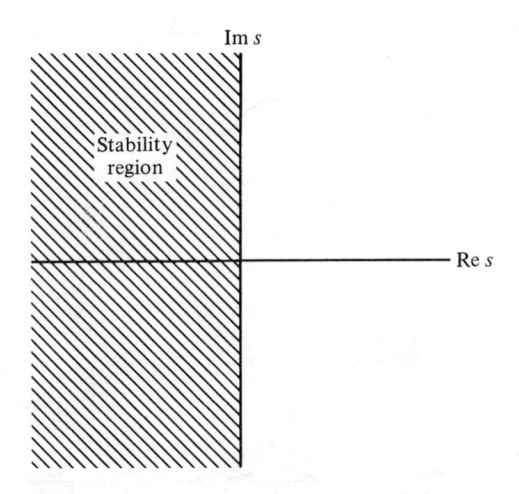

Figure 2.19 Complex s-plane showing the stability region.

where k is the integer part of t and τ lies between 0 and 1. Let M be an upper bound on the maximum magnitude of all the entries of $e^{\mathbf{A}\tau}$ over the interval $0 \leq \tau \leq 1$. Using the basic definition of matrix multiplication we may bound the magnitude of each element of $e^{\mathbf{A}t}$ as

$$|(e^{\mathbf{A}t})_{ij}| \leq Mn \max_{i,j}|(e^{\mathbf{A}k})_{ij}| \tag{2.65}$$

This results from bounding the expression for the magnitude of the elements of the product matrix, $e^{\mathbf{A}t}$, using the upper bound M on the magnitudes of the elements of the first factor and a similar upper bound on the magnitudes of the elements of the second factor, remembering that there are n terms in the sum that determines each element of the product. Taking the limit as $t \to \infty$ on the left side results in taking the limit as the integer variable $k \to \infty$ on the right side. For the largest magnitude element of $e^{\mathbf{A}k}$ to tend to zero, this whole matrix must tend to the zero matrix, so we require that

$$\lim_{k \to \infty} e^{\mathbf{A}k} = \lim_{k \to \infty} (e^{\mathbf{A}})^k = \mathbf{0} \tag{2.66}$$

In the first equality we have used the fact that $e^{\mathbf{A}k} = e^{(\mathbf{A}+\mathbf{A}+\cdots+\mathbf{A})} = (e^{\mathbf{A}})^k$. Recall that the limits of powers of matrices arose in our convergence analysis of iterative solutions of linear equations, equation (1.86), and in our discussion of the matrix version of the geometric series, equation (1.112); the limit in (2.66) holds if and only if the eigenvalues of $e^{\mathbf{A}}$ are less than 1 in magnitude. But since the eigenvalues of $e^{\mathbf{A}}$ are the complex numbers $\{e^{\lambda_i}\}$, where the $\{\lambda_i\}$ are the eigenvalues of **A**, this condition holds if and only if **A** has all of its eigenvalues in the open left half of the complex plane.

Characteristic Polynomial Conditions for Stability

The preceding analysis is summarized for practical purposes by saying that asymptotic stability of linear systems is strictly an algebraic property, being determined by the eigenvalues of the **A** matrix of the state equations. It is also true that asymptotic stability is determined by the characteristic polynomial of **A**, $p(\lambda) = \det(\lambda \mathbf{I} - \mathbf{A})$, since the eigenvalues are its zeros. In fact, it is not necessary to know the eigenvalues at all because it is possible to determine whether or not all of the zeros of a polynomial lie in the open left half of the complex plane by testing some inequalities involving easily computed functions of the polynomial's coefficients. The result is a stability test known as the Routh-Hurwitz test. The importance of this test is worth reemphasizing: Unlike the eigenvalues, which are the roots of $p(\lambda)$ and thus which are not expressible as algebraic functions of the coefficients of $p(\lambda)$, asymptotic stability can be determined from certain rational functions of these coefficients which are easily computable.

Now let us present the explicit form of the Routh-Hurwitz test. For the polynomial $p(\lambda) = \lambda^n + a_1 \lambda^{n-1} + a_2 \lambda^{n-2} + \cdots + a_n$, the first step is to examine the coefficients and verify that they are all positive; this is a necessary condition for the roots of $p(\lambda)$ to lie in the open left half of the complex plane. The second step requires computation of the *Routh array*. This array is composed of a sequence of rows of numbers that are computed recursively. The first two rows are formed from the coefficients of $p(\lambda)$ according to the following pattern:

$$\begin{array}{ccccc} 1 & a_2 & a_4 & \cdots & a_{n-1} \\ a_1 & a_3 & a_5 & \cdots & a_n \end{array} \qquad (2.67)$$

(This is the pattern for n odd; for n even, the last entry of the first row of the array is a_n, and the last entry of the second row is set equal to zero.) A simple calculation is used to generate the third row of the array from the first two. For notational convenience, let $a_0 = 1$; then take

$$b_1 = \frac{a_1 a_2 - a_0 a_3}{a_1} \, , \quad b_2 = \frac{a_1 a_4 - a_0 a_5}{a_1} \, , \quad b_3 = \frac{a_1 a_6 - a_0 a_7}{a_1} \, , \quad \ldots \qquad (2.68)$$

giving the three-row partial array

$$\begin{array}{ccccc} 1 & a_2 & a_4 & \cdots & a_{n-1} \\ a_1 & a_3 & a_5 & \cdots & a_n \\ b_1 & b_2 & b_3 & \cdots & \end{array} \qquad (2.69)$$

The remaining rows of the array are computed by following the same pattern of calculations, always using the last two rows to calculate the next row to be added. For example, the computation of the fourth row's elements uses the second and third rows in the previously described sequence of calculations:

$$c_1 = \frac{b_1 a_3 - a_1 b_2}{b_1} \, , \quad c_2 = \frac{b_1 a_5 - a_1 b_3}{b_1} \, , \quad c_3 = \frac{b_1 a_7 - a_1 b_3}{b_1} \, , \quad \ldots \qquad (2.70)$$

The Routh array for an nth-degree polynomial consists of $n+1$ rows, and the calculations above can be carried to completion provided that no zero entry arises in the first column of the array. Once the array is obtained, the array is examined to determine if the entries in its first column are all positive. If so, the polynomial $p(\lambda)$ has all of its zeros in the open left half of the complex plane. (If not, the number of sign changes going from top to bottom of the first column gives the number of zeros of $p(\lambda)$ lying in the right half of the complex plane.)

Examples

The Routh-Hurwitz test gives easy-to-apply tests for low-order polynomials in general form. For the second-degree polynomial $\lambda^2 + a_1 \lambda + a_2$, the Routh array is

$$\begin{matrix} 1 & a_2 \\ a_1 & \\ a_2 & \end{matrix}$$

from which the conditions $a_1 > 0$ and $a_2 > 0$ are obtained as the necessary and sufficient conditions for the zeros to lie in the open left half of the complex plane. These conditions can easily be checked by looking at the "quadratic formula" for the zeros. For the case of a third-degree polynomial, the formula for the zeros becomes rather cumbersome, but the Routh-Hurwitz test is easy to use and gives a simple result. The Routh array is

$$\begin{matrix} 1 & a_2 \\ a_1 & a_3 \\ \dfrac{a_1 a_2 - a_3}{a_1} & \\ a_3 & \end{matrix}$$

so in addition to the positivity of the coefficients, one additional inequality is needed to make all of the entries of the first column of the array positive: $a_1 a_2 > a_3$.

For systems described in state equation form, it is necessary to find the characteristic polynomial of the **A** matrix in order to apply the Routh-Hurwitz test as a test for asymptotic stability. Consider the airplane pitch dynamics model introduced earlier.

$$p(\lambda) = \det \begin{bmatrix} \lambda - m_q & -m_\alpha \\ -1 & \lambda - k_\alpha \end{bmatrix} = \lambda^2 - (m_q + k_\alpha)\lambda + (m_q k_\alpha - m_\alpha)$$

The conditions for asymptotic stability are $(m_q + k_\alpha) < 0$ and $(m_q k_\alpha - m_\alpha) > 0$. Ordinarily, airplanes are designed to have asymptotically stable pitch dynamics. However, high-performance military fighter aircraft are sometimes designed to be a bit unstable, relying on control forces to counteract their natural tendency to oscillate with increasing amplitude in response to forces such as wind gusts that create sudden changes in pitch rate.

Discussion

The characterization of stable systems that are only marginally stable is more complicated than a simple analysis of the diagonal case suggests; the complication arises from the need to consider multiple eigenvalues more carefully when they occur on the imaginary axis of the complex plane. If **A** has all of its eigenvalues in the open left half-plane except for some distinct eigenvalues whose real part is zero, then the system is

Sec. 2.3 Stability

marginally stable. The general necessary and sufficient condition for stability is that **A** have no eigenvalues in the open right half-plane and that the number of linearly independent eigenvectors corresponding to eigenvalues with zero real part equal the number of such eigenvalues (with multiplicities determined by $p(\lambda)$).

Asymptotic stability is often a highly desirable qualitative property because it assures that the effects of a system's initial conditions die out and are "forgotten" as time goes on. For a system such as a telecommunications satellite in geosynchronous orbit, a fixed orientation is desirable to keep its antennas pointed at particular locations on the Earth's surface. A nonzero initial condition, corresponding to an error in orientation or its derivative, can be used to model the effect of a micrometeorite impact. If a system is to be controlled by using the input signal to overcome the effects of "wrong" initial conditions, it is intuitively reasonable that small amounts of control effort will suffice for an asymptotically stable system because of its natural tendency to return to its rest state. An unstable system requires substantial control effort to counteract its natural tendency to drift away from its rest state at an exponentially growing rate. For uncontrolled systems and for systems where control system failures occur, this can lead to disaster; the Tacoma Narrows Bridge collapse and the Chernobyl nuclear reactor accident are familiar examples of such catastrophic events. Intuitively speaking again, it is reasonable to expect that inherently unstable systems are capable of fast response to changing inputs, and this idea is exploited in the design of highly maneuverable military aircraft. (Recall the stability analysis of the airplane pitch dynamics model.)

Input-Output Stability

As suggested in the discussion above, system stability has important consequences for systems with inputs, and we will close this section with a brief look at how such issues have been incorporated into a mathematical model suitable for various applications. Besides assuring that $\mathbf{y}(t) \to \mathbf{0}$ for large t when $\mathbf{u}(t) \equiv \mathbf{0}$, asymptotic stability guarantees some important relationships between $\mathbf{y}(t)$ and $\mathbf{u}(t)$ in other cases. There are various notions of *input-output stability* for systems that have been developed to capture a certain kind of "nice" qualitative characteristic of a system's input-output behavior. The intuitive idea is that "small" inputs lead to "small" outputs, where the measures used to describe the sizes of the input and output are chosen to reflect various physically meaningful criteria. Two of the most common formulations lead to the notions of *bounded-input, bounded-output stability*, often abbreviated as BIBO stability, and L_2 *stability*.

To keep notation simple, we will assume that the system has a single input and a single output so that $\mathbf{u}(t)$ and $\mathbf{y}(t)$ are scalar functions. BIBO stability is the property that a bounded input signal produces a bounded output signal; zero initial conditions, $\mathbf{x}(0) = \mathbf{0}$, are assumed. It is fairly easy to see that an asymptotically stable linear system is BIBO stable, since we have the explicit formula for the output signal to analyze. For zero initial conditions,

$$\mathbf{y}(t) = \int_0^t \mathbf{C}e^{\mathbf{A}(t-\tau)}\mathbf{B}\mathbf{u}(\tau)\,d\tau \tag{2.71}$$

If $|\mathbf{u}(t)|$ is bounded, let $\|\mathbf{u}(t)\|_\infty$ denote its supremum (the least upper bound) over all t.

Then

$$|\mathbf{y}(t)| \leq \|\mathbf{u}(t)\|_\infty \int_0^t |\mathbf{C}e^{\mathbf{A}(t-\tau)}\mathbf{B}|\,d\tau \leq \|\mathbf{u}(t)\|_\infty \int_0^\infty |\mathbf{C}e^{\mathbf{A}\tau}\mathbf{B}|\,d\tau \qquad (2.72)$$

Asymptotic stability assures that the latter integral is finite. (A careful argument about this is left for the reader to construct. Basically, $|\mathbf{C}e^{\mathbf{A}t}\mathbf{B}|$ is the magnitude of a sum of exponentially damped terms.) Thus we have the bound

$$|\mathbf{y}(t)| \leq K_1 \|\mathbf{u}(t)\|_\infty \qquad (2.73)$$

where K_1 is a finite positive constant depending on the system, but not on the input $\mathbf{u}(t)$. Since this bound holds for all t we may conclude that the least upper bound over t is similarly bounded:

$$\|\mathbf{y}(t)\|_\infty \leq K_1 \|\mathbf{u}(t)\|_\infty \qquad (2.74)$$

The second variety of input-output stability, L_2 stability, is the property that finite energy inputs result in finite energy outputs. (It is again assumed that the system's initial conditions are zero.) This is expressed mathematically as

$$\int_0^\infty \mathbf{y}^2(t)\,dt \leq K_2 \int_0^\infty \mathbf{u}^2(t)\,dt \qquad (2.75)$$

whenever the integral involving $\mathbf{u}$ is finite, where K_2 is a finite positive constant that depends on the system but not on the input $\mathbf{u}(t)$. The verification that asymptotic stability assures L_2 stability requires some details that will not be needed elsewhere in this book (the Cauchy-Schwarz inequality or Parseval's theorem for Fourier transforms), so we will not concern ourselves with it.

Input-output stability is a very natural topic to study in the context of interconnected systems, where an understanding of the stabilizing and destabilizing influences of one system upon another are of crucial importance. Feedback interconnections are frequently employed to transform an unstable system into a stable one. As a simple example, consider the system

$$\dot{\mathbf{y}}(t) + a\mathbf{y}(t) = \mathbf{u}(t) \qquad (2.76)$$

When a is a negative number, the system is not BIBO stable. (For this simple system, a state space description is obtained simply by choosing the output as a state (i.e., with $\mathbf{x}(t) = \mathbf{y}(t)$). Asymptotic stability holds if and only if $a > 0$. It is also easy to see that when $a < 0$, the output resulting from a constant input signal is exponentially growing.) Connecting this system in a feedback configuration with a gain k amounts to choosing the input $\mathbf{u}$ according to

$$\mathbf{u}(t) = -k\mathbf{y}(t) + \mathbf{v}(t) \qquad (2.77)$$

where $\mathbf{v}$ is the new label for the external input to the interconnected system (Fig. 2.20). These equations may easily be combined to obtain

$$\dot{\mathbf{y}}(t) + (a+k)\mathbf{y}(t) = \mathbf{v}(t) \qquad (2.78)$$

and the interconnected system is BIBO stable provided that $(a+k) > 0$. For feedback interconnections of more complicated systems, the conditions for BIBO stability are

Sec. 2.4　Frequency-Domain Characteristics of Linear Systems

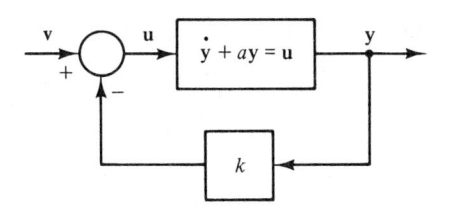

Figure 2.20 Stability of a feedback interconnection.

most easily described in terms of frequency-domain quantities. This brings us to the topic of the next section.

2.4 FREQUENCY-DOMAIN CHARACTERISTICS OF LINEAR SYSTEMS

As we have now seen, asymptotic stability of a system assures that the contribution to a system's output due to initial conditions fades away for large t; this leaves only the contribution due to system inputs, which can be analyzed in more detail. Indeed, the BIBO input-output stability property of asymptotically stable systems suggests that the family of phasors, inputs of the form $e^{j\omega t}$, is a "natural" class of input signals to investigate. Of course we can't really use phasors as inputs to any "real" system because this would require starting the input in the infinite past; but again because of asymptotic stability, we may simply interpret the phasor response of an asymptotically stable linear system (or linear differential equation) as giving an expression valid for times t large enough so that transients from initial conditions at $t = 0$ have died out.

The Transfer Function of a System

We start with the analysis of a single-input, single-output linear system described in input-output form:

$$\mathbf{y}^{(n)}(t) + a_1 \mathbf{y}^{(n-1)}(t) + \cdots + a_{n-1} \mathbf{y}^{(1)}(t) + a_n \mathbf{y}(t)$$
$$= b_1 \mathbf{u}^{(n-1)}(t) + b_2 \mathbf{u}^{(n-2)}(t) + \cdots + b_{n-1} \mathbf{u}^{(1)}(t) + b_n \mathbf{u}(t) \quad (2.79)$$

We examine solutions of this equation when the input is the complex exponential function $\mathbf{u}(t) = e^{st}$, where s is a complex number: $s = \sigma + j\omega$ in Cartesian form or $s = Re^{j\theta}$ in polar form. It is easy to see that the solution of the differential equation also takes the form of a complex exponential: $\mathbf{y}(t) = H(s)e^{st}$, where $H(s)$ is a complex number that depends on s (but does not vary with time t). Indeed, assuming this form for $\mathbf{y}$, plugging into the equation, and solving for $H(s)$ gives

$$H(s) = \frac{b_1 s^{n-1} + b_2 s^{n-2} + \cdots + b_{n-1} s + b_n}{s^n + a_1 s^{n-1} + \cdots + a_{n-1} s + a_n} \quad (2.80)$$

For the particular case of phasor inputs, $s = j\omega$. However, we see that the form of the output is the same for all complex exponential inputs. The effect of multiplication of the input, e^{st}, by the complex number $H(s)$ is most readily analyzed by representing both of the terms in the product in polar form: We already have $e^{st} = e^{\sigma t}e^{j\omega t}$ and we will define $H(s) = \rho(s)e^{j\phi(s)}$. In both products, the first factor is a nonnegative real number, the magnitude of e^{st} or $H(s)$, respectively. In the product, the magnitudes multiply and the angles add, giving

$$\mathbf{y}(t) = \rho(s)e^{\sigma t}e^{j(\omega t + \phi(s))} \tag{2.81}$$

In words, the effect of the system is to scale the magnitude of the input by a factor of $\rho(s)$ and to add a phase angle of $\phi(s)$ with respect to the phase of the input.

Viewed as a function of s, the function $H(s)$ plays an important role in characterizing a linear system, and it is commonly called the system's *transfer function*. For later use it will be important to have an expression for the transfer function in terms of the system state equations. Just as was the case in analyzing the input-output differential equations, we can verify that with a suitable choice of initial state and with the input $\mathbf{u} = e^{st}$, the solution of the state equations is a vector of phasors, say $\mathbf{x} = \mathbf{X}(s)e^{st}$. By substituting these choices of $\mathbf{x}$ and $\mathbf{u}$ in the state equations and solving, we find

$$\mathbf{x}(t) = (s\mathbf{I} - \mathbf{A})^{-1}\mathbf{B}e^{st} \tag{2.82}$$

so that

$$\mathbf{y}(t) = \mathbf{C}(s\mathbf{I} - \mathbf{A})^{-1}\mathbf{B}e^{st} \tag{2.83}$$

from which we may conclude that

$$H(s) = \mathbf{C}(s\mathbf{I} - \mathbf{A})^{-1}\mathbf{B} \tag{2.84}$$

Thus the transfer function of a system is easily obtained from whichever description is originally used as a model.

Because of the superposition property of the response of linear systems, we know that any input signal that can be written as a linear combination of complex exponentials will produce a corresponding linear combination of transfer function-weighted complex exponentials as an output. The analysis of this phenomenon lies at the heart of the theory of Laplace transforms. This powerful analytic method will probably be at least partly familiar to many readers; an in-depth study of this topic will not be pursued in this book. We will simply mention one fact that serves to tie together the time- and frequency-domain analyses of linear systems: The transfer function $H(s)$ is the Laplace transform of the weighting function $h(t)$.

The Frequency Response Function

For our purposes, phasor inputs will suffice. We will call $H(j\omega)$ the *frequency response function* of the system. $H(j\omega)$ is a complex number whose magnitude, $\rho(j\omega)$ and phase $\phi(j\omega)$ provide a useful characterization of the frequency dependence of the phasor response of the system. Indeed, for many system design problems, the frequency response characteristics of the system provide the most natural framework in which to specify the desired properties of the system; a wide range of examples may be found in the frequency-selective filters used in telecommunications systems and audio equipment.

Sec. 2.4   Frequency-Domain Characteristics of Linear Systems

The most common graphical representation of a frequency response function is the *Bode plot*, which actually consists of two graphs, the first being a plot of the magnitude $\rho(j\omega)$ of a frequency response function $H(j\omega)$ versus the frequency ω, plotted on a log-log scale. (It is customary to plot $20\log_{10}(\rho(j\omega))$ versus $\log_{10}\omega$. The scale of the vertical axis is then in decibels (dB) when the frequency response function represents a circuit voltage response.) The second graph of a Bode plot shows the phase function $\phi(j\omega)$ plotted against the log of the frequency ω. A little analysis shows why these particular choices of plots are made. First, if we write

$$H(j\omega) = \frac{b(j\omega)}{a(j\omega)} \qquad (2.85)$$

where we have used obvious notation for the numerator and denominator polynomials of H which are written out in terms of coefficients above, then

$$\log|H(j\omega)| = \log|b(j\omega)| - \log|a(j\omega)| \qquad (2.86)$$

so at each frequency the contributions of the numerator and denominator terms are easily combined to obtain the log magnitude of the frequency response function. (Similarly, the phase angles combine directly, without taking logs, in the same way.)

Even better, it is clear that the log magnitude of a polynomial can be obtained by adding up the log magnitudes of its polynomial factors. Since we may write

$$a(s) = \prod_{k=1}^{n}(s - \alpha_k) \qquad (2.87)$$

and

$$b(s) = b_1 \prod_{m=1}^{n-1}(s - \beta_m) \qquad (2.88)$$

by using the roots of the polynomials $a(s)$ and $b(s)$, $\{\alpha_k\}$ and $\{\beta_m\}$, respectively, we get the following simplified expression:

$$\log|H(j\omega)| = \log|b_1| + \sum_{m=1}^{n-1}\log|j\omega - \beta_m| - \sum_{k=1}^{n}\log|j\omega - \alpha_k| \qquad (2.89)$$

This expression admits a nice graphical interpretation since the term $|j\omega - s_0|$ is the length of the line joining the point s_0 to the point $j\omega$ in the complex plane. So for each frequency ω, the log magnitude expression is determined by the sum and difference of the logs of various distances that are easily visualized from a plot of the location of the "poles" and "zeros" of the function $H(s)$ (Fig. 2.21). (We use the standard terminology of complex function theory: The poles are the points where the function $H(s)$ is unbounded, namely the zeros of its denominator $a(s)$, which we have denoted by the $\{\alpha_k\}$; the zeros are the points where $H(s)$ has the value 0, namely the zeros of its numerator $b(s)$, which we have denoted by the $\{\beta_m\}$.)

A similar decomposition of the phase into the sum and difference of the angles of the complex numbers represented by the lines joining the poles and zeros to the point

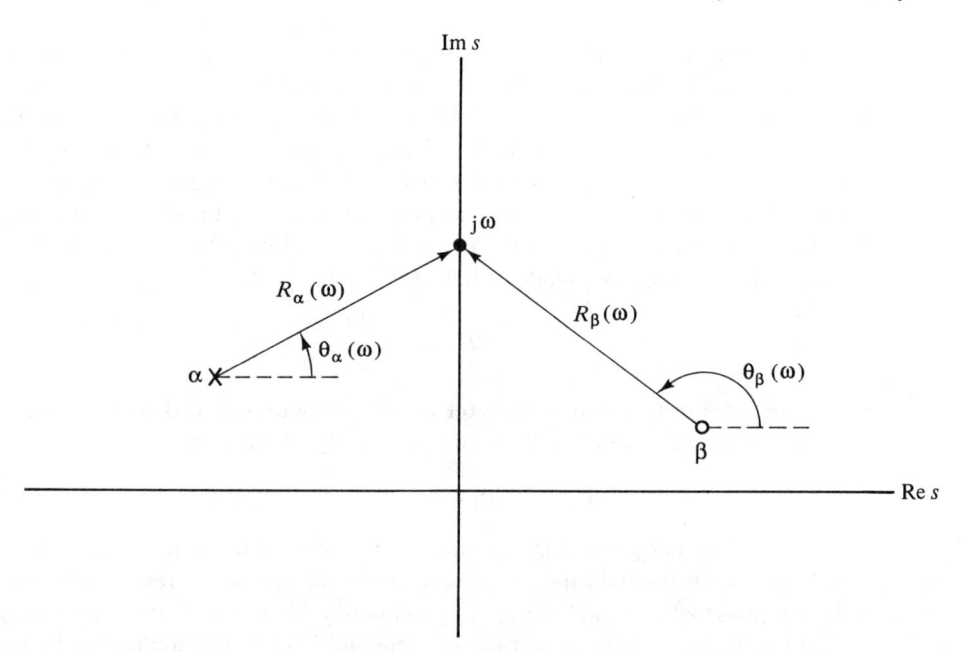

Figure 2.21 Complex s-plane plot showing relevant angles and magnitudes. $j\omega - \alpha = R_\alpha(\omega)e^{j\theta_\alpha(\omega)}$ and $j\omega - \beta = R_\beta(\omega)e^{j\theta_\beta(\omega)}$.

$j\omega$ (plus a contribution of π if b_1 is negative) can be obtained. The details are left to the student to work out.

What remains to be explained is why Bode plots involve the log of the independent variable ω. A little thought shows that this is the most convenient choice for obtaining a graph with a simple asymptotic form. Indeed, for large (i.e., "huge") ω any term of the form $|s_0 - j\omega|$ grows linearly with ω, so each term in the expression for the log magnitude of the frequency response function grows like $\log \omega$ for large ω. Thus, by using a log-log plot, the asymptotic form of a Bode magnitude plot for a frequency response function of the form studied here is linear; no other choice of scaling of the ω axis will give such a simple asymptotic form.

Frequency Response for First- and Second-Order Systems

We will give two examples that illustrate typical Bode plot features. These are representative cases that should be studied and understood in detail because of their importance in analyzing Bode plots for more complex systems in terms of simpler pieces. First we will investigate the simplest first-order system,

$$\dot{y}(t) + ay(t) = u(t) \tag{2.90}$$

whose transfer function is

$$H_1(s) = \frac{1}{s+a} \tag{2.91}$$

Sec. 2.4 Frequency-Domain Characteristics of Linear Systems

We assume that $a > 0$ so that the system is asymptotically stable (Fig. 2.22). To obtain the Bode magnitude plot we write

$$\log|H_1(j\omega)| = -\log a - \log|j(\omega/a) + 1| \tag{2.92}$$

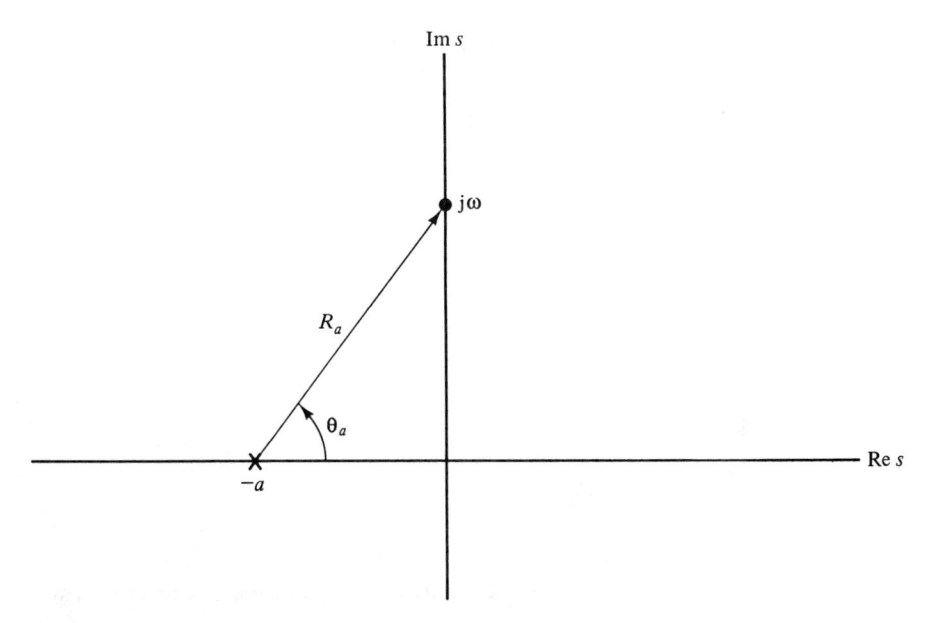

Figure 2.22 Complex s-plane showing single real pole and Bode plot quantities. $j\omega + a = R_a e^{j\theta_a}$, where magnitude R_a and phase angle θ_a vary with frequency ω.

Then it is clear that apart from a constant level determined by the $-\log a$ term, the plot is a fixed curve with respect to the normalized frequency variable (ω/a). For $\omega \ll a$, the frequency-dependent term is negligible and the plot is a horizontal straight line. For $\omega \gg a$, the frequency-dependent term is the dominant one, and its approximate value is $-\log(\omega/a)$. Thus the Bode magnitude plot is a straight line with slope -1 for very large ω. It is also easy to see that $\log|H_1(j\omega)|$ is monotonically decreasing as ω goes from 0 to ∞.

The analysis above shows that the Bode magnitude plot may be approximated as a piecewise linear function consisting of two straight-line segments that intersect at the value $(\omega/a) = 1$. Of course, the actual plot is a smooth curve that approaches the two straight lines asymptotically as $\omega \to 0$ and $\omega \to \infty$. It is easy to calculate that the true value of the magnitude at $(\omega/a) = 1$ is smaller than its asymptotic limit for small ω by the amount $\log\sqrt{2}$. This means that the magnitude of $H_1(ja)$ is a factor of $\sqrt{2}$ smaller than the magnitude of $H_1(0)$ (Fig. 2.23). For obvious reasons, we say that the system acts as a low-pass filter whose *bandwidth* is a (radians per second).

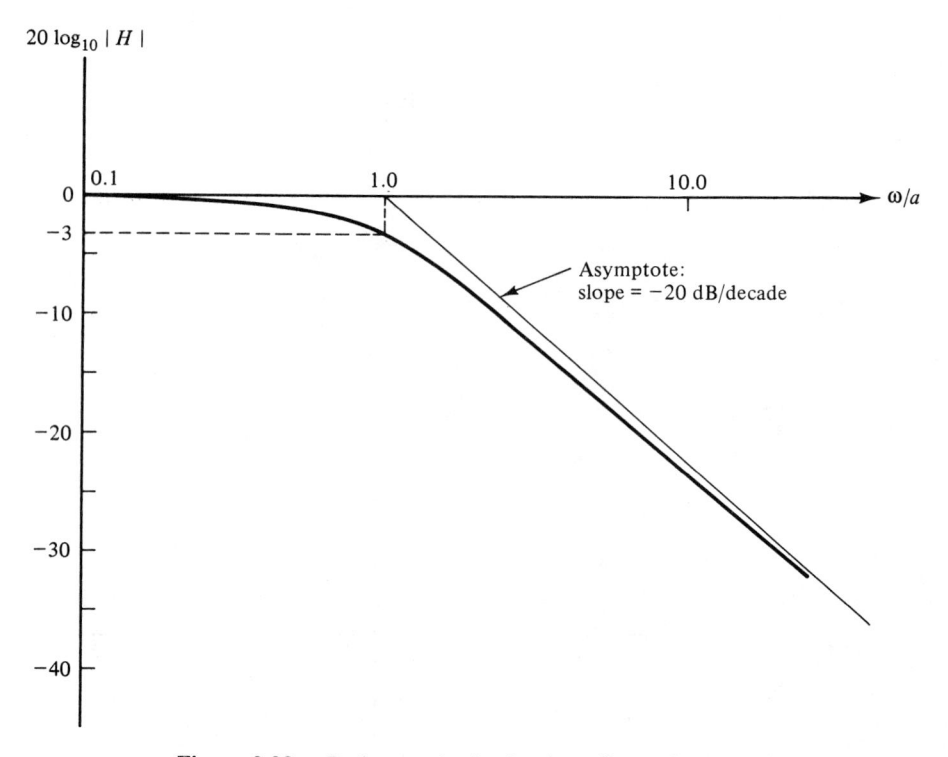

Figure 2.23 Bode magnitude plot for a first-order system.

From Fig. 2.22 the phase function has asymptotes 0 and $-\pi/2$ for very small and very large ω, respectively, and it is montonically decreasing (Fig. 2.24).

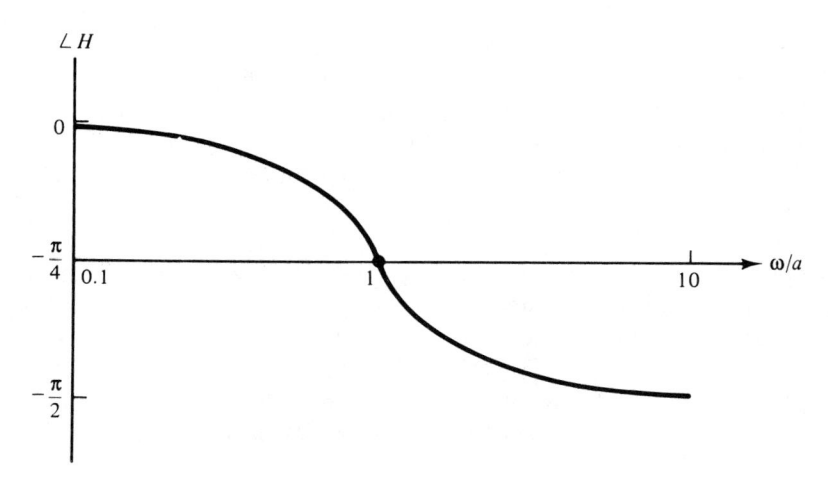

Figure 2.24 Bode phase plot for a first-order system.

Sec. 2.4 Frequency-Domain Characteristics of Linear Systems

The simplest second-order system is described by the equation

$$\ddot{y}(t) + a_1\dot{y}(t) + a_2 y(t) = u(t) \tag{2.93}$$

We again assume that the system is asymptotically stable, and in addition we will restrict attention to the case where the poles of the associated transfer function are not purely real numbers, since this case can be analyzed as the product of two first-order transfer functions of the kind just considered. A bit of simple algebra leads us to the choices $a_1 = 2\zeta\omega_n$ and $a_2 = \omega_n^2$. The resulting transfer function is

$$H_2(s) = \frac{1}{s^2 + 2\zeta\omega_n s + \omega_n^2} \tag{2.94}$$

This form arises from insisting that the denominator of $H_2(s)$ be factorable in the form

$$s^2 + 2\zeta\omega_n s + \omega_n^2 = (s - Re^{j\theta})(s - Re^{-j\theta}) \tag{2.95}$$

where $|R| > 0$ and $\pi/2 < \theta < \pi$, thus imposing the stability and complex-pole constraints desired (Fig. 2.25). These conditions on R and θ correspond to the following conditions on ζ and ω_n: $0 < \zeta < 1$ and $\omega_n > 0$. The parameter ζ is known as the system's *damping ratio*, while ω_n is the system's *natural frequency*.

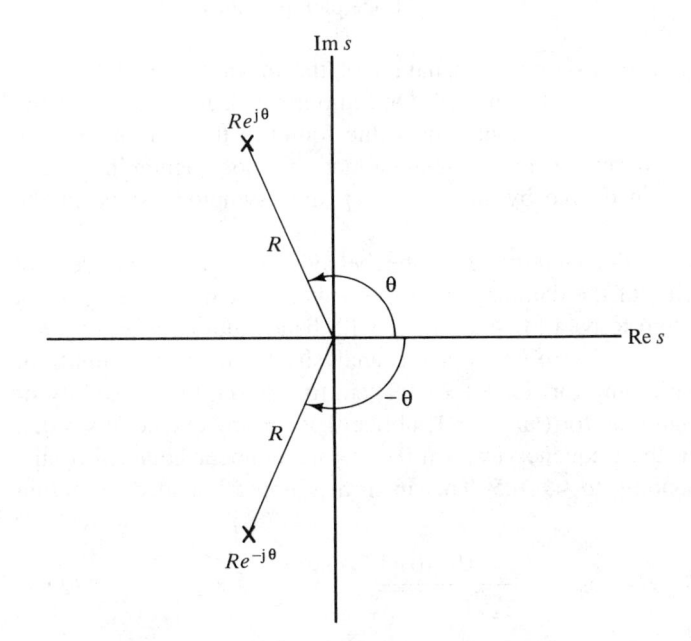

Figure 2.25 Complex s-plane showing a complex-conjugate pole pair.

Fig. 2.26 shows the magnitudes and angles that determine the Bode plot at a typical frequency ω. Rewriting the transfer function in a normalized form again facilitates an analysis of the system's Bode magnitude plot, so we write

$$H_2(j\omega) = \frac{1/\omega_n^2}{(j\omega/\omega_n)^2 + 2\zeta(j\omega/\omega_n) + 1} \tag{2.96}$$

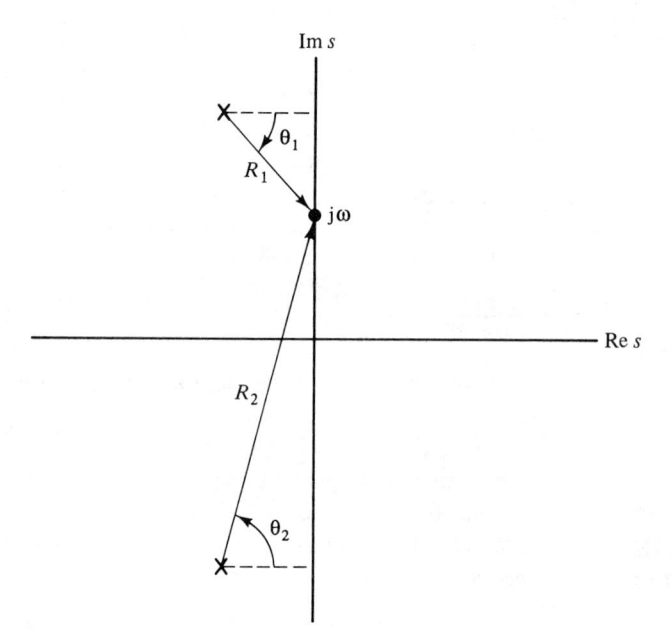

Figure 2.26 Complex s-plane showing a complex-conjugate pole pair and Bode plot quantities.

As in the first-order example, the asymptotic behavior of the magnitude plot for very small and very large (ω/ω_n) is easily determined. One important difference is that the slope of the plot for large (ω/ω_n) is -2, twice the value found in the first-order case. (And in general for a system described by a transfer function whose denominator and numerator polynomials differ in degree by an integer n_d, the asymptotic slope of the Bode plot is $-n_d$.)

For values of (ω/ω_n) in the vicinity of 1, the behavior of the magnitude plot depends critically on the value of the damping ratio ζ. For heavy damping, say $\zeta > 0.5$, the Bode plot shows the system to be a low-pass filter with bandwidth $\omega_r = \omega_n\sqrt{1-\zeta^2}$ (Figs. 2.27 and 2.28). This can be verified by some analysis. First, the magnitude of $H_2(j\omega)$ is monotonically decreasing for (ω/ω_r) > 1. Then, to rule out the possibility of a significant peak in the magnitude for (ω/ω_r) < 1, notice that for any choice of $\zeta > 0.5$, the magnitude of $H_2(j\omega)$ on the frequency interval $0 < \omega < \omega_r$ is upper bounded by the value of $|H_2(j\omega_r)|$ corresponding to $\zeta = 0.5$. This in turn is upper bounded by noting that for any ζ,

$$|H_2(j\omega_r)| = \frac{|H_2(0)|}{2\zeta\sqrt{1-0.75\zeta^2}} \tag{2.97}$$

and for $\zeta = 0.5$ the right side is bounded above by $1.1095|H_2(0)|$.

Light damping, corresponding to small values of ζ, say $\zeta < 0.25$, produces a pronounced peak in the Bode magnitude plot roughly centered on the point (ω/ω_r) = 1 (Fig. 2.27). In the limit of small damping, the system acts as a narrow bandpass filter. The width of the peak (taking the filter passband to be those frequencies where the magnitude is smaller than its peak value by no less than a $\sqrt{2}$ factor) is approximately

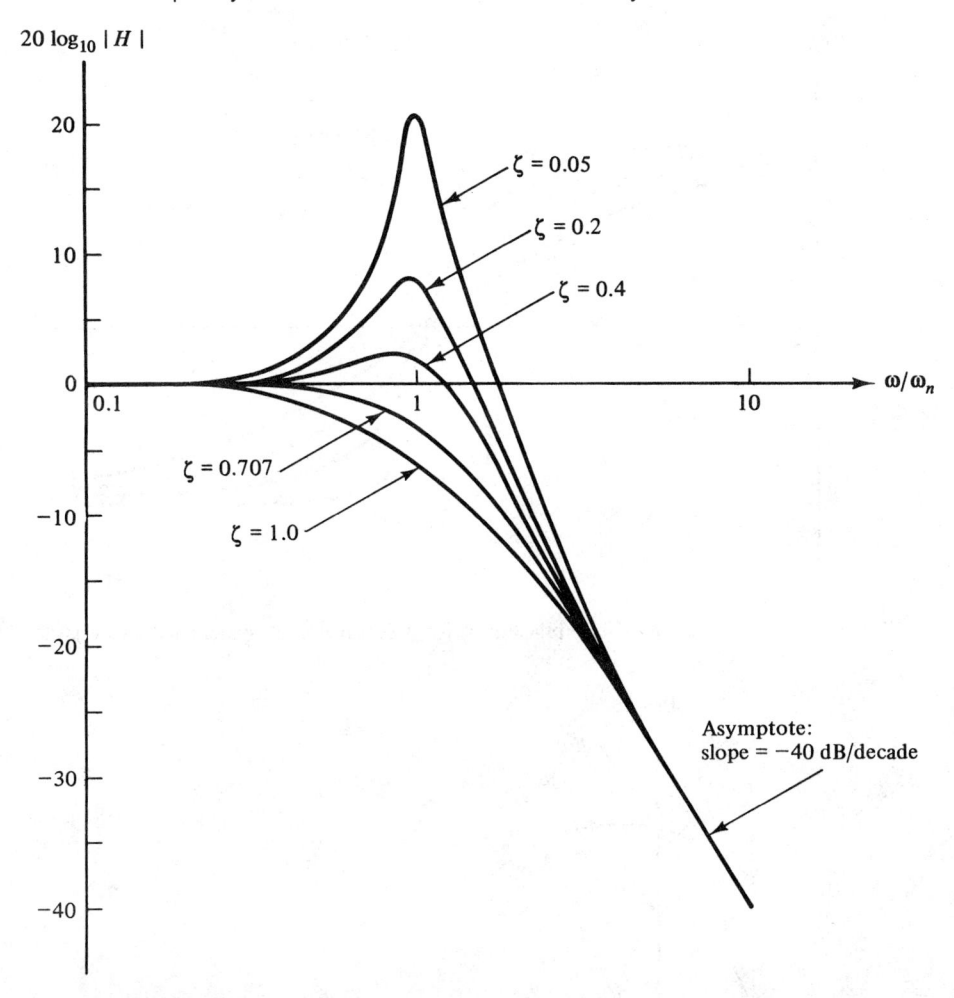

Sec. 2.4 Frequency-Domain Characteristics of Linear Systems

Figure 2.27 Bode magnitude plot for a second-order system with various damping values.

$2\zeta\omega_n$, so the relative bandwidth of the filter, given by the ratio of bandwidth to center frequency, is approximately 2ζ. (The approximations for small ζ result from the usual process of retaining only the lowest power of ζ in suitable power series expansions for the center frequency and peak width expressions.) A nice geometric construction for the peak frequency and half-power frequency is available (Fig. 2.29).

Example

The *Butterworth filters* comprise a class of systems with low-pass filter characteristics. For any order $n > 0$, the transfer function of the nth-order normalized Butterworth filter is given by the expression

$$H_n(s) = \frac{\overline{H}}{(s - \lambda_1)(s - \lambda_2) \cdots (s - \lambda_n)}$$

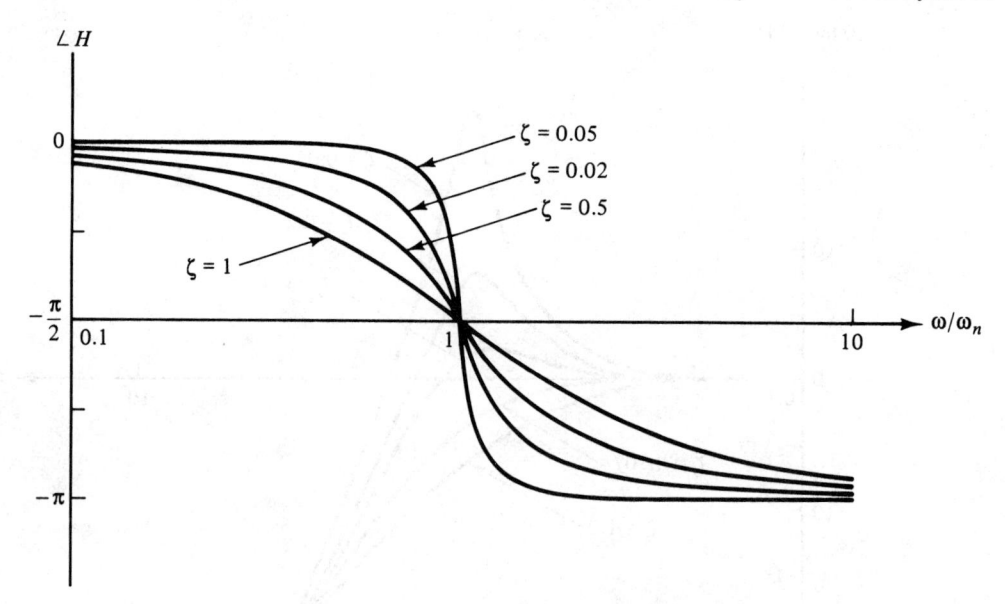

Figure 2.28 Bode phase plot for a second-order system with varous damping values.

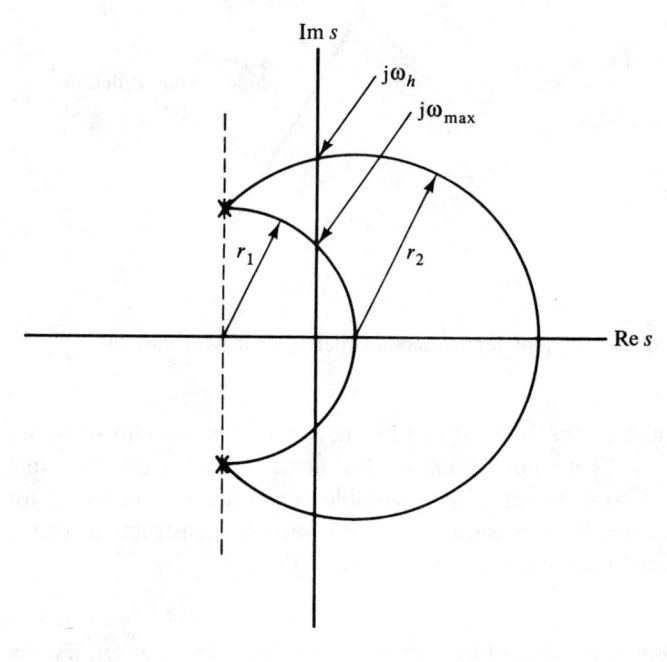

Figure 2.29 Complex s-plane showing a complex-conjugate pole pair and construction for ω_{max}, Bode magnitude peak frequency, and ω_h, Bode magnitude half-power frequency.

where

$$\lambda_k = e^{j\left(\frac{\pi}{2} + \frac{(2k-1)\pi}{2n}\right)} \quad \text{for } 1 \leq k \leq n$$

and $\bar{H}$ is simply an amplitude scaling factor used to adjust the gain at zero frequency (the "dc gain"). The poles of the nth-order normalized Butterworth filter lie in the left half of

Sec. 2.4 Frequency-Domain Characteristics of Linear Systems

the complex plane, equally spaced in angle along the unit circle. As n increases, the filters become "sharper." Bode magnitude plots of various normalized Butterworth filters are shown in Fig. 2.30. These plots are conveniently obtained by using the fact that the squared magnitude of a complex number is the magnitude of the product of the number and its complex conjugate, so that

$$|H_n(j\omega)|^2 = |H_n(j\omega)H_n(-j\omega)|$$

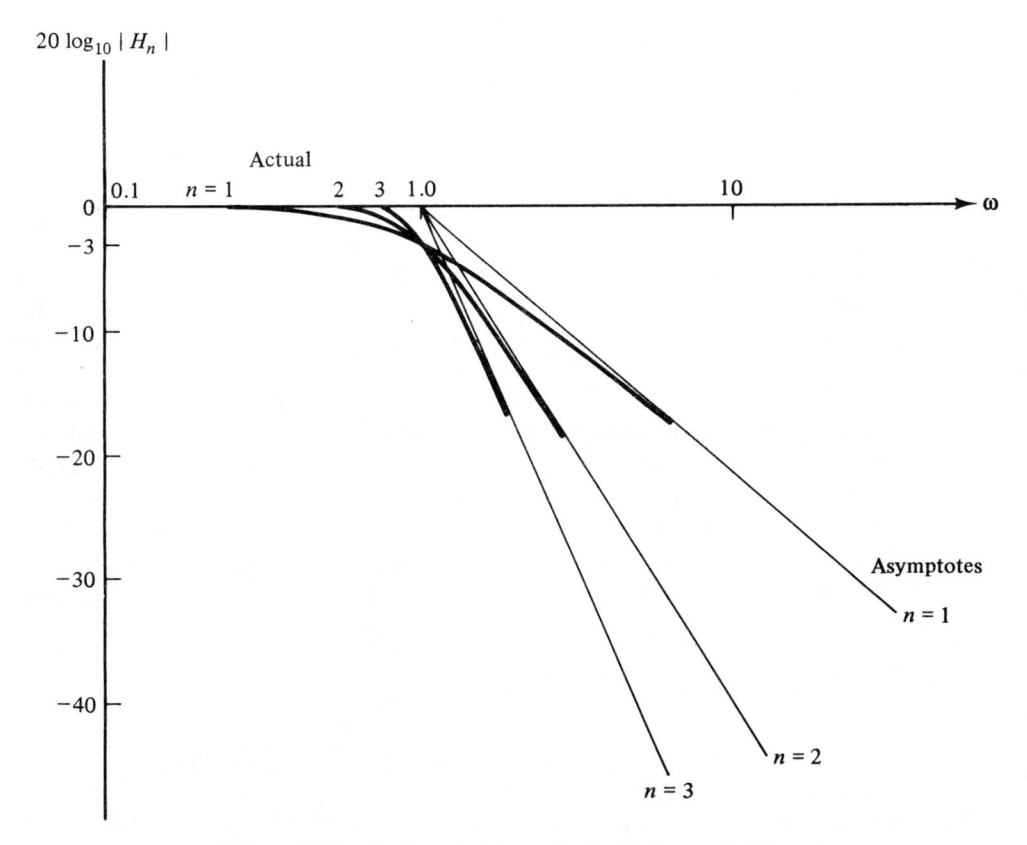

Figure 2.30 Bode magnitude plot for Butterworth filters.

The right side of this equation is the magnitude of $H_n(s)H_n(-s)$ when $s = j\omega$. Notice that the poles of $H_n(-s)$ are the negatives of the poles of $H_n(s)$ and that the $2n$th power of each pole of $H_n(s)$ (e.g., λ_k^{2n}) equals $(-1)^{n-1}$. Hence the poles of $H_n(s)H_n(-s)$ are all of the $2n$th roots of $(-1)^{n-1}$. This gives

$$|H_n(j\omega)|^2 = \frac{|\overline{H}|^2}{|(j\omega)^{2n} - (-1)^{n-1}|} = \frac{|\overline{H}|^2}{\omega^{2n} + 1}$$

From this expression it is easy to see, using the same kind of analysis applied earlier to the study of first- and second-order systems, that the magnitude of $H_n(j\omega)$ is monotonically decreasing in ω and that the Bode magnitude plot is well-approximated by two straight lines that intersect at the value $\omega = 1$, the filter bandwidth (Fig. 2.30).

In applications, the design of a Butterworth-type filter involves two choices; the order n is determined by the required sharpness of the filter cutoff, and the desired bandwidth ω_o, expressed in units of radians per second, serves as a frequency scaling parameter that specifies the filter transfer function as $H_n(s/\omega_o)$. For example, suppose that $n=3$ is selected and that a bandwidth of 1 kHz (i.e., $\omega_o = 2000\pi$ rad/s) is desired. The resulting Butterworth filter transfer function is

$$H_3(s) = \frac{\overline{H}}{((s/\omega_o)+1)((s/\omega_o)^2+(s/\omega_o)+1)}$$

Choosing $\overline{H} = 1$ will give unity dc gain.

2.5 TIME-DOMAIN CHARACTERISTICS OF LINEAR SYSTEMS

As indicated above, frequency-domain properties are very helpful in the study of steady-state responses of linear systems to inputs such as phasors (and to general inputs through the use of Laplace transforms). The steady-state response represents the asymptotic behavior of a system for times large enough that initial conditions have had a chance to die out, under the assumption that the system is stable. Here we will briefly introduce one qualitative measure of a system's *transient response*, by which we mean, intuitively speaking, how the system reacts to sudden changes in its input. Clearly, transient response characteristics are key features of many systems; think of the power steering system of an automobile that must be designed to provide fast response to a driver's commands. Automobile suspensions are another case where frequency response characteristics are important for a comfortable ride in steady-state conditions, yet good transient response characteristics are important in providing safe handling in the event that road hazards (potholes, tire blowouts, etc.) are encountered.

System Step Response and Time Constant

As a characteristic input for exhibiting transient response, the unit step function is commonly adopted. We have already introduced the notation $\mathbf{1}(t)$ for the unit step function, the function that is 0 for $t < 0$ and 1 for $t \geq 0$. Clearly, this is a reasonable mathematical representation for an abruptly changing input. For the state space system

$$\dot{\mathbf{x}}(t) = \mathbf{A}\mathbf{x}(t) + \mathbf{B}\mathbf{u}(t) \tag{2.98}$$

$$\mathbf{y}(t) = \mathbf{C}\mathbf{x}(t) \tag{2.99}$$

we may write down the solution explicitly when the input function $\mathbf{u}(t) = \mathbf{1}(t)$. We assume that the initial conditions are $\mathbf{0}$ at time $t = 0$ since we are interested in response to the abrupt change in input; you can think of this as meaning that the system has reached steady-state as the result of all of its past inputs. In any event, the step response is

$$\mathbf{x}_s(t) = \int_0^t e^{\mathbf{A}(t-\tau)}\mathbf{B}\,d\tau = \mathbf{A}^{-1}(e^{\mathbf{A}t} - \mathbf{I})\mathbf{B} \tag{2.100}$$

Sec. 2.5 Time-Domain Characteristics of Linear Systems

From this expression it is clear that the step response is determined by the matrix exponential function, and that for asymptotically stable systems the step response tends to the constant value $-\mathbf{A}^{-1}\mathbf{B}$. We have made the tacit assumption that $\mathbf{A}$ is invertible in order to evaluate the integral above; this is always true for asymptotically stable systems, whose eigenvalues all lie in the left half of the complex plane. Of course, it is always possible to determine the step response by finding the specific form of the matrix exponential $e^{\mathbf{A}t}$, multiplying by $\mathbf{B}$, and evaluating the integral (element by element).

As we know from our previous discussion, all of the elements of the matrix exponential are linear combinations of terms of the form $e^{\lambda_i t}$, where λ_i is an eigenvalue of the matrix $\mathbf{A}$. (There may be terms of this form multiplied by polynomials in t if $\mathbf{A}$ has repeated eigenvalues.) This suggests using the eigenvalue of $\mathbf{A}$ whose real part is least negative to provide a measure of the speed of response. To see how this is done, notice that if

$$e^{\lambda t} = e^{\lambda_R t} e^{j\lambda_I t} \tag{2.101}$$

where $\lambda = \lambda_R + j\lambda_I$ is a (complex) eigenvalue of $\mathbf{A}$, with $\lambda_R < 0$ from the assumption of asymptotic stability, we can bound the magnitude of the exponential by

$$|e^{\lambda t}| \le e^{-t/\tau} \tag{2.102}$$

where the "time constant" $\tau = |1/\lambda_R|$. Then it is clear that the "slowest" eigenvalue of $\mathbf{A}$ is the one with the longest time constant, and we take as the time constant of the matrix $\mathbf{A}$

$$\tau_\mathbf{A} = \max \left| \frac{1}{\mathrm{Re}(\lambda_i)} \right| \tag{2.103}$$

where the maximum is taken over all eigenvalues λ_i of $\mathbf{A}$.

It is useful to follow up on the two examples introduced earlier in connection with frequency response and Bode plots in order to see how time- and frequency-domain characteristics of systems are related. In looking at the first- and second-order examples, keep in mind that our analysis of system time constants will allow us to interpret the results for more general systems, namely those whose slowest eigenvalue (or complex conjugate pair of eigenvalues) are "well separated" from the remaining, "fast" eigenvalues. For such systems, the appropriate first- or second-order example provides a good estimate for the dominant, slow-response mode.

For the first-order system,

$$\dot{y}(t) + ay(t) = u(t); \qquad y(0) = 0 \tag{2.104}$$

an analytical solution for the step response is easy to find:

$$y(t) = \frac{1}{a}(1 - e^{-at})\mathbf{1}(t) \tag{2.105}$$

Fig. 2.31 is a graph of this expression. The system time constant $1/a$ is the inverse of the bandwidth of this low-pass filter.

For the second-order system,

$$\ddot{y}(t) + 2\zeta\omega_n \dot{y}(t) + \omega_n^2 y(t) = \omega_n^2 u(t); \qquad y(0) = \dot{y}(0) = 0 \tag{2.106}$$

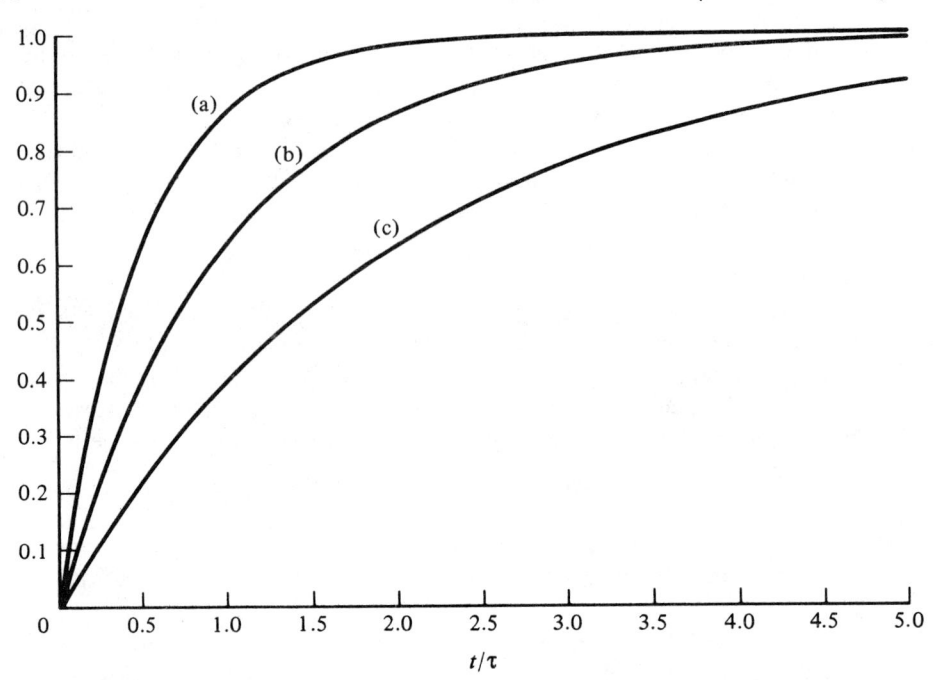

Figure 2.31 Step response for a first-order system. Curves: (a) $1/2\,\tau_{nom}$; (b) τ_{nom}; (c) $2\tau_{nom}$.

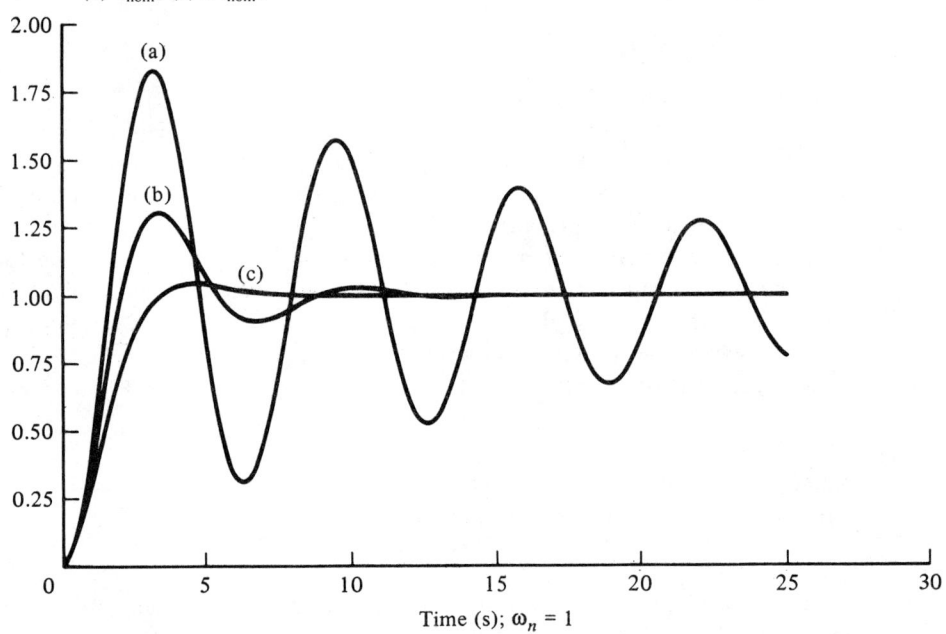

Figure 2.32 Step response for a second-order system: $\omega_n = 1$; $\tau_{nom} = 2\sqrt{2}$. Curves: (a) $6\tau_{nom}$; (b) τ_{nom}; (c) $1/2\,\tau_{nom}$.

Sec. 2.6 Discrete-Time Linear Systems

the step response involves substantially more calculation, but it may be verified to be

$$y(t) = \left(1 + \frac{\omega_n}{\omega_r} e^{-\zeta\omega_n t} \sin(\omega_r t - \phi)\right) \mathbf{1}(t) \quad (2.107)$$

where $\omega_r = \omega_n\sqrt{1-\zeta^2}$; the phase angle ϕ can be expressed in terms of ω_n and ζ also, but we will not need an explicit form for our purposes. Figs. 2.32 and 2.33 show families of second-order system step responses for fixed values of natural frequency and damping ratio, respectively. We note that the ratio ω_n/ω_r is approximately equal to the "overshoot" in the solution. Based on our previous analysis, we again find that the system bandwidth determines the system time constant, $1/\zeta\omega_n$. For a heavily damped system, this follows from a suitable approximation of ω_r for values of ζ close to 1. For a lightly damped system, the system bandwidth is associated with the bandpass filter interpretation obtained from the earlier Bode plot analysis.

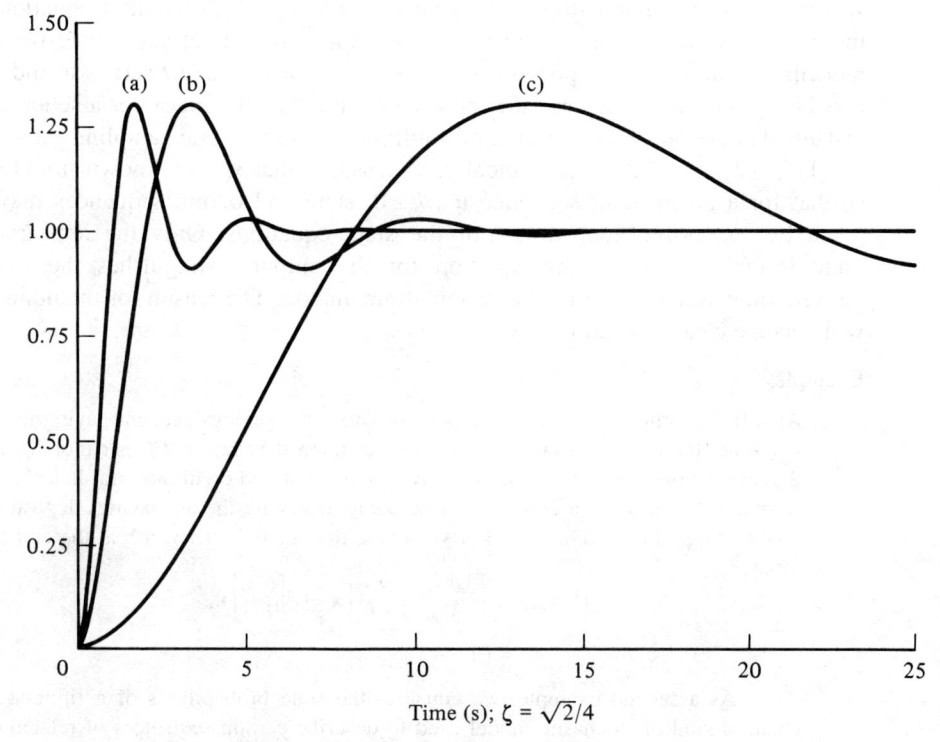

Figure 2.33 Step response for a second-order system: $\zeta = \sqrt{2}/4$; $\tau_{nom} = 2\sqrt{2}$. Curves: (a) $1/2\,\tau_{nom}$; (b) τ_{nom}; (c) $4\tau_{nom}$.

2.6 DISCRETE-TIME LINEAR SYSTEMS

Essentially all of the material we have developed in this chapter can be adapted to the case of discrete-time linear systems. Differential equations are replaced by difference equations, the notions of stability are suitably changed, and the frequency domain is modified appropriately. In this section we will give a very brief overview, motivated by the importance of discrete-time systems for applications where digital computation is involved, such as system simulation and computer-based control and signal processing. These applications will be covered in considerable detail in Chapter 3. A few other examples will be introduced shortly to provide some additional motivation.

State Space and Input-Output Models

The basic state equation description of a linear, time-invariant, discrete-time system is the following:

$$\mathbf{x}_{k+1} = \mathbf{F}\mathbf{x}_k + \mathbf{G}\mathbf{u}_k \qquad (2.108)$$

$$\mathbf{y}_k = \mathbf{H}\mathbf{x}_k + \mathbf{J}\mathbf{u}_k \qquad (2.109)$$

Just as in the continuous-time case where an analogous differential equation provides the model, we think of $\mathbf{x}$, $\mathbf{u}$, and $\mathbf{y}$ as state, input, and output vectors, respectively. For modeling of time-varying phenomena, where the real variable t serves to index the signals being considered, we use the integer variable k as the index for a sequence of time instants that are chosen to be integral multiples of some basic sampling interval T_s: ..., $(k-1)T_s$, kT_s, $(k+1)T_s$, ... Typically, we assume that $\mathbf{x}_0$ is a known initial condition, so that for a given input sequence $\mathbf{u}_k$, $k \geq 0$, state and output sequences may be computed by successive applications of the state equations. Only the appearance of an input-dependent term in the equation for the output distinguishes the form of the discrete-time system from its continuous-time analog. The reason for including this term will become clear in what follows.

Examples

As a first example, we consider a person who has a savings account paying interest at rate i_s (% per month) and a credit card charging interest at rate i_c (% per month). Let $\mathbf{x}_k$ be a 2-vector whose components are the savings account and credit account balances at the end of month k. Let $\mathbf{u}_k$ be a 2-vector whose components are the net savings deposit and the net credit charge during month k. Let $\mathbf{y}_k$ denote the person's net worth at the end of month k. Then

$$\mathbf{x}_{k+1} = \begin{bmatrix} 1+i_s & 0 \\ 0 & 1+i_c \end{bmatrix} \mathbf{x}_k + \begin{bmatrix} 1 & 0 \\ 0 & 1 \end{bmatrix} \mathbf{u}_k$$

$$\mathbf{y}_k = [\,1 \;\; -1\,]\,\mathbf{x}_k$$

As a second example, we consider the state probabilities of a finite-state Markov chain, a kind of stochastic model used to describe certain sequences of related events. Let $\mathbf{x}_0$ be an n-vector of *nonnegative* components whose sum is 1; thus the components of $\mathbf{x}_0$ can be viewed as the probabilities of a set of n mutually exclusive and exhaustive outcomes of an event at time instant 0. Let $\mathbf{F}$ be an $n \times n$ matrix whose entries are all nonnegative and whose column-sums are all 1, that is,

$$\sum_{i=1}^{n} (\mathbf{F})_{ij} = 1$$

The matrix element $(\mathbf{F})_{ij}$ is the conditional probability of the ith outcome following immediately after the jth outcome. For $k > 0$, $\mathbf{x}_k$ is a vector whose components give the

Sec. 2.6 Discrete-Time Linear Systems

probabilities of the n outcomes at time instant k. The formula for the probabilities is most conveniently expressed in the form of a discrete-time linear system:

$$\mathbf{x}_{k+1} = \mathbf{F}\mathbf{x}_k$$

As a particular case, suppose that the outcomes of a coin-flipping experiment are classified as H or T (head or tail). We suppose that there is no dependence between tosses of the coin at different time instants and that H and T are equally likely outcomes at each toss. Let the outcome of event E_k take values 1 or 2 according to the number of different outcomes resulting from the flips at time instants k, $k-1$, and $k-2$. For this case

$$\mathbf{F} = \begin{bmatrix} \frac{1}{2} & \frac{5}{6} \\ \frac{1}{2} & \frac{1}{6} \end{bmatrix}$$

and the state vector $\mathbf{x}_k$ has two components. Notice that

$$\mathbf{x}_e = \begin{bmatrix} \frac{5}{8} \\ \frac{3}{8} \end{bmatrix}$$

is an equilibrium solution; if $\mathbf{x}_0 = \mathbf{x}_e$, then $\mathbf{x}_k = \mathbf{x}_e$ for all $k > 0$. This is called the *stationary distribution* for the Markov chain.

As a third example, recall from Chapter 1 the discussion of splitting methods for solving linear equations iteratively. It was shown that solving

$$\mathbf{A}\mathbf{x} = \mathbf{y}$$

could be accomplished, under suitable conditions, by an iterative method given in equation (1.84),

$$\mathbf{A}_1 \mathbf{x}_{k+1} = \mathbf{A}_2 \mathbf{x}_k + \mathbf{y}$$

Taking $\mathbf{F} = \mathbf{A}_1^{-1}\mathbf{A}_2$, $\mathbf{G} = \mathbf{A}_1^{-1}$, and $\mathbf{u}_k = \mathbf{y}$ produces linear discrete-time state equations of the form (2.108).

As in the continuous-time case, the general solution to the state equations can be written as the sum of a homogeneous solution and a particular solution. However, the form of the discrete-time solution is much more elementary, amounting only to a very simple formula that describes the result of successive applications of the state equations:

$$\mathbf{x}_k = \mathbf{F}^k \mathbf{x}_0 + \sum_{i=0}^{k-1} \mathbf{F}^{k-1-i} \mathbf{G} \mathbf{u}_i \tag{2.110}$$

(where the summation term is zero for $k=0$). This formula is the discrete-time counterpart to the expression for the solution of a continuous-time system in terms of a matrix exponential function and a convolution integral, equation (2.26).

Clearly, the discrete-time case is much more elementary in form. Still, some of the ideas and methods introduced in the study of continuous-time systems can be used advantageously in the discrete-time case also. For example, computing $\mathbf{F}^k$ can be done with the aid of the eigenvector-eigenvalue factorization of $\mathbf{F}$, just as computation of $e^{\mathbf{A}t}$ can be computed with the aid of the eigenvector-eigenvalue factorization of $\mathbf{A}$. Indeed, exactly this trick was used in our proof of the Cayley-Hamilton theorem; see equation (1.35).

The input-output description of a discrete-time linear system may be described as a convolution sum,

$$\mathbf{y}_k = \sum_{i=0}^{k} g_{k-i} \mathbf{u}_i \qquad (2.111)$$

again in analogy with the convolution integral representation in the continuous-time case, equation (2.31), where the weighting function plays a role corresponding to the sequence $\{g_l\}$ that appears in this formula.

The sequence $\{g_l\}$ has an interesting interpretation; it is the *unit pulse response* of the discrete-time system. To see this, we use the solution formula for the state equations, (2.110), along with the output equation for the system, equation (2.109), to obtain an expression for the output sequence as a function of the input sequence, assuming that the initial state $\mathbf{x}_0 = \mathbf{0}$:

$$\mathbf{y}_k = \mathbf{J}\mathbf{u}_k + \sum_{i=0}^{k-1} \mathbf{H}\mathbf{F}^{k-1-i}\mathbf{G}\mathbf{u}_i, \qquad k > 0 \qquad (2.112)$$

If we define g_l as follows:

$$g_l = \begin{cases} \mathbf{J}, & l = 0 \\ \mathbf{H}\mathbf{F}^{l-1}\mathbf{G}, & l > 0 \end{cases} \qquad (2.113)$$

then we obtain the convolution sum formula for the zero-state output response, equation (2.111). Notice also that the formula for g_l in (2.113) is precisely the output that is obtained when the unit pulse sequence

$$\delta_k = \begin{cases} 1, & k = 0 \\ 0, & k > 0 \end{cases} \qquad (2.114)$$

is the system input. This is the origin of the name "unit pulse response." If $\mathbf{J} = 0$, so that the state equations have no "direct feedthrough" from input to output, then the unit pulse response is zero at time instant zero. To allow for nonzero values is the reason that the $\mathbf{J}$ term is included in the discrete-time state equation model in the first place.

Next we turn to the general form of an input-output model for discrete-time linear systems. You will notice that the derivative term of the continuous-time model was replaced by a "unit time shift" operation to obtain the discrete-time state equations, and this leads by analogy to the following high-order difference equation model:

$$\mathbf{y}_{k+n} + a_1 \mathbf{y}_{k+n-1} + \cdots + a_{n-1} \mathbf{y}_{k+1} + a_n \mathbf{y}_k$$
$$= b_0 \mathbf{u}_{k+n} + b_1 \mathbf{u}_{k+n-1} + b_2 \mathbf{u}_{k+n-2} + \cdots + b_{n-1} \mathbf{u}_{k+1} + b_n \mathbf{u}_k \qquad (2.115)$$

This is the discrete-time analog of equation (2.37). Notice the first term on the right side of the equation, which corresponds to inclusion of the $\mathbf{J}$ term in the state equations.

Frequency-Domain Analysis and the Sampling Theorem

We now turn to the topic of frequency-domain analysis. As with continuous-time systems, we can use complex exponential signals as a means of characterization. We choose to conform with usage that has become "standard" and define the discrete-time

Sec. 2.6 Discrete-Time Linear Systems

complex exponential function $\mathbf{u}_k = z^k$, where z is a complex number; this specializes to the sampled values of a discrete-time phasor when $z = e^{jv}$, so that $\mathbf{u}_k = e^{jvk}$.

Since the variable k serves only as a counter or index, it is necessary to use the sampling time interval T_s to make an interpretation of the v variable as a frequency: the quantity $\omega_p = v/T_s$ is the radian frequency of the phasor whose samples make up the discrete signal e^{jvk} when the sampling time interval is T_s. Because of the periodicity of the complex exponential function, $e^{jv} = e^{j(v+2\pi)}$, there is an ambiguity in representing discrete-time phasors; this is customarily resolved by limiting the admissible range of v to an interval of length 2π. We adopt the convention that $-\pi < v < \pi$. Using the interpretation of v in terms of phasor frequency, this inequality may be rewritten as

$$|\omega_p| < \frac{\pi}{T_s} \tag{2.116}$$

This is the mathematical statement of the celebrated *Sampling Theorem* for phasors:

In order to reconstruct a phasor $e^{j\omega t}$ from a set of uniformly spaced samples with a sampling interval T_s seconds (i.e., a sampling frequency $f_s = 1/T_s$ hertz), the frequency of the phasor, $f_p = \omega_p/2\pi$ hertz, must be less than twice the sampling frequency.

Actually, the Sampling Theorem may be proved for a very general class of signals composed of superpositions of component phasors (using the Fourier transform to make this notion precise) with the following result:

In order to reconstruct a signal $s(t)$ from a set of uniformly spaced samples taken as a sampling frequency of f_s hertz, the signal must contain only components whose frequencies are less than $f_s/2$.

The Sampling Theorem involves a fundamental limitation on the use of digital computing methods for processing continuous-time signals; the details of its implications, both theoretical and practical, are left for a course in digital signal processing.

Continuing with the analysis of the input-output difference equation, (2.115), when the input is $\mathbf{u}_k = z^k$, it is easily verified that the solution takes the form $\mathbf{y}_k = G(z)z^k$, where $G(z)$ is a complex number that is characteristic of the particular system:

$$G(z) = \frac{b_0 z^n + b_1 z^{n-1} + b_2 z^{n-2} + \cdots + b_{n-1} z + b_n}{z^n + a_1 z^{n-1} + \cdots + a_{n-1} z + a_n} \tag{2.117}$$

As in the continuous-time case (see equations (2.80, 2.81)), we may express the complex number $G(z)$ in polar form, $G(z) = |G(z)|e^{j\psi(z)}$, where $\psi(z)$ is the angle or argument of $G(z)$. Then the output corresponding to the input $\mathbf{u}_k = z^k$ is

$$\mathbf{y}_k = |G(z)|e^{j\psi(z)}z^k \tag{2.118}$$

and the system has the effect of magnitude scaling (by $|G(z)|$) and addition of a phase angle $\psi(z)$. The frequency response function for a discrete-time system is $G(e^{jv})$, which corresponds to the choice of phasor inputs. As a function of z, $G(z)$ is called the transfer function of the system. An analysis of the state equations when the system input is z^k and the initial state is zero leads to the following expression for the transfer function:

$$G(z) = \mathbf{J} + \mathbf{H}(z\mathbf{I} - \mathbf{F})^{-1}\mathbf{G} \qquad (2.119)$$

The continuous-time counterpart is equation (2.84).

$G(e^{j\nu})$ is a periodic function of the variable ν. It is customary to plot $\log|G(e^{j\nu})|$ and $\psi(e^{j\nu})$ as a function of ν on the interval $-\pi < \nu < \pi$ as a means of displaying frequency response characteristics graphically.

As mentioned in our earlier discussion concerning continuous-time signals, the Laplace transform enables frequency-domain techniques to be extended from complex exponential signals to more general signals. There is also a transform for discrete-time signals, the *z-transform*, which plays the same role. The transfer function turns out to be the z-transform of the unit pulse response sequence, in analogy with the relation between the transfer function and weighting function for continuous-time systems.

Stability of Discrete-Time Linear Systems

We close this section with a discussion of stability for linear discrete-time systems. Returning to the state equations, (2.108, 2.109),

$$\mathbf{x}_{k+1} = \mathbf{F}\mathbf{x}_k + \mathbf{G}\mathbf{u}_k$$

$$\mathbf{y}_k = \mathbf{H}\mathbf{x}_k + \mathbf{J}\mathbf{u}_k$$

we introduce some definitions that parallel those made earlier (see Section 2.3). We say that the system is *stable* if every solution $\mathbf{x}_k$ corresponding to $\mathbf{u}_k \equiv \mathbf{0}$ is bounded for all k. If a system is stable and in addition every solution $\mathbf{x}_k$ corresponding to $\mathbf{u}_k \equiv \mathbf{0}$ tends to $\mathbf{0}$ as $k \to \infty$, we say that the system is *asymptotically stable*. A stable system that is not asymptotically stable is said to be *marginally stable*.

The asymptotic behavior of $\mathbf{x}_k$ determines the asymptotic behavior of $\mathbf{y}_k$. Since the solution of the state equations for $\mathbf{u}_k \equiv \mathbf{0}$ is given by $\mathbf{x}_k = \mathbf{F}^k\mathbf{x}_0$, the fact that stability and asymptotic stability are determined by the matrix $\mathbf{F}$ in the state equations is clear. As before, consideration of the diagonal case goes a long way toward showing exactly how stability depends on the eigenvalues of $\mathbf{F}$. A linear discrete-time system is asymptotically stable if and only if all of the eigenvalues of $\mathbf{F}$ lie inside the unit circle in the complex plane, or in other words, $|\lambda_i| < 1$ for every eigenvalue, λ_i, of $\mathbf{F}$, (Fig. 2.34).

The condition for asymptotic stability can be checked by operations performed on the coefficients of the characteristic polynomial $p(\lambda) = \det(\lambda\mathbf{I} - \mathbf{F})$, using what is called the Schur-Cohn test, the discrete-time counterpart of the Routh-Hurwitz test mentioned earlier. A sequence of monic polynomials

$$\{p_k(\lambda) : 1 \leq k \leq n, \deg p_k(\lambda) = k\} \qquad (2.120)$$

is defined by the recursion

$$p_k(\lambda) = \lambda p_{k-1}(\lambda) + \rho_k \lambda^{k-1} p_{k-1}(\lambda^{-1}) \qquad (2.121)$$

Sec. 2.6 Discrete-Time Linear Systems

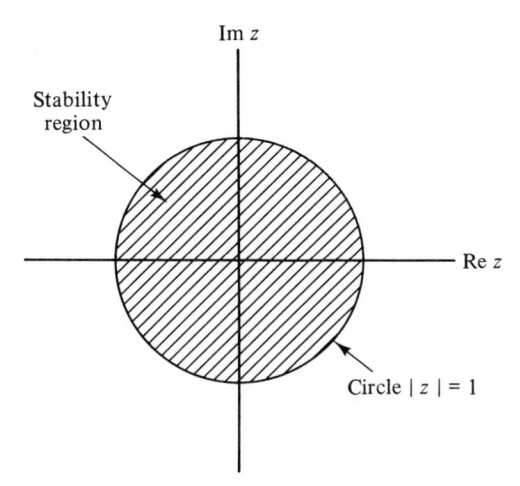

Figure 2.34 Complex z-plane showing the stability region.

starting with $k=n$, $p_n(\lambda) = p(\lambda)$, solving for $p_{n-1}(\lambda)$, and repeating to solve for $p_{n-2}(\lambda)$, and so on. Notice that $\rho_k = p_k(0)$ and that $\lambda^{k-1}p_{k-1}(\lambda^{-1})$ is the polynomial of degree no greater than $k-1$ formed by reversing the order of the coefficients of $p_{k-1}(\lambda)$. The polynomials obtained from $p(\lambda)$ by this recursion provide the following convenient set of necessary and sufficient conditions for the zeros of $p(\lambda)$ to lie inside the unit circle in the complex plane:

$$|\rho_k| < 1, \quad 1 \leq k \leq n \tag{2.122}$$

Examples

One convenient implementation of the Schur-Cohn test will be described by means of an example. Suppose that

$$p_4(\lambda) = p(\lambda) = \lambda^4 + a_1\lambda^3 + a_2\lambda^2 + a_3\lambda + a_4$$

The last coefficient gives $p_4(0)$, so

$$\rho_4 = a_4$$

Form the matrix

$$\mathbf{U}_4 = \begin{bmatrix} (1-\rho_4^2)^{-1} & -\rho_4(1-\rho_4^2)^{-1} \end{bmatrix}$$

Omit the first and last coefficients of $p(\lambda)$ and take the remaining ones in natural and reverse order to form the matrix

$$\mathbf{V}_4 = \begin{bmatrix} a_1 & a_2 & a_3 \\ a_3 & a_2 & a_1 \end{bmatrix}$$

Then the matrix product

$$\mathbf{U}_4 \mathbf{V}_4 = \begin{bmatrix} b_1 & b_2 & b_3 \end{bmatrix}$$

is a row vector whose entries are the coefficients of the next polynomial in the Schur-Cohn sequence,

$$p_3(\lambda) = \lambda^3 + b_1\lambda^2 + b_2\lambda + b_3$$

Thus the last term, b_3, is identified as ρ_3; then the matrices $\mathbf{U}_3$ and $\mathbf{V}_3$ are formed and multiplied to find the coefficients of $p_2(\lambda)$, and similarly for $p_1(\lambda)$. As each successive polynomial is obtained, its last coefficient provides the quantity to be used in the stability test. The results obtained are:

$$\rho_3 = b_3$$
$$\mathbf{U}_3 = \begin{bmatrix} (1-\rho_3^2)^{-1} & -\rho_3(1-\rho_3^2)^{-1} \end{bmatrix}$$
$$\mathbf{V}_3 = \begin{bmatrix} b_1 & b_2 \\ b_2 & b_1 \end{bmatrix}$$
$$\mathbf{U}_3\mathbf{V}_3 = \begin{bmatrix} c_1 & c_2 \end{bmatrix}$$
$$\rho_2 = c_2$$
$$\mathbf{U}_2 = \begin{bmatrix} (1-\rho_2^2)^{-1} & -\rho_2(1-\rho_2^2)^{-1} \end{bmatrix}$$
$$\mathbf{V}_2 = \begin{bmatrix} c_1 \\ c_1 \end{bmatrix}$$
$$\mathbf{U}_2\mathbf{V}_2 = \begin{bmatrix} d_1 \end{bmatrix}$$
$$\rho_1 = d_1$$

For second-order systems, the Schur-Cohn test provides very explicit conditions. Notice that the last two steps of the process carried out for a fourth-degree polynomial give the test for a second-degree polynomial. Using the same notation as above, $p_2(\lambda) = \lambda^2 + c_1\lambda + c_2$. Then

$$\rho_2 = c_2$$

and the coefficient d_1 in $p_1(\lambda) = \lambda + d_1$ can be given explicitly in terms of c_1 and c_2, so that

$$\rho_1 = d_1 = \frac{c_1}{1+c_2}$$

The set of pairs of values that correspond to stability is

$$\{ (c_1, c_2) : |c_2| < 1 \text{ and } |c_1| < |1+c_2| \}$$

This is the interior of the triangular region in the (c_1, c_2)-plane defined by the points $(2,1)$, $(-2,1)$, and $(0,-1)$ (Fig. 2.35).

For stable systems that are only marginally stable, the case is a bit more complicated, as would be expected. When eigenvalues of $\mathbf{F}$ occur on the unit circle, a more careful analysis is required. The general necessary and sufficient condition for stability is that $\mathbf{F}$ have no eigenvalues outside the unit circle and that the number of linearly independent eigenvectors corresponding to eigenvalues with unit magnitude equal the number of such eigenvalues (with multiplicities determined by $p(\lambda)$).

2.7 NOTES AND REFERENCES

Our discussion of linear systems differs from the "traditional" approach in one important respect. We emphasize time-domain methods and do not give a comprehensive treatment of Fourier and Laplace transform methods. Most, if not all, students will have been introduced to transform methods in courses on circuits and signals, and a first course in differential equations usually includes some such material also. We have

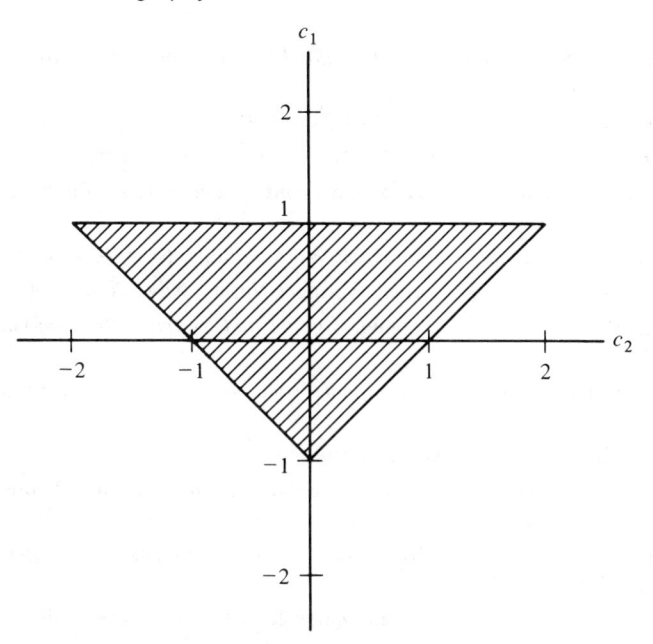

Figure 2.35 Stability region for discrete-time characteristic polynomial $p_2(\lambda) = \lambda^2 + c_1\lambda + c_2$.

chosen this approach to make room for the inclusion of material on nonlinear systems and optimization (Chapters 4 and 5), and because time-domain methods are of primary importance in large-scale, real-world applications of system analysis and design where numerical methods are used to obtain solutions.

There are many books available for complementary and supplementary reading. Books for a traditional linear systems course at a comparable level of sophistication include those by Kamen [5], Kwakernaak and Sivan [6], Oppenheim, Willsky, and Young [9], Siebert [11], Soliman and Srinath [12], Ziemer, Tranter, and Fannin [16]. Also of some interest are the books by Cannon [1] and McClamroch [8]; these provide a broader perspective on modeling of dynamic systems. For a mathematical perspective, the book by Hirsch and Smale [3] is recommended. For senior/graduate-level treatments of linear systems, see Delchamps [2], Kailath [4], Luenberger [7], and Sontag [13].

Both continuous-time systems and discrete-time systems are now usually covered in introductory linear systems books (cf. [5], [6], [9], [11]). The pioneering book by Steiglitz [14] opened the way for introductory-level courses involving discrete-time systems; Strum and Kirk [15] is a second-generation introductory book of this kind. These books provide the background for a follow-on course in digital signal processing, using Oppenheim and Schafer [10], for example; the material also complements the topics covered in a circuits and signals course as a lead-in to the material covered in this book.

BIBLIOGRAPHY

[1] R. H. Cannon, *Dynamics of Physical Systems*, McGraw-Hill, New York, 1967.
[2] D.F. Delchamps, *State Space and Input-Output Linear Systems*, Springer-Verlag, New York, 1988.

[3] M.W. Hirsch and S. Smale, *Differential Equations, Dynamical Systems, and Linear Algebra*, Academic Press, New York, 1974.
[4] T. Kailath, *Linear Systems*, Prentice-Hall, Englewood Cliffs, NJ, 1980.
[5] E.W. Kamen, *Introduction to Signals and Systems*, 2nd ed., Macmillan, New York, 1990.
[6] H. Kwakernaak and R. Sivan, *Modern Signals and Systems*, Prentice-Hall, Englewood Cliffs, NJ, 1991.
[7] D.G. Luenberger, *Introduction to Dynamic Systems*, John Wiley & Sons, New York, 1979.
[8] N.H. McClamroch, *State Models of Dynamic Systems*, Springer-Verlag, New York, 1980.
[9] A.V. Oppenheim and A.S. Willsky, with I.T. Young, *Signals and Systems*, Prentice-Hall, Englewood Cliffs, NJ, 1983.
[10] A.V. Oppenheim and R.W. Schafer, *Discrete-Time Signal Processing*, Prentice-Hall, Englewood Cliffs, NJ, 1989.
[11] W.M. Siebert, *Circuits, Signals, and Systems*, McGraw-Hill, New York, 1986.
[12] S.S. Soliman and M.D. Srinath, *Continuous and Discrete Signals and Systems*, Prentice-Hall, Englewood Cliffs, NJ, 1990.
[13] E.D. Sontag, *Mathematical Control Theory: Deterministic Finite Dimensional Systems*, Springer-Verlag, New York, 1990.
[14] K. Steiglitz, *An Introduction to Discrete Systems*, John Wiley & Sons, New York, 1974.
[15] R.D. Strum and D.E. Kirk, *First Principles of Discrete Systems and Digital Signal Processing*, Addison-Wesley, Reading, MA, 1988.
[16] R.E. Ziemer, W.H. Tranter, and D.R. Fannin, *Signals and Systems*, 2nd ed., Macmillan, New York, 1989.

PROBLEMS

1. Use the power series definition to evaluate $e^{\mathbf{A}t}$ when
$$\mathbf{A} = \begin{bmatrix} 0 & 1 & 1 \\ 0 & 0 & 1 \\ 0 & 0 & 0 \end{bmatrix}$$

2. Show that $\mathbf{A} e^{\mathbf{A}t} = e^{\mathbf{A}t} \mathbf{A}$ (i.e., that the matrices $\mathbf{A}$ and $e^{\mathbf{A}t}$ commute for all times t).

3. Suppose that $\mathbf{A}_1 \mathbf{A}_2 = \mathbf{A}_2 \mathbf{A}_1$. Show that
$$e^{\mathbf{A}_1 t} e^{\mathbf{A}_2 t} = e^{(\mathbf{A}_1 + \mathbf{A}_2)t} = e^{\mathbf{A}_2 t} e^{\mathbf{A}_1 t}$$
Use this property to evaluate $e^{(\lambda \mathbf{I} + \mathbf{A})t}$, where $\mathbf{A}$ is the matrix given in Problem 1.

4. The solution of linear matrix differential equations can be written in a form that includes the case of state space equations as a particular application. Show by direct verification that for $n_1 \times n_1$ and $n_2 \times n_2$ matrices $\mathbf{A}_1$ and $\mathbf{A}_2$, respectively, the $n_1 \times n_2$ matrix $\mathbf{X}(t)$ given by
$$\mathbf{X}(t) = e^{\mathbf{A}_1 t} \mathbf{X}_0 e^{\mathbf{A}_2 t} + \int_0^t e^{\mathbf{A}_1 (t-\tau)} \mathbf{U}(\tau) e^{\mathbf{A}_2 (t-\tau)} d\tau$$
where $\mathbf{X}_0$ is a constant $n_1 \times n_2$ matrix, is the solution to the matrix differential equation
$$\dot{\mathbf{X}}(t) = \mathbf{A}_1 \mathbf{X}(t) + \mathbf{X}(t) \mathbf{A}_2 + \mathbf{U}(t)$$
with initial condition $\mathbf{X}(0) = \mathbf{X}_0$. (As usual, the dot represents the time derivative of the matrix.)

Problems

5. For the system

$$\dot{x}(t) = \begin{bmatrix} -7 & 1 \\ -12 & 0 \end{bmatrix} x(t) + \begin{bmatrix} 0 \\ 1 \end{bmatrix} u(t)$$

$$y(t) = [\,1 \quad 0\,]\,x(t)$$

Find the solution $x(t)$ when $x(0) = \begin{bmatrix} 1 \\ 0 \end{bmatrix}$ and $u(t) = 1(t)$, the unit step function. (You should obtain explicit expressions for the components of $x(t)$ as functions of time.)

6. For the state equations

$$\dot{x}(t) = Ax(t) + Bu(t); \quad A = \begin{bmatrix} 0 & -2 \\ 1 & -3 \end{bmatrix}, \quad B = \begin{bmatrix} 1 \\ 0 \end{bmatrix}$$

with $x(0) = \begin{bmatrix} 0 \\ 1 \end{bmatrix}$ and $u(t) = 1(t-2)$, find an analytical expression for the solution vector $x(t)$ for $t > 0$.

7. Write down a set of state equations in the form

$$\dot{x}(t) = Ax(t) + Bu(t)$$

$$y = Cx(t)$$

for a system whose input-output relationship is given by

$$\ddot{y}(t) + 6\dot{y}(t) + 5y(t) = 4\dot{u}(t) + 2u(t)$$

Find the transfer function from u to y for this system.

8. Find the input-output differential equation corresponding to the following state space model:

$$\dot{x}(t) = \begin{bmatrix} 1 & 0 \\ 2 & 3 \end{bmatrix} x(t) + \begin{bmatrix} 1 \\ 0 \end{bmatrix} u$$

$$y = [\,4 \quad 1\,]\,x(t)$$

9. For the state equations

$$\dot{x}(t) = \begin{bmatrix} -2 & 1 \\ -3 & 0 \end{bmatrix} x(t) + \begin{bmatrix} 4 \\ 5 \end{bmatrix} u(t)$$

$$y(t) = [\,1 \quad 0\,]\,x(t)$$

find the input-output differential equation for y in terms of u. Find all sets of initial conditions for y, $\dot{y}$, and u at $t = 0$, so that the solution of the input-output differential equation coincides with the solution obtained from the state equations when

$$x(0) = \begin{bmatrix} 1 \\ 1 \end{bmatrix}$$

10. For the input-output differential equation

$$\ddot{y}(t) + a_1 \dot{y}(t) + a_2 y(t) = u(t)$$

write down a corresponding set of state equations in the form

$$\dot{x}(t) = Ax(t) + Bu(t)$$

$$y(t) = Cx(t)$$

11. Give a state space model whose input-output relationship is the following differential equation:

$$\ddot{y}(t) + 2\dot{y}(t) + 2y(t) = \dot{u}(t) + 3u(t)$$

Find the initial conditions for the state vector when the input-output description has $\dot{y}(0) = 1$, $y(0) = 3$, and $u(0) = 4$. Find the response $y(t)$ to a unit step input when all initial conditions are zero.

12. For the system in Problem 5, determine whether or not the system is asymptotically stable; explain your conclusions.

13. Show that the Routh-Hurwitz test gives the expected results when applied to testing the stability of a fourth-degree polynomial formed from the product of two second-degree polynomials.

14. Give a rough sketch and also graph more carefully the Bode plot for the transfer function formed from the ratio of the two monic polynomials obtained by choosing the zeros to be $\{-2+3j, -2-3j, 1.5\}$ and the poles to be $\{-0.5, -1-j, -1+j, -0.25+4j, -0.25-4j\}$.

15. For the second-order system whose transfer function is given in equation (2.96), use computational experiments to obtain graphs of the peak value of the Bode plot and the peak location as a function of frequency ω over the range $0 < \zeta < 1/4$. Also determine the width of the peak as a function of ζ for the same values of ζ. (You will need to use a convenient, consistent definition of the width of a peak.)

16. For the system with transfer function

$$H(s) = \frac{s+1}{s^2 + 2s + 2}$$

find a good approximation of the output response to the input $\mathbf{u}(t) = \cos(\omega t + \pi/4)$ when the input frequency, ω, is very large. (Assume that the input was first applied to the system in the remote past so that "initial condition" terms can be ignored.)

17. For the motion of a point with mass m moving in the x direction under the influence of spring and damping forces and an externally applied force $u(t)$, Newton's laws lead to a second-order differential equation of the form

$$m\ddot{x}(t) + f\dot{x}(t) + k x(t) = u(t)$$

The equations for a large class of interconnected systems of r point masses can be collected into a vector second-order differential equation

$$\mathbf{M}\ddot{\mathbf{x}}(t) + \mathbf{F}\dot{\mathbf{x}}(t) + \mathbf{K}\mathbf{x}(t) = \mathbf{u}(t)$$

where $\mathbf{x}(t)$ is an r-vector of displacements, $\mathbf{u}(t)$ is an r-vector of external forces, and $\mathbf{M}$, $\mathbf{F}$, and $\mathbf{K}$ are $r \times r$ real symmetric matrices; both $\mathbf{M}$ and $\mathbf{K}$ also have positive eigenvalues.

(a) Use displacements and velocities as states to obtain a $2r$-dimensional state space equation for this system.

(b) When the external forces are zero, determine the condition on the matrix $\mathbf{F}$ which assures asymptotic stability of the state $\mathbf{0}$. *Hint:* Investigate the total energy.

(c) For the undamped case, $\mathbf{F} = \mathbf{0}$, the equations describe a set of coupled linear oscillators. Assume that solutions of the form $\mathbf{x}(t) = \mathbf{X}(\omega)\exp(j\omega t)$ may be found; this means that all of the point masses are oscillating at a single frequency. Verify this assumption by using the vector second-order differential equations, showing how the "mode shape vector" $\mathbf{X}(\omega)$ and the corresponding natural frequency ω are related to eigenvectors and eigenvalues of the matrix $\mathbf{M}^{-1}\mathbf{K}$. What are corresponding eigenvectors and eigenvalues of the state equations?

18. Continuing with Problem 17, investigate the behavior of a linear chain of r identical point masses connected by identical springs. Assume that the chain is terminated at both ends by infinitely massive "walls." Use computational experiments to determine the natural frequencies and mode shapes for an increasing sequence of values of r, say $r = 2, 3, 4, 5$. Describe any patterns that you find.

19. The Fibonacci numbers are defined by the difference equation

$$y_{k+2} - y_{k+1} - y_k = 0, \quad k \geq 0$$

with initial conditions $y_0 = 1$ and $y_1 = 2$. Choose the state vector

$$\mathbf{x}_k = \begin{bmatrix} y_{k+1} \\ y_k \end{bmatrix}$$

and write state equations that produce the Fibonacci numbers as outputs at time instants 0, 1, 2, ... By solving these state equations, obtain an analytical expression for the ith Fibonacci number.

20. Find explicit inequalities involving the polynomial coefficients obtained from the Schur-Cohn stability test applied to the third-degree polynomial $p(\lambda) = \lambda^3 + a_1\lambda^2 + a_2\lambda + a_3$.

CHAPTER 3

Discretization of Continuous-Time Systems

The importance of discretization stems from two kinds of applications. The first involves analysis and simulation of systems using digital computers and numerical methods. As a tool for studying the behavior of complex systems, computer simulation is indispensable. It is an ideal method for studying effects of a hazardous phenomenon such as earthquake-induced structural failures and the efficacy of preventive measures that might be considered for reducing associated risks. Simulation is a versatile tool being used to great advantage in applications ranging from VLSI system design to aircraft design to chemical reactor design. Whether the problem scale is molecular, as in the modeling of quantum semiconductor device behavior, global, as in the modeling of weather and climate processes, or astronomical, as in the trajectory planning for the *Voyager* spacecraft missions, computer simulation can offer an economical (or even feasible!) basis for "experimental" studies of various processes.

In order to represent general signals for manipulation by computer, some kind of discrete representation of waveforms must be employed. (By "general" signals we mean that we want the capability to use signals that cannot be represented in terms of analytical expressions as a function of time.) The simplest such discrete representation is to replace a signal by a sequence of sample values taken at multiples of a fixed intersample time unit, denoted h in what follows. (In the previous chapter's discussion of discrete-time systems, the variable T_s was used instead of h, but the conventional notation for discretization in the numerical analysis literature is h, and we choose to conform to this choice.) So instead of a time function, say $z(t)$, we deal with the sequence $z(kh)$, where k runs over some consecutive subset of the integers. We use the terms *sequence* and *discrete-time signal* interchangeably.

The Sampling Theorem, introduced in the preceding chapter, provides substantial justification for uniform sampling of signals under reasonable conditions. In particular, for signals containing no frequency components higher than W hertz, the sequence of samples obtained by uniform sampling at any rate higher than $2W$ samples/second, the so-called *Nyquist sampling rate*, suffices to allow perfect reconstruction of the signal. Furthermore, errors introduced when the assumptions of an exact band limit and perfect accuracy of the sample values are really only engineering approximations made for simplicity do not turn out to be catastrophic in most cases of interest. As will be seen in this chapter, the effects of uniform sampling on various properties of linear systems can be analyzed in great detail; as a result, a good understanding of various issues related to system discretization is available.

Discretization is an important topic for another set of reasons involving what might be called "real-time applications." The widespread (and still growing) use of digital hardware as a part of an operational system has been based to a large extent on using discretization as a means by which continuous-time processes and signals can be interfaced to discrete-time digital processors. Both general-purpose, programmable digital hardware and "hard-wired" special-purpose digital systems are being used; microprocessor-based systems for use in control systems, e.g., automobile engine control systems, provide examples of the former, while custom VLSI systems for telecommunications signal processing functions (e.g., echo cancellation on satellite telephone channels) provide examples of the latter. Compact disk (CD) technology provides another example in the consumer electronics field, where the tremendous versatility of digital signal processing systems is only beginning to be exploited commercially. It is often important for the design of a digital signal processing system to be based on an appropriate discrete-time model for the continuous-time signals that make up its operating environment.

3.1 BASIC DISCRETIZATION METHODS

Having said a few words about why there is substantial interest in the topic of discretization, we now turn to a discussion of various techniques that can be employed.

Sample-and-Hold Discretization

The first approach to discretization is the one involving the most assumptions; it applies only to linear systems of the usual form:

$$\dot{\mathbf{x}}(t) = \mathbf{A}\mathbf{x}(t) + \mathbf{B}\mathbf{u}(t) \tag{3.1}$$

$$\mathbf{y}(t) = \mathbf{C}\mathbf{x}(t) \tag{3.2}$$

We will call the method *sample-and-hold* discretization since it is based on the further assumption that the system input stays constant between sampling instants (i.e., $\mathbf{u}(t) = \mathbf{u}(kh)$, $kh \leq t < (k+1)h$), and we imagine that the sampled values of the state and output also serve as good approximations over the same range of times between sampling instants. (We expect that if the input is "almost" constant between sampling times, the discretization remains a good approximation.)

To derive the discretized system, we apply the known form of the solution to the state equations, see (2.29), to write the solution at time $t = (k+1)h$ in terms of the state and input at time $t = kh$, giving

$$\mathbf{x}((k+1)h) = e^{\mathbf{A}h}\mathbf{x}(kh) + \int_0^h e^{\mathbf{A}(h-\tau)}\mathbf{B}\mathbf{u}(kh+\tau)\,d\tau \tag{3.3}$$

which simplifies due to the assumption of a constant input to

$$\mathbf{x}((k+1)h) = e^{\mathbf{A}h}\mathbf{x}(kh) + \left(\int_0^h e^{\mathbf{A}\tau}\mathbf{B}\,d\tau\right)\mathbf{u}(kh) \tag{3.4}$$

We define $\mathbf{F} = e^{\mathbf{A}h}$ and $\mathbf{G} = \int_0^h e^{\mathbf{A}\tau}\mathbf{B}\,d\tau$. Then we may rewrite these equations in a form that corresponds to a discrete-time linear system in state space form as introduced in the preceding chapter:

$$\mathbf{x}_d((k+1)h) = \mathbf{F}\mathbf{x}_d(kh) + \mathbf{G}\mathbf{u}(kh) \tag{3.5}$$

$$\mathbf{y}_d(kh) = \mathbf{C}\mathbf{x}_d(kh) \tag{3.6}$$

(The differences in notation should be obvious: We avoid the use of subscripts for sequence indices because we want to emphasize the connection with the "time" variable of the underlying continuous-time system. The output equation also has $\mathbf{H} = \mathbf{C}$ and $\mathbf{J} = \mathbf{0}$ in the notation of the general linear discrete-time system model introduced earlier, equations (2.108, 2.109).) We use the d subscript in this section to denote a discrete approximation; here the only approximation introduced is that of a constant input over each intersample interval. Recall that we encountered $\int_0^h e^{\mathbf{A}\tau}\,d\tau$ in a discussion of step response; see equation (2.100). We found that if $\mathbf{A}$ is invertible, which will be the case if the continuous-time system is asymptotically stable, then we may simplify the expression for $\mathbf{G}$ to $\mathbf{G} = \mathbf{A}^{-1}(\mathbf{F} - \mathbf{I})\mathbf{B}$.

An obvious application of sample-and-hold discretization arises in digital control systems. If the input to a linear system is being determined by a digital processor (e.g., a microprocessor) that operates on observations processed by an analog-to-digital (A-D) converter, then the input to the system can reasonably be expected to remain constant over a basic time interval during which a new input value is being computed and processed by a digital-to-analog (D-A) converter into a suitable system input.

Example

The airplane pitch response example will be used to illustrate how the sample-and-hold discretization procedure is carried out. We refer to Chapter 2, where the following equations were given as the mathematical model for this system:

$$\begin{bmatrix} \dot{q}(t) \\ \dot{\alpha}(t) \end{bmatrix} = \begin{bmatrix} m_q & m_\alpha \\ 1 & k_\alpha \end{bmatrix} \begin{bmatrix} q(t) \\ \alpha(t) \end{bmatrix} + \begin{bmatrix} m_e & m_f \\ k_e & k_f \end{bmatrix} \begin{bmatrix} e(t) \\ f(t) \end{bmatrix}$$

As a representative problem, suppose that $m_\alpha = 18$, $m_q = -0.6$, $k_\alpha = -1.2$, and consider the system when only elevator control inputs are used, so that $f(t) = 0$. The state equations thus reduce to

$$\begin{bmatrix} \dot{q}(t) \\ \dot{\alpha}(t) \end{bmatrix} = \begin{bmatrix} -0.6 & 18 \\ 1 & -1.2 \end{bmatrix} \begin{bmatrix} q(t) \\ \alpha(t) \end{bmatrix} + \begin{bmatrix} m_e \\ k_e \end{bmatrix} e(t)$$

The eigenvalue-eigenvector factorization can be used to compute the matrix exponential function (all numbers rounded to two decimal places):

Sec. 3.1 Basic Discretization Methods

$$\mathbf{A} = \begin{bmatrix} -0.6 & 18 \\ 1 & -1.2 \end{bmatrix} = \begin{bmatrix} 1 & 1 \\ 0.22 & -0.25 \end{bmatrix} \begin{bmatrix} 3.35 & 0 \\ 0 & -5.15 \end{bmatrix} \begin{bmatrix} 0.54 & 2.12 \\ 0.46 & -2.12 \end{bmatrix}$$

so

$$\mathbf{F} = e^{\mathbf{A}h} = \begin{bmatrix} 1 & 1 \\ 0.22 & -0.25 \end{bmatrix} \begin{bmatrix} e^{3.35h} & 0 \\ 0 & e^{-5.15h} \end{bmatrix} \begin{bmatrix} 0.54 & 2.12 \\ 0.46 & -2.12 \end{bmatrix}$$

and

$$\mathbf{G} = \begin{bmatrix} 0.07 & 1.04 \\ 0.06 & 0.03 \end{bmatrix} (\mathbf{F} - \mathbf{I}) \begin{bmatrix} m_e \\ k_e \end{bmatrix}$$

Notice that the original system has one positive ("unstable") eigenvalue and one negative ("stable") eigenvalue. The sample-and-hold discretization of the system shares this property for every (positive) choice of the sampling interval h; recall that for discrete-time systems, a "stable" eigenvalue is one with magnitude less than 1 and an "unstable" eigenvalue has magnitude greater than 1.

Forward and Backward Difference Methods

A second class of discretization methods involve making approximations directly in the state equations, eliminating the derivative by replacing it with some kind of difference quotient. An advantage of such methods is that they may be applied to time varying and nonlinear systems, where there are no analytical solutions available for use in obtaining a discretization. The two most obvious approximations to the derivative $\dot{\mathbf{x}}(t)$ are the forward difference

$$\dot{\mathbf{x}}(kh) \approx \frac{\mathbf{x}((k+1)h) - \mathbf{x}(kh)}{h} \tag{3.7}$$

and the backward difference

$$\dot{\mathbf{x}}(kh) \approx \frac{\mathbf{x}(kh) - \mathbf{x}((k-1)h)}{h} \tag{3.8}$$

which are both exact in the limit $h \to 0$ (Fig. 3.1). Thus for small h, we expect these approximations to be good ones. For the forward difference discretization method (also called the Euler method), (3.7) is used in (3.1) and the discretized system is

$$\mathbf{x}_d((k+1)h) = (\mathbf{I} + h\mathbf{A})\mathbf{x}_d(kh) + h\mathbf{B}\mathbf{u}(kh) \tag{3.9}$$

$$\mathbf{y}_d(kh) = \mathbf{C}\mathbf{x}_d(kh) \tag{3.10}$$

For the backward difference discretization method, (3.8) is used in (3.1) and the discretized system is

$$\mathbf{x}_d(kh) = (\mathbf{I} - h\mathbf{A})^{-1} \mathbf{x}_d((k-1)h) + h(\mathbf{I} - h\mathbf{A})^{-1} \mathbf{B}\mathbf{u}(kh) \tag{3.11}$$

$$\mathbf{y}_d(kh) = \mathbf{C}\mathbf{x}_d(kh) \tag{3.12}$$

Example

The calculation of the forward and backward difference discretizations of the aircraft example described earlier is straightforward and is left to the reader. Attention should be paid to what conditions must be imposed on the size of h to ensure that the asymptotic

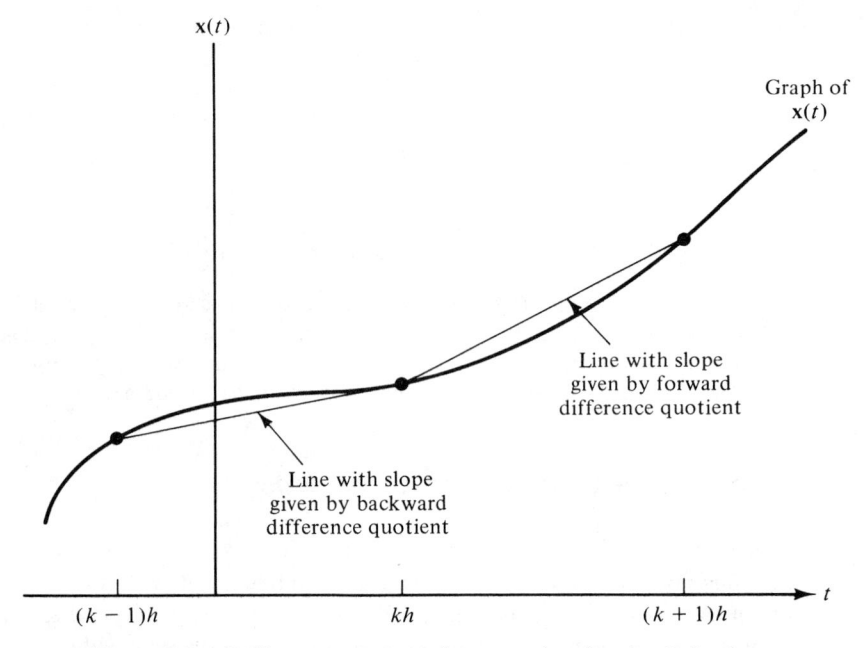

Figure 3.1 Difference quotients for approximating the derivative.

behavior of the discretized systems match that of the original continuous-time system (i.e., one "stable eigenvalue" and one "unstable eigenvalue").

Discretization as Approximate Integration

There is another way of looking at discretization that provides some useful insights, particularly because it applies to systems of the very general form

$$\dot{\mathbf{x}}(t) = \mathbf{f}(\mathbf{x}(t), \mathbf{u}(t)) \tag{3.13}$$

Nonlinear systems such as (3.13) will be studied in Chapter 4. For the present discussion, we integrate both sides of this equation from $t = kh$ to $t = (k+1)h$ and solve for $\mathbf{x}((k+1)h)$ to obtain

$$\mathbf{x}((k+1)h) = \mathbf{x}(kh) + \int_{kh}^{(k+1)h} \mathbf{f}(\mathbf{x}(\tau), \mathbf{u}(\tau))\, d\tau \tag{3.14}$$

In order to obtain an expression depending only on sampled values of the signals, we must make an approximation to the integral on the right side of this equation. Since h is small, the easiest approximation to make is that the integrand is a constant; obvious choices for the value of $\mathbf{f}(\mathbf{x}(\tau), \mathbf{u}(\tau))$ are

(a) $\mathbf{f}(\mathbf{x}(\tau), \mathbf{u}(\tau)) \approx \mathbf{f}(\mathbf{x}(kh), \mathbf{u}(kh))$
(b) $\mathbf{f}(\mathbf{x}(\tau), \mathbf{u}(\tau)) \approx \mathbf{f}(\mathbf{x}((k+1)h), \mathbf{u}((k+1)h))$
(c) $\mathbf{f}(\mathbf{x}(\tau), \mathbf{u}(\tau)) \approx [\mathbf{f}(\mathbf{x}(kh), \mathbf{u}(kh)) + \mathbf{f}(\mathbf{x}((k+1)h), \mathbf{u}((k+1)h))]/2$

The first two of these choices correspond to Euler and backward difference approximations, respectively. Agreement with our previous work is easily checked by substituting

Sec. 3.1 Basic Discretization Methods 129

$\mathbf{A}\mathbf{x}(t) + \mathbf{B}\mathbf{u}(t)$ for $\mathbf{f}(\mathbf{x}(t),\mathbf{u}(t))$. The third method, known as *trapezoidal discretization* (Fig. 3.2), gives the following result when applied to linear systems:

$$\mathbf{x}_d((k+1)h) = \left(\mathbf{I} + \frac{h\mathbf{A}}{2}\right)\left(\mathbf{I} - \frac{h\mathbf{A}}{2}\right)^{-1}\mathbf{x}_d(kh)$$
$$+ \frac{1}{2}\left(\mathbf{I} - \frac{h\mathbf{A}}{2}\right)^{-1}\mathbf{B}h(\mathbf{u}((k+1)h) + \mathbf{u}(kh)) \qquad (3.15)$$

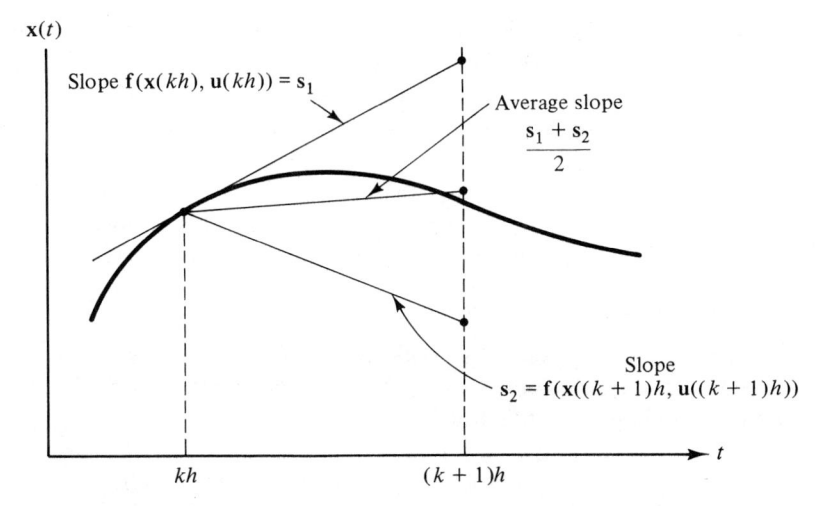

Figure 3.2 Trapezoidal method uses the average of slopes at the end points.

Because there is nothing more to say about it, we won't bother to repeat the output equation for the remaining methods that will be described.

Explicit and Implicit Methods

An important distinction between various discretization methods, especially for applications to nonlinear systems, needs to be made. There are two quite different classes of discretization methods, the *explicit* ones and the *implicit* ones. The situation is easily illustrated by considering how the nonlinear system (3.13) would be discretized. A key difference between the Euler method and the other two methods (backward difference and trapezoidal) arises from the use of the "future" value $\mathbf{x}((k+1)h)$ as an argument of the function $\mathbf{f}$ in the approximations used. When $\mathbf{f}$ is a nonlinear function, this means that it will generally not be possible to "solve" for $\mathbf{x}((k+1)h)$ explicitly (either analytically or with a noniterative numerical method) in terms of $\mathbf{x}(kh)$ and sampled values of $\mathbf{u}$.

Example

This discussion may be illustrated with a simple one-dimensional example:

$$\dot{x}(t) = \cos(x(t)) + u(t)$$

Using backward difference discretization leads to

$$x_d((k+1)h) = x_d(kh) + h\cos(x_d((k+1)h)) + h\,u((k+1)h)$$

which cannot be solved analytically for $x_d((k+1)h)$.

To use implicit discretization methods for nonlinear systems requires that some numerical technique be available for solving nonlinear equations of the form

$$\mathbf{x}_d((k+1)h) = \mathbf{F}(\mathbf{x}_d(kh), \mathbf{u}(kh), \mathbf{x}_d((k+1)h), \mathbf{u}((k+1)h))) \tag{3.16}$$

One popular technique is Newton's method, which will be discussed in Chapter 4, after we have introduced derivatives and linearization. At this point, however, it is worth pointing out that since discretization amounts to an approximation itself, it is not necessary that an exact solution of the nonlinear equations be found, only "sufficiently accurate" ones.

The Euler discretization is the simplest example of an explicit method. Since $\mathbf{x}_d((k+1)h)$ is not used as an argument of the function $\mathbf{f}$ in this method, an explicit formula for obtaining successive discretized values is available, namely

$$\mathbf{x}_d((k+1)h) = \mathbf{x}_d(kh) + h\,\mathbf{f}(\mathbf{x}_d(kh), \mathbf{u}(kh)) \tag{3.17}$$

Another explicit method is second-order Runge-Kutta discretization, which may be viewed as a way of approximating trapezoidal discretization by replacing its implicit part by an Euler-type explicit one:

$$\mathbf{x}_d((k+1)h) = \mathbf{x}_d(kh) + \frac{h}{2}[\mathbf{f}_1 + \mathbf{f}_2] \tag{3.18}$$

where

$$\mathbf{f}_1 = \mathbf{f}(\mathbf{x}_d(kh), \mathbf{u}(kh)) \tag{3.19}$$

$$\mathbf{f}_2 = \mathbf{f}(\mathbf{x}_d(kh) + h\,\mathbf{f}_1, \mathbf{u}((k+1)h)) \tag{3.20}$$

Another variation, with the advantage of not requiring inputs at two different sampling instants, is the second-order Ralston-Runge-Kutta method:

$$\mathbf{x}_d((k+1)h) = \mathbf{x}_d(kh) + h\left[\frac{1}{3}\mathbf{f}_1^* + \frac{2}{3}\mathbf{f}_2^*\right] \tag{3.21}$$

where we take

$$\mathbf{f}_1^* = \mathbf{f}(\mathbf{x}_d(kh), \mathbf{u}(kh)) \tag{3.22}$$

$$\mathbf{f}_2^* = \mathbf{f}(\mathbf{x}_d(kh) + \frac{3}{4}h\,\mathbf{f}_1^*, \mathbf{u}(kh)) \tag{3.23}$$

When discretization techniques are used to obtain numerical solutions of differential equations (i.e., when they are used for system simulation), it is not really necessary to restrict the points at which the input function $\mathbf{u}(t)$ is evaluated as part of the process of approximating the right side of the differential equation, $\mathbf{f}(\mathbf{x}(t), \mathbf{u}(t))$. In particular it is often advantageous to use input samples taken on a finer grid of points than the grid chosen for computing solution, and many such methods can be found in the numerical

Sec. 3.2 Analysis of Discretization Techniques   131

analysis literature. One popular technique, due to its relative simplicity, is the fourth-order Runge-Kutta method, which uses evaluations of **f** corresponding to input samples taken at instants $t = kh/2$. It is an explicit method, given by the following:

$$\mathbf{x}_d((k+1)h) = \mathbf{x}_d(kh) + \frac{h}{6}[\mathbf{f}_1 + 2\mathbf{f}_2 + 2\mathbf{f}_3 + \mathbf{f}_4] \qquad (3.24)$$

where

$$\mathbf{f}_1 = \mathbf{f}(\mathbf{x}_d(kh), \mathbf{u}(kh)) \qquad (3.25)$$

$$\mathbf{f}_2 = \mathbf{f}(\mathbf{x}_d(kh) + \frac{h}{2}\mathbf{f}_1, \mathbf{u}((k+\frac{1}{2})h)) \qquad (3.26)$$

$$\mathbf{f}_3 = \mathbf{f}(\mathbf{x}_d(kh) + \frac{h}{2}\mathbf{f}_2, \mathbf{u}((k+\frac{1}{2})h)) \qquad (3.27)$$

$$\mathbf{f}_4 = \mathbf{f}(\mathbf{x}_d(kh) + h\,\mathbf{f}_3, \mathbf{u}((k+1)h)) \qquad (3.28)$$

The fourth-order Runge-Kutta method is widely used because it offers an excellent "price-performance" trade-off. For the "cost" of four function evaluations, highly accurate discretization is achieved.

A discussion of the wide range of more sophisticated methods for numerical integration of differential equations is best left for a book in numerical analysis. We will provide a brief discussion of discretization for system models consisting of partial (rather than ordinary) differential equations in the last section of this chapter. We now turn to some analysis of the discretization methods already introduced; this is aimed at providing an understanding of the issues involved in selecting a discretization method from among the many possibilities and in choosing a discretization step size h.

3.2 ANALYSIS OF DISCRETIZATION TECHNIQUES

Several basic properties may be associated with every discretization method; *accuracy*, *stability*, and *computational requirements* are three important ones. In selecting a discretization method for a particular application, performance comparisons must be made and the trade-offs considered. As will be seen below, stability considerations can preclude the use of explicit methods; the added computational burden imposed by an implicit method is often well worth bearing because the stability limits of explicit methods can only be overcome by a drastic reduction in the discretization step size, which in turn imposes an even greater computational load.

Accuracy of Discretization Methods

To start, we consider some accuracy analyses for discretization of linear systems. Of primary interest will be the quality of the methods, judged in comparison with the sample-and-hold discretization method, which is exact when the system input remains constant between sampling times. The notion of *accuracy* is intended to provide an indication of the ultimate performance of a discretization method in the limit of vanishing discretization step size.

First consider the Euler, or forward difference, discretization. In the state equations for the Euler method, the matrix $(\mathbf{I}+h\mathbf{A})$ is an approximation to $\mathbf{F} = e^{\mathbf{A}h}$, the corresponding term in the state equations for sample-and-hold discretization, accurate to first order in h. That is, since we are interested in small h for accurate discretization, if we use the power series definition of $e^{\mathbf{A}h}$, we see that it differs from $(\mathbf{I}+h\mathbf{A})$ only in terms involving $h^2, h^3, \ldots$, terms that tend to zero much faster than h as $h \to 0$; this is the meaning of the expression "accurate to first order in h." Similarly, $h\mathbf{B}$ is an approximation to $\mathbf{G}$ in the sample-and-hold discretization, accurate to first order in h. Thus for small h, we would expect the Euler discretization and the sample-and-hold discretization to give quite similar results.

We can make the same kind of analysis for the backward difference discretization. We note that $(\mathbf{I}-h\mathbf{A})^{-1}$ is also an approximation to $e^{\mathbf{A}h}$ accurate to first order in h. This follows from the "matrix version" of the geometric series formula

$$(\mathbf{I}-h\mathbf{A})^{-1} = \mathbf{I} + h\mathbf{A} + (h\mathbf{A})^2 + \cdots \qquad (3.29)$$

which was developed in Chapter 1; see (1.116). The infinite series converges for h small enough: $1/h$ must be larger than the magnitudes of all of the eigenvalues of $\mathbf{A}$. Another argument suggesting that $(\mathbf{I}-h\mathbf{A})^{-1}$ provides a first-order approximation is to start with $e^{-\mathbf{A}h} \approx (\mathbf{I}-h\mathbf{A})$ and then invert both sides of the approximation; to make this argument precise involves some manipulation of matrix norms that is left to the reader's ingenuity.

Stability of Discretization Methods

Next we turn to a consideration of stability. The sample-and-hold discretization has an important property: Asymptotically stable continuous-time systems have discretizations that are also asymptotically stable. This may be seen by recalling that in continuous-time systems, asymptotically stable systems are those having the eigenvalues of $\mathbf{A}$ in the left half of the complex plane. For discrete-time systems, asymptotically stable systems are those having the eigenvalues of $\mathbf{F}$ inside the unit circle in the complex plane. Since $\mathbf{F} = e^{\mathbf{A}h}$ for the sample-and-hold discretization, the eigenvalues of $\mathbf{F}$ are the quantities $\{e^{\lambda h} : \lambda$ an eigenvalue of $\mathbf{A}\}$. All of these have magnitude less than 1, and the desired conclusion follows.

One approach to studying the effects of various discretization methods on stability employs the system input-output response. We start with a look at sample-and-hold discretization. Just as we may express the transfer function of the original continuous-time system, (3.1, 3.2) in terms of the matrices of the state equations as $H(s) = \mathbf{C}(s\mathbf{I}-\mathbf{A})^{-1}\mathbf{B}$ (cf. (2.84)), we may also express the transfer function of the discrete-time system arising from discretization, (3.5, 3.6). The result is $G(z) = \mathbf{C}(z\mathbf{I}-\mathbf{F})^{-1}\mathbf{G}$ (cf. (2.119)). There is no simple relationship that holds between the transfer functions $H(s)$ and $G(z)$. However, it is easy to investigate the relationship for any particular case using numerical calculations. Low-order examples can easily be worked out by hand.

In contrast to the sample-and-hold case, a particularly nice feature of the Euler and backward difference discretization methods is that both of them lead to simple relationships between the transfer function of the continuous-time system, $H(s) = \mathbf{C}(s\mathbf{I} - \mathbf{A})^{-1}\mathbf{B}$, and the transfer function of the discretized version. For the Euler method we compute the z-transform transfer function of (3.9, 3.10) as

$$G_E(z) = \mathbf{C}(z\mathbf{I} - (\mathbf{I} + h\mathbf{A}))^{-1} h\mathbf{B} \qquad (3.30)$$

which may be rearranged using some algebra into the form

$$\begin{aligned} G_E(z) &= \mathbf{C}\left(\frac{z-1}{h}\mathbf{I} - \mathbf{A}\right)^{-1}\mathbf{B} \\ &= H\left(\frac{z-1}{h}\right) \end{aligned} \qquad (3.31)$$

Thus the z-transform transfer function of the discrete-time system obtained as the Euler discretization of the continuous-time system whose transfer function is $H(s)$ can be obtained simply by setting $s = (z-1)/h$. This corresponds to taking the forward difference quotient $(z-1)/h$ as an approximation to the differentiation operator, s, of Laplace transforms. Written the other way around, the unit delay operator z^{-1} is approximated as $1/(sh+1)$ a low-pass filter that can be recognized as the transfer function of an RC voltage divider. In a similar way, the analysis of the backward difference approximation gives $s = (z-1)/(zh)$, leading to the transfer function relationship

$$G_B(z) = H\left(\frac{z-1}{zh}\right) \qquad (3.32)$$

These simple relationships between transfer functions in the s and z domains are valuable for studying various effects of discretization. For the case of the Euler method, the mapping between s and z has one undesirable property with important practical consequences. As will soon be shown, the region of the complex s-plane corresponding to asymptotic stability, namely the region $\operatorname{Re} s < 0$, is *not* mapped into the region of the z-plane corresponding to stability for discrete-time systems, namely $|z| < 1$. Thus some stable continuous-time systems will give rise to unstable Euler discretizations, a clearly undesirable result from a qualitative point of view, but also one that leads to catastrophic numerical errors when solution of the discretized system is attempted using a digital computer with its inherent limited numerical accuracy.

For the Euler method, $\operatorname{Re} s < 0$ holds when $\operatorname{Re} z < 1$. To determine systems that admit stable discretizations, we need to restrict the eigenvalues of $\mathbf{A}$ to lie in that subset of $\operatorname{Re} s < 0$ corresponding to $|z| < 1$. Substituting for z in terms of s, this is the set of s satisfying the inequality $|1 + sh| < 1$, which is the interior of the circle of radius $1/h$ centered at $s = -1/h$ (Fig. 3.3). We obtain an explicit constraint on the choice of the sampling time: h must be chosen small enough so that all of the eigenvalues of $\mathbf{A}$ lie within this circle (whose radius increases with decreasing h). There is a unfortunate aspect of this stability condition: the "most stable" eigenvalue of $\mathbf{A}$ (i.e., the one with the most negative real part) can play the determining role in the selection of h small enough to guarantee stability of the Euler discretization. For a "stiff system," one whose eigenvalues are widely spread out in the complex plane, a small discretization step size is required for stability reasons that may have nothing to do with accuracy requirements. The following contrived example illustrates this point quite clearly.

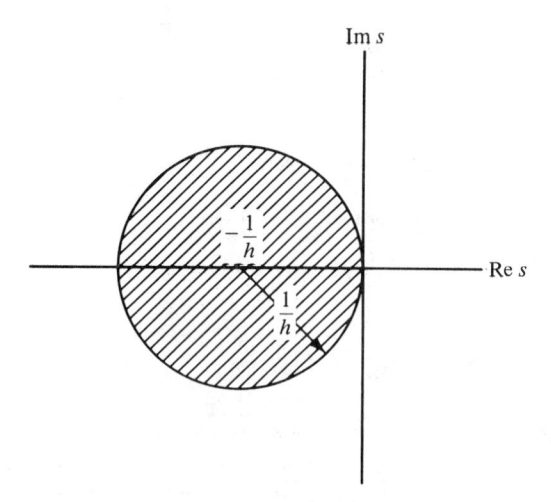

Figure 3.3 Complex s-plane showing the region mapped into $|z| < 1$ by the mapping $s = (z - 1)/h$.

Example

For the stiff system

$$\dot{x}_1(t) = -99x_1(t) + x_2(t)$$
$$\dot{x}_2(t) = -x_2(t)$$

the choice of $h = 0.2$ would seem to be reasonable on the basis of accuracy considerations. After all, initial conditions on the "fast" variable, x_1, certainly disappear from the solution in a time much shorter than a single discretization step. However, for stability h must be about 10 times smaller (e.g., $h = 0.02$), which corresponds to a 10-fold increase in computational effort. A numerical experiment to verify that stability is required to obtain a valid solution to the Euler discretized system is left for the reader.

For the backward difference discretization method, the situation is somewhat improved. All stable systems have stable backward difference discretizations. (The terminology "A-stable," for *absolutely stable*, is sometimes used for such a method.) There is still a potential problem for the discretization of a continuous-time system that is unstable, because the instability region (i.e., the region in the s-plane leading to unstable discretizations) is given by the set of complex s satisfying the inequality $|1 - sh| < 1$, which is the interior of the circle of radius $1/h$ centered at the point $1/h$ on the real axis in the complex s-plane (Fig. 3.4). This circle lies completely within the right half of the complex s-plane, so we see that the backward difference method of discretization (an implicit method) always preserves stability, allowing the discretization step size to be chosen according to accuracy requirements. Still, caution must be used when the backward difference discretization is used on unstable continuous-time systems; instability may not be preserved unless the discretization time step is taken to be very small.

It turns out that all explicit discretization methods suffer to some extent from stability limitations (none are A-stable), while some implicit methods (but not all of them) avoid such limitations; at least in the case of the backward difference method (and the trapezoidal method to be discussed next) the added computational requirements can be worthwhile, especially in applications to "stiff" systems.

Sec. 3.2 Analysis of Discretization Techniques

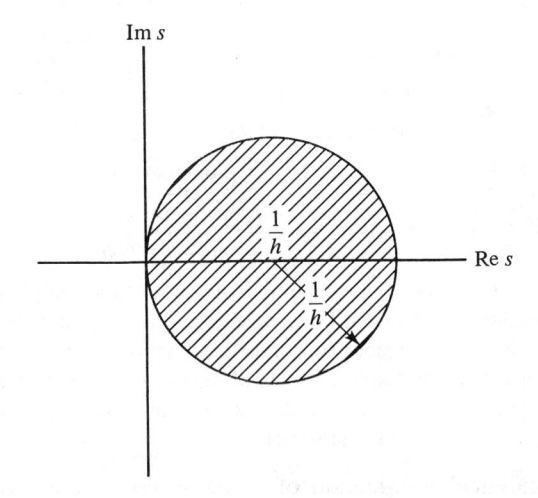

Figure 3.4 Complex s-plane showing the region mapped into $|z| > 1$ by the mapping $s = (z-1)/zh$.

Analysis of the Trapezoidal Discretization Method

Now we turn our attention to the trapezoidal method of discretization. It has the desirable property of providing second-order (in h) accuracy, as we will show later. First we will explore the corresponding transfer function relations. To find the desired answer, we use an intuitive argument first. Recall that the matrices $(\mathbf{I}+h\mathbf{A})$ and $(\mathbf{I}-h\mathbf{A})^{-1}$ provide first-order (in h) accuracy as approximations to $e^{\mathbf{A}h}$. An easy verification shows that multiplicative combination $(\mathbf{I}+h\mathbf{A}/2)(\mathbf{I}-h\mathbf{A}/2)^{-1}$ results in an approximation of $e^{\mathbf{A}h}$ accurate to second order in h. By analogy with the two previous methods, this suggests the relation

$$z = \frac{1+sh/2}{1-sh/2} \tag{3.33}$$

from which we solve for s to obtain

$$s = \frac{2}{h}\frac{z-1}{z+1} \tag{3.34}$$

So we are led to the (correct!) determination of the transfer function relation satisfied by the trapezoidal discretization:

$$G_T(z) = H\left(\frac{2}{h}\frac{z-1}{z+1}\right) \tag{3.35}$$

Once we prove this formula we will have justified another name for trapezoidal discretization that is commonly used: the *bilinear transformation method*. A third name is also used, especially in the control systems field, where the technique is known as Tustin's method.

Example

We return to the airplane example of the preceding section, taking the angle-of-attack, α, to be the output of the system. Thus the state equation description of the system is

$$\begin{bmatrix} \dot{q}(t) \\ \dot{\alpha}(t) \end{bmatrix} = \begin{bmatrix} -0.6 & 18 \\ 1 & -1.2 \end{bmatrix} \begin{bmatrix} q(t) \\ \alpha(t) \end{bmatrix} + \begin{bmatrix} m_e \\ k_e \end{bmatrix} e(t)$$

$$y(t) = \begin{bmatrix} 0 & 1 \end{bmatrix} \begin{bmatrix} q(t) \\ \alpha(t) \end{bmatrix}$$

The transfer function is easily found to be

$$H(s) = \frac{k_e s + (m_e + 0.6 k_e)}{s^2 + 1.8 s - 17.28}$$

The transfer function of the Euler discretization, to take one case, is

$$G(z) = H\left(\frac{z-1}{h}\right) = \frac{h k_3 (z-1) + h^2 (m_e + 0.6 k_e)}{(z-1)^2 + 1.8 h (z-1) - 17.28 h^2}$$

which would ordinarily be simplified after the numerical value of the sampling interval h is selected. The transfer functions for backward difference discretization and Tustin's discretization can be obtained using the corresponding transfer function relations. Notice that to determine the transfer function of the sample-and-hold discretization, obtained in an example in the preceding section, you must compute the transfer function of the associated state equations for the discretized model from first principles.

Before we give an analytical verification of this transfer function relation, it is worth pointing out a (perhaps unexpected) benefit of the Tustin method: It preserves both stability (it is A-stable) and instability! To see this involves showing that whenever $\mathrm{Re}\, s < 0$, then $|z| < 1$, where

$$z = \frac{1 + sh/2}{1 - sh/2} \tag{3.36}$$

This calculation is a tedious one, and an indirect approach turns out to be much easier. First, notice that the mapping between the variables z and s is bijective: If

$$\frac{1 + s_1 h/2}{1 - s_1 h/2} = \frac{1 + s_2 h/2}{1 - s_2 h/2} \tag{3.37}$$

then cross-multiplying to eliminate fractions gives

$$1 + \frac{s_1 h}{2} - \frac{s_2 h}{2} - \frac{s_1 s_2 h^2}{4} = 1 + \frac{s_2 h}{2} - \frac{s_1 h}{2} - \frac{s_1 s_2 h^2}{4} \tag{3.38}$$

from which we see that $s_1 = s_2$, so distinct values of s lead to distinct values of z. The same argument holds with the roles of s and z interchanged. So we can show that stability is preserved by showing that $\mathrm{Re}\, s < 0$ whenever $|z| < 1$. Take any such z, written in the form $z = Re^{j\theta}$. Then the corresponding s is

$$s = \frac{2}{h} \frac{Re^{j\theta} - 1}{Re^{j\theta} + 1} \tag{3.39}$$

Multiplying numerator and denominator by the conjugate of the denominator and expressing the result in terms of real and imaginary parts, we obtain

$$\mathrm{Re}\, s = \frac{R^2 - 1}{|Re^{j\theta} + 1|^2} \tag{3.40}$$

So for $|z| = R < 1$, $\mathrm{Re}\, s < 0$, as was to be shown.

Now we return to verifying that trapezoidal discretization corresponds to the bilinear transformation between the s and z variables in the transfer function domain. We will solve this problem by investigating the form of the transfer function that results

Sec. 3.2 Analysis of Discretization Techniques

from substituting for s in terms of z, manipulating the expression until it is in the form where we can identify it as the z-transform of the discrete-time state equations obtained by trapezoidal discretization. We start with the substitution of s in terms of z, giving

$$G(z) = \mathbf{C}\left(\frac{2}{h}\frac{z-1}{z+1}\mathbf{I} - \mathbf{A}\right)^{-1}\mathbf{B} = \mathbf{C}\left(\frac{z-1}{z+1}\mathbf{I} - \frac{h\mathbf{A}}{2}\right)^{-1}\mathbf{B}\frac{h}{2} \qquad (3.41)$$

For notational simplicity, we will study the expression

$$\begin{aligned}\left(\frac{z-1}{z+1}\mathbf{I} - \mathbf{X}\right)^{-1} &= ((z-1)\mathbf{I} - (z+1)\mathbf{X})^{-1}(z+1)\\ &= (z(\mathbf{I}-\mathbf{X}) - (\mathbf{I}+\mathbf{X}))^{-1}(z+1)\\ &= \left[(\mathbf{I}-\mathbf{X})(z\mathbf{I} - (\mathbf{I}-\mathbf{X})^{-1}(\mathbf{I}+\mathbf{X}))\right]^{-1}(z+1)\\ &= \left[z\mathbf{I} - (\mathbf{I}-\mathbf{X})^{-1}(\mathbf{I}+\mathbf{X})\right]^{-1}(\mathbf{I}-\mathbf{X})^{-1}(z+1) \qquad (3.42)\end{aligned}$$

To get our answer to come out in the desired form, we will use the identity

$$(\mathbf{I}-\mathbf{X})^{-1}(\mathbf{I}+\mathbf{X}) = (\mathbf{I}+\mathbf{X})(\mathbf{I}-\mathbf{X})^{-1} \qquad (3.43)$$

which is easily checked by cross-multiplying to get rid of the inverses. Now we set $\mathbf{X} = h\mathbf{A}/2$ to obtain

$$G(z) = \mathbf{C}\left[z\mathbf{I} - \left(\mathbf{I}+\frac{h\mathbf{A}}{2}\right)\left(\mathbf{I}-\frac{h\mathbf{A}}{2}\right)^{-1}\right]^{-1}\left(\mathbf{I}-\frac{h\mathbf{A}}{2}\right)^{-1}\mathbf{B}\frac{h}{2}(z+1) \qquad (3.44)$$

which we identify as the transfer function of the discrete-time state equations

$$\mathbf{x}_d((k+1)h) = \left(\mathbf{I}+\frac{h\mathbf{A}}{2}\right)\left(\mathbf{I}-\frac{h\mathbf{A}}{2}\right)^{-1}\mathbf{x}_d(kh)$$
$$+ \left(\mathbf{I}-\frac{h\mathbf{A}}{2}\right)^{-1}\mathbf{B}\frac{h}{2}(\mathbf{u}((k+1)h) + \mathbf{u}(kh)) \qquad (3.45)$$

$$\mathbf{y}_d(kh) = \mathbf{C}\mathbf{x}_d(kh) \qquad (3.46)$$

Of course, this is the trapezoidal discretization; cf. (3.15).

It is interesting to see how trapezoidal discretization provides second-order (in h) accuracy. For this, we start with the exact solution, equation (3.3), rewritten here for convenience:

$$\mathbf{x}((k+1)h) = e^{\mathbf{A}h}\mathbf{x}(kh) + \int_0^h e^{\mathbf{A}(h-\tau)}\mathbf{B}\mathbf{u}(kh+\tau)\,d\tau$$

We already know that $(\mathbf{I}+h\mathbf{A}/2)(\mathbf{I}-h\mathbf{A}/2)^{-1}$ provides an approximation for $e^{\mathbf{A}h}$, accurate to second order (in h), so we need to examine the term involving the integral. Using the power series for the matrix exponential,

$$\int_0^h e^{\mathbf{A}(h-\tau)}\mathbf{B}\mathbf{u}(kh+\tau)\,d\tau \approx \int_0^h \mathbf{B}\mathbf{u}(kh+\tau)\,d\tau + \int_0^h (h-\tau)\mathbf{A}\mathbf{B}\mathbf{u}(kh+\tau)\,d\tau \qquad (3.47)$$

For small h, any smooth input $\mathbf{u}$ can be approximated by the Mean Value Theorem as

$$\mathbf{u}(kh+\tau) \approx \mathbf{u}(kh) + \left(\frac{\mathbf{u}((k+1)h) - \mathbf{u}(kh)}{h}\right)\tau, \quad 0 \leq \tau \leq h \qquad (3.48)$$

Putting this expression into the integrals in equation (3.47) and simplifying gives

$$\int_0^h e^{\mathbf{A}(h-\tau)}\mathbf{B}\mathbf{u}(kh+\tau)\,d\tau \approx h(\mathbf{B}+\frac{h}{2}\mathbf{A}\mathbf{B})\left[\frac{\mathbf{u}(kh)+\mathbf{u}((k+1)h)}{2}\right]$$
$$-h^2\mathbf{A}\mathbf{B}\left[\frac{\mathbf{u}((k+1)h)-\mathbf{u}(kh)}{12}\right] \quad (3.49)$$

The last term can be ignored when considering terms up to second order in h because of the Mean Value Theorem, so we find that

$$\int_0^h e^{\mathbf{A}(h-\tau)}\mathbf{B}\mathbf{u}(kh+\tau)\,d\tau \approx \left(\mathbf{I}+\frac{h\mathbf{A}}{2}\right)\mathbf{B}\frac{h}{2}(\mathbf{u}((k+1)h)+\mathbf{u}(kh))$$
$$\approx \left(\mathbf{I}-\frac{h\mathbf{A}}{2}\right)^{-1}\mathbf{B}\frac{h}{2}(\mathbf{u}((k+1)h)+\mathbf{u}(kh)) \quad (3.50)$$

where the approximations are accurate to second order in h as was to be shown.

Further Discussion

The accuracy of the two Runge-Kutta methods for the case of linear systems may be analyzed using a direct approach (although this is a bit laborious for the fourth-order method); the results justify the labels of second-order and fourth-order that have already been used. To determine the stability limitations, a direct approach must be used because neither of the Runge-Kutta discretization methods provides a simple relationship between the transfer functions of the original continuous-time system and its discretized version.

Example

A stability analysis for the second-order Runge-Kutta method, equations (3.18–3.20), serves to illustrate how a direct approach is carried out. For the system

$$\dot{\mathbf{x}}(t) = \mathbf{A}\mathbf{x}(t)$$

the discretization is

$$\mathbf{x}_d((k+1)h) = \mathbf{x}(kh) + \frac{h}{2}[\mathbf{A}\mathbf{x}_d(kh) + \mathbf{A}(\mathbf{x}_d(kh)+h\mathbf{A}\mathbf{x}_d(kh))]$$
$$= (\mathbf{I}+h\mathbf{A}+0.5h^2\mathbf{A}^2)\mathbf{x}_d(kh)$$

(Notice that this shows the method's second-order accuracy in the case of linear systems with no input.) For stability of the discretized system, the eigenvalues of $(\mathbf{I}+h\mathbf{A}+0.5h^2\mathbf{A}^2)$ must have magnitudes less than unity; but these eigenvalues are simply the quantities $(1+h\lambda+0.5h^2\lambda^2)$, where λ varies over the eigenvalues of $\mathbf{A}$. Thus the stability condition is that the stable eigenvalues of $\mathbf{A}$ lie in the subset of the complex plane defined by the inequalities

$$|1+hs+0.5h^2s^2| < 1 \quad \text{and} \quad \text{Re } s < 0$$

The stability and accuracy analyses carried out above carry over to "well-behaved" nonlinear systems as well. For accuracy analysis, it suffices to think of the

Sec. 3.2   Analysis of Discretization Techniques   139

solutions of (well-behaved) nonlinear systems as being described by power series, even when there is no "closed-form" expression for the solutions analogous to the one provided by the matrix exponential function arising in the solution of linear systems. The notion of accuracy is then quite simply related to the matching of low-order terms in the power series solutions. Stability analysis follows directly from the results for linear systems using the idea of *linearization* of nonlinear systems, which will be covered in Chapter 4.

Frequency Response Properties

So far, our analysis of discretization methods has concentrated on accuracy and stability properties. For many applications, frequency response characteristics are also important. For example, analog circuits for audio and video signal processing applications are often designed to meet certain frequency response specifications (e.g., as low-pass or bandpass filters), so it is natural to investigate how the frequency response of a continuous-time system is transformed by discretization. A second important motivation is that the hardware implementation of discretized systems constitutes one of the principal methods for digital filter synthesis. For brevity we will limit our discussion of frequency response properties to the Tustin (trapezoidal) method, since we have seen that it has other desirable properties; the application to digital filter design will be illustrated in the next section.

We choose to measure frequency in units of radians per second. This is the usual custom when setting $s = j\omega_c$ (subscript denotes "continuous") and working with complex phasors of the form $e^{j\omega_c t}$. For the discrete-time case we must be careful to use the appropriate frequency variable, ω_d (subscript denotes "discrete"); if ν represents the "normalized" frequency variable whose units are radians per sample, then $\omega_d = \nu/h$ when the sampling interval is h seconds. Thus we set $z = e^{j\omega_d h}$, and the discretized, or sampled, phasors take the form $e^{j\omega_d h k}$.

Recall from (3.34) that the Tustin discretization produces the relationship

$$s = \frac{2}{h} \frac{z-1}{z+1}$$

which takes $s = j\omega_c$ into $z = e^{j\omega_d h}$, where

$$\omega_c = \frac{2}{h} \tan \frac{\omega_d h}{2} \qquad (3.51)$$

The inverse of this relationship between the frequency variables is

$$\omega_d = \frac{2}{h} \tan^{-1} \frac{\omega_c h}{2} \qquad (3.52)$$

Thus the Tustin discretization of a system introduces a nonlinear "warping" of the frequency variable ω_c into ω_d (Fig. 3.5). From the graph of the warping function it is clear that for small values of $\omega_c h$, there is a roughly linear relationship between the two frequency variables. This can be seen analytically by using the Taylor series for the arctangent function:

$$\omega_d = \frac{2}{h}\tan^{-1}\frac{\omega_c h}{2} \approx \omega_c - \frac{\omega_c^3 h^2}{12} = \omega_c\left(1 - \frac{(\omega_c h)^2}{12}\right) \tag{3.53}$$

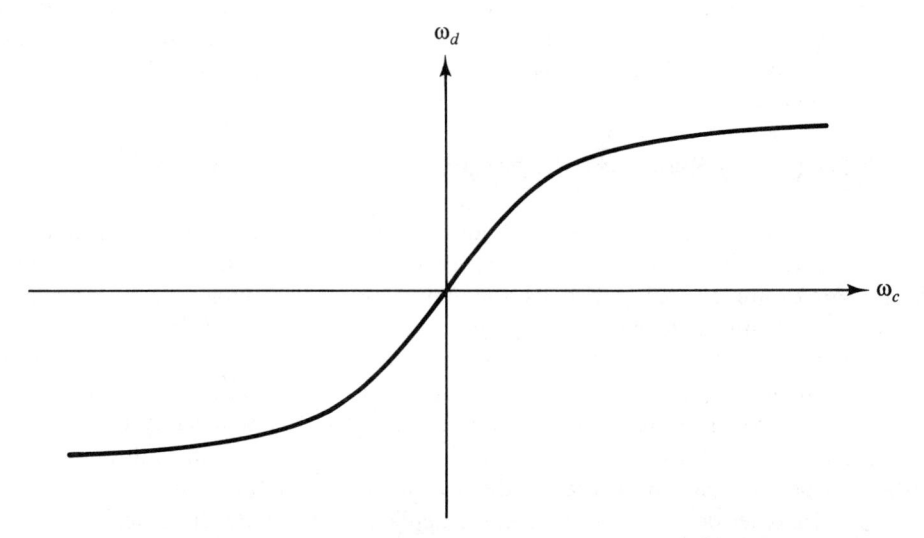

Figure 3.5 Frequency warping function: $\omega_d = (2/h)\tan^{-1}(\omega_c h/2)$.

From elementary properties of the arctangent function it also follows that for all positive values of ω_c, $\omega_d < \omega_c$, and ω_d is a monotonically increasing function of ω_c.

From (3.35), the frequency response of the Tustin discretization of a continuous-time system with frequency response function $H(j\omega_c)$ is given by

$$G_T(e^{jh\omega_d}) = H(j\omega_c) \tag{3.54}$$

Let $M_{G_T}(\omega_d)$ denote the magnitude of G_T as a function of ω_d, and similarly for $M_H(\omega_c)$; then the chain rule for differentiation gives

$$\frac{dM_H}{d\omega_c} = \frac{dM_{G_T}}{d\omega_d}\frac{d\omega_d}{d\omega_c} \tag{3.55}$$

Since $d\omega_d/d\omega_c$ is positive, this shows that regions of increasing magnitude M_H are transformed into regions of increasing magnitude M_{G_T}, and that the regions of decreasing magnitude also correspond. In practical terms, this means that the discretization of a low-pass filter H is a low-pass filter G_T, and G_T will be a bandpass filter if H is a bandpass filter.

A simple modification of the Tustin discretization can be employed to normalize the transformation of the frequency response function $H(j\omega_c)$ at a particular frequency, say ω_c^*. For example, ω_c^* could be chosen as the bandwidth in the case of a low-pass filter, or as the center frequency in the case of a bandpass filter. Notice that by suitably increasing the slope of the warping function relating ω_d to ω_c, making it greater than 1 at the origin, it would be possible to make any particular value of ω_c, say ω_c^*, equal to a

Sec. 3.3   Discretization and Digital Filter Design

given value of ω_d, ω_d^*. (An example is given in the next section.) The simplest way of scaling the slope of the warping function is to multiply ω_c by a factor less than 1; this simple linear scaling is sometimes called a "prewarping" transformation, since it allows for compensation of the nonlinear frequency warping at a prespecified pair of frequencies, (ω_c^*, ω_d^*).

The modified Tustin discretization method incorporating prewarping is accomplished by replacing the bilinear transformation (3.34), with a differently scaled version:

$$s = \alpha \frac{z-1}{z+1} \tag{3.56}$$

The relationship between ω_c and ω_d for this modified Tustin discretization method is

$$\omega_c = \alpha \tan \frac{\omega_d h}{2} \tag{3.57}$$

Thus choosing α as

$$\alpha = \frac{\omega_c^*}{\tan(\omega_d^* h/2)} \tag{3.58}$$

will give

$$G_{T(\alpha)}(e^{j\omega_d^* h}) = H(j\omega_c^*) \tag{3.59}$$

where we have indicated the dependence of the discretized frequency response function on α explicitly.

The discretized state equations corresponding to the modified Tustin discretization are easily obtained by substituting the new scaling factor α for $2/h$; the discretized state equations are

$$\mathbf{x}_d((k+1)h) = \left(\mathbf{I} + \frac{\mathbf{A}}{\alpha}\right)\left(\mathbf{I} - \frac{\mathbf{A}}{\alpha}\right)^{-1} \mathbf{x}_d(kh)$$
$$+ \frac{1}{\alpha}\left(\mathbf{I} - \frac{\mathbf{A}}{\alpha}\right)^{-1} \mathbf{B}(\mathbf{u}((k+1)h) + \mathbf{u}(kh)) \tag{3.60}$$

$$\mathbf{y}_d(kh) = \mathbf{C}\mathbf{x}_d(kh) \tag{3.61}$$

The property of second-order accuracy (in h) of the discretization is preserved under prewarping, thanks to the second-order accuracy (in h) of the expression $1/\alpha$ as an approximation to $h/2$.

3.3 DISCRETIZATION AND DIGITAL FILTER DESIGN

Our focus in the first two sections of this chapter has been on the discretization of a continuous-time state equation model of a system. We found that the Tustin method offers a number of advantages such as second-order accuracy, preservation of stability, and an easily expressed transformation of the frequency response function. The latter property is frequently exploited in certain digital filter design problems. (Here we use the term *digital filter* to mean simply a linear discrete-time system.) In this section we will explore two digital filter design methods related to discretization, using a case study for illustration and as the basis for some comparisons.

As analyzed at the end of the last section, the bilinear transformation between the s and z transform variables imposed by the Tustin method produces a nonlinear, but monotonic relationship between the frequency variables. As a consequence, continuous-time filters with certain frequency-domain characteristics (e.g., low-pass, bandpass, and high-pass filters) are mapped into digital filters with the same characteristics. Hence for a wide range of applications of digital filtering techniques where the desired signal processing operations require such frequency-selective filters, the Tustin method provides a means for taking advantage of the wealth of available knowledge about continuous-time filter design that has been built up by decades of work on passive and active electrical circuits.

As an alternative to the Tustin method, we will also describe another method of going from continuous-time filters to digital filters, the *impulse invariance design method*. This method is also widely used, and it does not suffer from the nonlinear frequency warping effect arising in the Tustin method. However, it is subject to aliasing effects such as arise in connection with the Sampling Theorem, so the method must be used with careful attention to the choice of sampling rate. The impulse invariance method is similar in spirit to the sample-and-hold discretization method, as will become clear in our discussion.

We choose a simple, but representative, low-pass digital filter design problem as an example. A continuous-time signal is sampled at a sampling frequency of 8 kHz and it is desired to have a low-pass filter that passes the samples of signal components in the range from 0 to 1 kHz. Both of the design methods start with a continuous-time low-pass filter whose performance is judged to be adequate. For the purpose of this example, we use as a *prototype* continuous-time filter the three-pole Butterworth filter whose transfer function is

$$H_p(s) = \frac{1}{(s+1)(s^2+s+1)} \tag{3.62}$$

This prototype has a bandwidth of 1 radian per second, or $(2\pi)^{-1}$ hertz, and its magnitude is a monotonically decreasing function of ω. Associated with this prototype is the *nominal* continuous-time filter with bandwidth 1 kHz, whose transfer function is obtained by scaling the prototype filter to make its cutoff frequency 1 kHz, corresponding to $\omega_{co} = 2000\pi$, namely

$$H_n(s) = H_p\left(\frac{s}{\omega_{co}}\right) = H_p\left(\frac{s}{2000\pi}\right) \tag{3.63}$$

Filter Design Using Tustin Discretization

Now consider the design of a digital filter by applying Tustin discretization. As a first approach, we will compensate for the nonlinear frequency warping of the discretization by scaling the bandwidth of the nominal filter H_n appropriately. The bandwidth of the

Sec. 3.3 Discretization and Digital Filter Design

required digital filter is 1 kHz, so $\omega_d = 2000\pi$ radians per second. At the sampling frequency of 8 kHz, this corresponds to $\pi/4$ radians per sample (i.e., $\omega_d h = \pi/4$). The required bandwidth for the continuous-time filter is obtained from the inverse of the warping function:

$$\omega_c = \frac{2}{h}\tan\frac{\omega_d h}{2} = 16{,}000\tan\frac{\pi}{8} \approx 6627.417 \qquad (3.64)$$

As expected, this is somewhat larger than the bandwidth of the nominal filter, $\omega_{co} = 2000\pi \approx 6283.185$. We denote the ratio by $\eta = 16{,}000\tan(\pi/8)/2000\pi$. The transfer function to be discretized is $H(s) = H_n(s/\eta)$. Using the bilinear transformation (3.35) gives the discrete-time transfer function for the Tustin discretization

$$G_T(z) = H\left(16{,}000\,\frac{z-1}{z+1}\right) \qquad (3.65)$$

As a second approach to this design problem we will use the modified Tustin discretization that incorporates prewarping. We start with the prototype transfer function, $H_p(s)$, whose bandwidth, 1 radian per second, becomes ω_c^*; the value of ω_d^* is 2000π, the desired filter bandwidth in radians per second. For an 8-kHz sampling rate, $h = 1/8000$, and the value of the scaling parameter is obtained from (3.58),

$$\alpha = \frac{1}{\tan(2000\pi/16{,}000)} = \cot\frac{\pi}{8} \qquad (3.66)$$

and thus

$$G_T(z) = H_p\left(\alpha\frac{z-1}{z+1}\right) \qquad (3.67)$$

It is easily seen that the two expressions for $G_T(z)$, (3.65) and (3.67), are equivalent.

Filter Design Based on Impulse Invariance

Now we turn to the impulse invariance method for digital filter design. This is the name given to a procedure based on matching the sampled values of the weighting function of a continuous-time system to the samples of the unit pulse response sequence of a discrete-time system. The motivation for this procedure is the observation that the convolution sum

$$y_k = \sum_{i=0}^{k} g_i u_{k-i} \qquad (3.68)$$

is a natural (Riemann sum) approximation to the convolution integral

$$y(t) = \int_0^t \mathbf{h}(\tau)\mathbf{u}(t-\tau)\,d\tau \qquad (3.69)$$

when we take $t = kT_s$, $\mathbf{h}(0)/2 = \mathbf{g}_0/T_s$, and $\mathbf{h}(iT_s) = \mathbf{g}_i/T_s$, $i > 0$. Since these two expressions give the output signals for continuous-time and discrete-time linear systems, respectively, it is natural to use the approximation as the basis for a discretization procedure.

To carry out the solution of this (infinite) set of equations, we turn to a frequency-domain method. As a means of motivating this approach, consider the simplest weighting function, $\mathbf{h}(t) = h_0 e^{\lambda t} \mathbf{1}(t)$. This weighting function corresponds to the first-order system

$$\dot{y}(t) - \lambda y(t) = h_0 u(t) \tag{3.70}$$

and so to the transfer function

$$H(s) = \frac{h_0}{s - \lambda} \tag{3.71}$$

Now it is easy to recognize the samples of this weighting function as the terms in a geometric sequence:

$$\mathbf{h}(iT_s) = h_0 e^{\lambda i T_s}, \quad i > 0 \tag{3.72}$$

so the prescribed choice of $\mathbf{g}_i$ corresponds to the unit pulse response sequence of the first-order discrete-time system

$$y_{k+1} - e^{\lambda T_s} y_k = \frac{T_s h_0}{2} (u_{k+1} + e^{\lambda T_s} u_k) \tag{3.73}$$

and so to the transfer function

$$G(z) = \frac{T_s h_0}{2} \frac{z + \xi}{z - \xi} \tag{3.74}$$

where $\xi = e^{\lambda T_s}$.

Now we turn to a more general case. Let $H(s)$ be the transfer function associated with the weighting function $\mathbf{h}(t)$. ($H(s)$ is the Laplace transform of $\mathbf{h}(t)$.) For convenience in exposition, we will assume here that $H(s)$ has distinct poles so that it may be expanded into a simple partial fraction expansion.

$$H(s) = \sum_{i=1}^{n} \frac{r_i}{s - \lambda_i} \tag{3.75}$$

(If $H(s)$ should have repeated poles, a more complicated partial fraction expansion would be required, and a corresponding discrete-time system transfer function for such terms would be used. We omit the details here.)

Now the transformation used for the simple example can be applied to each term of the summation, giving

$$G(z) = \sum_{i=1}^{n} \frac{T_s r_i}{2} \frac{z + \xi_i}{z - \xi_i} \tag{3.76}$$

where $\xi_i = e^{\lambda_i T_s}$. This is the transfer function for the impulse invariance method of discretization.

In thinking about the performance of this discretization method, the most important problem can be expected to arise from the approximation of the individual exponential terms of the weighting function by their samples taken at a sampling interval of T_s. Since the Sampling Theorem may not be applied to such exponential signals, aliasing of frequency components at frequencies higher than $2/T_s$ hertz will occur and contribute to the frequency response of the discrete-time system, $G(e^{j\omega_d})$. This effect can be reduced by choosing the sampling interval small enough so that the frequency response of the continuous-time system, $H(j\omega_c)$, is small at frequencies higher than $2/T_s$ hertz. The advantage of the impulse invariance method is that no nonlinear warping of the frequency scale is introduced by the discretization; this can be important for applications involving multipassband filters, for example.

Continuing with the example started earlier, we start with the nominal transfer function $H_n(s)$, (3.63), whose cutoff frequency is 1 kHz, corresponding to $\omega_{co} = 2000\pi$. We then obtain the partial fraction expansion of $H_n(s)$:

$$H_n(s) = \sum_{i=1}^{3} \frac{r_i}{s - \lambda_i} \qquad (3.77)$$

where $\lambda_1 = -\omega_{co}$, $\lambda_2 = \omega_{co}((j\sqrt{3} - 1)/2)$, $\lambda_3 = \omega_{co}((-j\sqrt{3} - 1)/2)$, $r_1 = \omega_{co}$, $r_2 = \omega_{co}^2/(\lambda_2 - \omega_{co})$, and $r_3 = \omega_{co}^2/(\lambda_3 - \omega_{co})$. The resulting discretized system is then easily calculated from the general expression.

The impulse invariance method is just one of a class of methods that can be obtained by insisting that the samples of a particular continuous-time system output are matched by a discrete-time system, it being implicitly assumed that the input signal of the continuous-time system is processed in some known way to obtain the input sequence for the discrete-time system. Put this way, we see immediately that sample-and-hold discretization could also be called the *step invariance method*. As one example of a property that sample-and-hold has in common with impulse invariance, notice that both preserve stability as the result of the exponential function mapping the left half of the complex plane into the unit circle in the complex plane.

A fundamentally different approach to discrete-time system (digital filter) design involves the formulation of a suitable performance criterion and the solution of the corresponding optimization problem. Some of the examples in Chapter 5 will illustrate this approach.

3.4 DISCRETIZATION FOR DISTRIBUTED PARAMETER SYSTEMS

Systems described by ordinary differential equations, while providing a large class of models for dynamic processes, do not suffice for modeling many of the natural phenomena arising in important technological applications; *partial differential equations* comprise a wide class of models used in fluid dynamics, elasticity, electromagnetism, thermodynamics, and quantum mechanics. We will use the term *distributed parameter system* to mean a system described by partial differential equations, although the term is often applied to even broader classes of system models. It is not the intention to give a comprehensive treatment of distributed parameter systems here. Rather, we will briefly discuss only a single aspect, discretization, the theme of this chapter, using examples to illustrate some of the important issues.

As we have already seen to some extent, discretization is a process of replacing analytical models (such as ordinary and partial differential equations) with models in a form suitable for solution by numerical computation; in particular, the domain of each independent variable needs to be replaced by a discrete set of points. When there are two or more independent variables involved in the description of a quantity, such as a temperature or a magnetic field that may vary in both temporal and spatial domains, partial differential equations models are used, and a wide variety of discretization techniques may be applied. The variety of techniques reflects the fact that from a computational point of view it is advantageous to exploit various properties, such as symmetry and smoothness, that arise in the solutions.

Before going further, we will describe two situations involving distributed parameter systems so that the reader can begin to think more concretely about such problems. First, consider the problem of determining the temperature distribution in material placed in contact with a set of heat sources. Such a problem would arise in the analysis and design of a cooler for keeping a six-pack of your favorite beverage cold; is a more specific description needed to convince the reader of the importance of this kind of example? Similar problems are involved in the design of automobile radiators, heat sinks for the power transistors in audio amplifiers, and so on. The temporal and spatial variation of temperature is modeled by a partial differential equation whose particular form will be given later; the heat sources provide boundary conditions for this equation.

As a second problem, consider the description of a magnetic field, a vector-valued quantity varying in space and time; this is an important problem in electromagnetic theory that is "solved" by using Maxwell's equations, a well-known partial differential model. The design of electrical devices such as transformers, motors, and generators relies on such models; so does the design of high-resolution cathode ray tube displays and magnetic plasma containment devices used for high-temperature fusion reactions.

Many, many other examples of problems where partial differential equation models are applied can be given, such as the lift and drag characteristics of an aircraft wing at various altitudes, speeds, and angles of attack, the displacement of points on a plucked guitar string, and the description of the electrons orbiting an atomic nucleus. Some explicit examples will be given later, although the origin of the models will be left for the interested reader to explore.

The importance of distributed parameter systems is not the point to be emphasized in this section. The point to be made here is that there is an important role played by numerical solution methods for such systems. While traditional introductory mathematical physics and engineering courses use analytically tractable examples (and analytical methods of approximate solution), such methods are generally limited to special situations involving idealizations of material properties and simplified spatial geometries. Realistic problems require numerical solution methods, and a wide variety of software is available for applications in many fields.

There are several factors that make the discretization of distributed parameter models considerably more complicated than the process of obtaining a simple difference equation model to replace an ordinary differential equation. Second- and higher-order partial derivatives are commonly encountered in applications; recall that it was always

Sec. 3.4   Discretization for Distributed Parameter Systems   **147**

assumed that state space models were given in the form of first-order differential equations for a state vector. As a result, it is necessary to employ suitable finite difference approximations. Even for first-order partial derivative operators, the best choice of discrete approximation may be different for different variables. The approaches described earlier in this chapter dealt with discretization when time was the independent variable; for spatial variables (or others) it may be appropriate to use finite difference approximations that do not make any distinction between "past" and "future," especially in regard to achieving a suitable counterpart to the notion of stability.

A second kind of complication arising in distributed parameter models relates to boundary conditions. In the state space models already considered, it has always been assumed that the solution values needed to specify a unique solution of the differential equation are all given at a single time instant (i.e., for a single value of the independent variable). This is already too restrictive an assumption for some systems that may be modeled using ordinary differential equations. For example, the periodic function $x(t) = \cos t$ is naturally specified by the differential equation

$$\ddot{x}(t) + x(t) = 0 \tag{3.78}$$

with the boundary conditions $x(0) = 1$ and $x(\pi/2) = 0$. For distributed parameter models, provisions for handling general kinds of boundary conditions must be incorporated as part of the discretization process, and we generally expect that discretizations will involve implicit, rather than explicit, equations to be solved. Handling irregular boundary geometries can be a tricky problem that affects the discretization process.

The two primary methods of discretization for distributed parameter models are *finite difference methods* and *finite element methods*. The former are extensions of the methods that have been covered in the first few sections of this chapter. Each partial differential operator is replaced by an approximation based on sampled values of the function such as a forward or backward difference; because most distributed parameter models involve partial derivative operators in more than one variable, this generally turns a partial differential equation model into a partial difference equation model.

Finite Difference Methods

Consider the description of the current-voltage relationships along a semi-infinite transmission line. Maxwell's equations lead to a distributed parameter model for this system described by the so-called *telegraph equations*:

$$\frac{\partial i(x,t)}{\partial x} = -C(x) \frac{\partial v(x,t)}{\partial t} \tag{3.79}$$

$$\frac{\partial v(x,t)}{\partial x} = -L(x) \frac{\partial i(x,t)}{\partial t} \tag{3.80}$$

where x is the distance coordinate along the transmission line, $0 \le x < \infty$, $L(x)$ is the line inductance per unit length at position x, and $C(x)$ is the line capacitance per unit length at position x; we will assume that the line is initially quiescent: Both $v(x, 0)$ and $i(x, 0)$ are zero. This system may be discretized by using forward differences to approximate the derivatives of both the space and time variables (Fig. 3.6). We introduce the simplifying notation $v(n\,\Delta x, m\,\Delta t) = v_{n,m}$ and $i(n\,\Delta x, m\,\Delta t) = i_{n,m}$. Then the discretized equations are

$$(i_{n+1,m} - i_{n,m})\,\Delta t = C(n\,\Delta x)(v_{n,m} - v_{n,m+1})\,\Delta x \tag{3.81}$$

$$(v_{n+1,m} - v_{n,m})\,\Delta t = L(n\,\Delta x)(i_{n,m} - i_{n,m+1})\,\Delta x \tag{3.82}$$

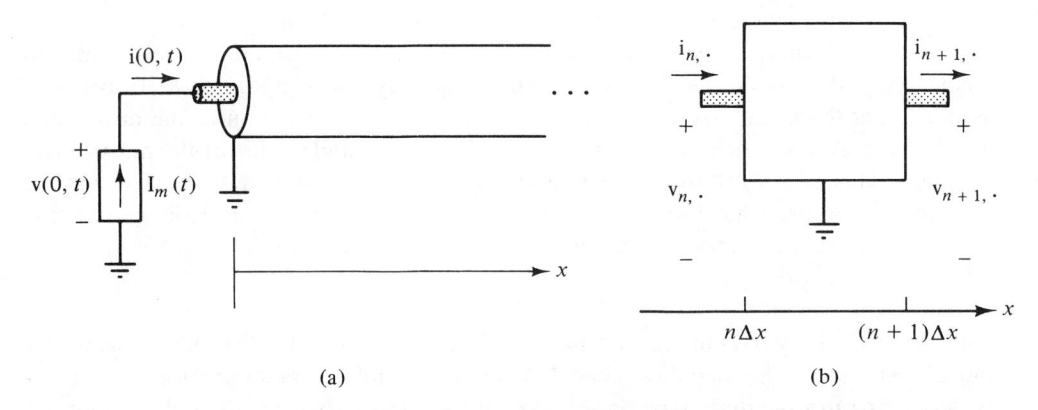

Figure 3.6 Semi-infinite transmission line; spatial discretization of a transmission line.

Rearranging the equations into a form that suggests how the discretized solution is actually computed, we have

$$v_{n,m+1} = v_{n,m} - \frac{1}{C_n}(i_{n,m} - i_{n+1,m}) \tag{3.83}$$

$$i_{n,m+1} = i_{n,m} - \frac{1}{L_n}(v_{n,m} - v_{n+1,m}) \tag{3.84}$$

where

$$L_n = L(n\,\Delta x)\frac{\Delta x}{\Delta t} \quad \text{and} \quad C_n = C(n\,\Delta x)\frac{\Delta x}{\Delta t} \tag{3.85}$$

Besides the boundary conditions for a quiescent line, $v_{n,0} = 0$ and $i_{n,0} = 0$, we assume that a driving input is applied to the line; for example, it is driven at the left end by an ideal current source so that

$$i_{0,m} = I_m \quad \text{with } I_0 = 0 \tag{3.86}$$

The "computational wavefront," which describes the sample points at which the partial difference equations for i and v may be calculated, is a line of slope -1 in (n,m)-index space (Fig. 3.7). Using this model, it is straightforward to compute various quantities of interest for characterizing the behavior of this transmission line as a two-terminal device such as its impulse response (voltage response to current impulse), step response, or complex impedance (voltage response to a unit amplitude complex phasor current input).

Sec. 3.4 Discretization for Distributed Parameter Systems

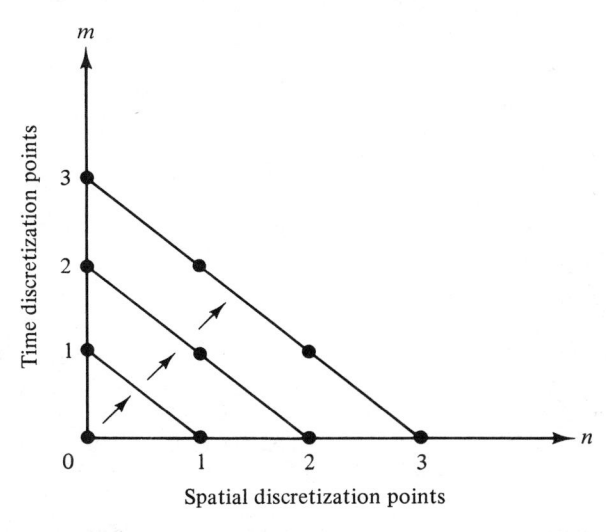

Figure 3.7 Computational wavefront for discretized telegraph equations.

In carrying out the discretization, the usual problem of selecting the discretization step sizes must be handled. For this problem, involving first derivatives in both space and time, it is almost obvious that there is a similar dependence on the spatial and temporal discretization step sizes. We expect that Δx and Δt must approach zero at the same rate in order to achieve a meaningful limit; in other words, we need $\Delta x / \Delta t \to c$ for some constant c as the two discretization steps sizes tend to zero. Based on earlier discussions in this chapter, stability of the discretization is also an important consideration.

It will be recognized that the two first-order equations above can be combined to obtain a second-order equation for one of the variables. For example, by differentiating the first equation with respect to x and the second with respect to t, the v dependence can be eliminated by combining the two resulting equations, giving

$$\frac{\partial^2 i(x,t)}{\partial x^2} = L(x)C(x) \frac{\partial^2 i(x,t)}{\partial t^2} \tag{3.87}$$

However, it should be noticed that the general solution of this second-order equation has terms corresponding to a "phantom" (nonphysical) first-order model with negative L and C terms. In solving the second-order equation it is thus important to avoid any difficulties that might be associated with the uninteresting solutions to the phantom model.

As a second example, consider the problem of determining the electrostatic potential in a "slab" of material, finite in extent in two spatial directions, but infinite in the third, with a spatial charge density ρ (Fig. 3.8). Denoting the potential by ϕ, simplification of Maxwell's equations leads to Poisson's equation:

$$\nabla^2 \phi = -\rho \tag{3.88}$$

where units have been chosen to normalize the equation. Choosing a rectangular coordinate system, with x and y being the spatial coordinates of interest, Poisson's equation takes the more concrete form

$$\frac{\partial^2 \phi(x,y)}{\partial x^2} + \frac{\partial^2 \phi(x,y)}{\partial y^2} = -\rho(x,y) \tag{3.89}$$

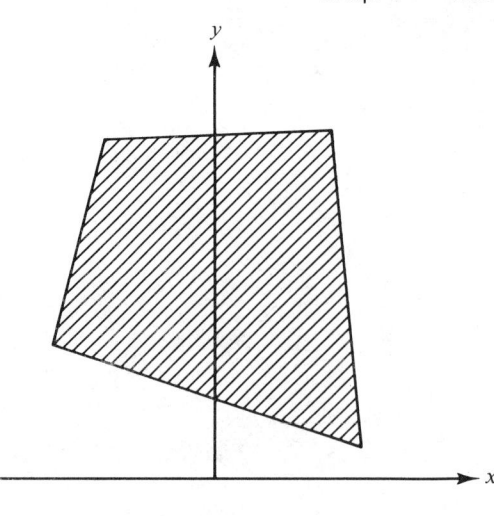

Figure 3.8 (x,y) cross section showing region of definition of potential function $\phi(x,y)$ and charge density function $\rho(x,y)$.

Centered second differences may be used to approximate the second derivative terms. For the x derivative, this amounts to the approximation

$$\frac{\partial^2 \phi}{\partial x^2} \approx \frac{\frac{\partial \phi(x+\Delta x/2,y)}{\partial x} - \frac{\partial \phi(x-\Delta x/2,y)}{\partial x}}{\Delta x}$$

$$\approx \frac{\phi(x+\Delta x,y) - 2\phi(x,y) + \phi(x-\Delta x,y)}{(\Delta x)^2} \tag{3.90}$$

where the second expression comes from using difference approximations to the first derivatives appearing in the previous expression. (One way to derive finite difference approximations for high-order derivatives is to require an exact fit for the corresponding polynomial functions. The second derivative of x^2 is the constant function 2; the second difference approximation above gives the value 2 at each sample point when applied to x^2.) A similar approximation is applied for the second derivative term in y. Choosing $\Delta x = \Delta y = \Delta$ and combining the results gives the discretized model

$$\phi(x+\Delta,y) + \phi(x,y+\Delta) + \phi(x-\Delta,y) - \phi(x,y-\Delta) - 4\phi(x,y) = -\Delta^2 \rho(x,y) \tag{3.91}$$

(Notice the remarkable property of this equation: On each five-point "neighborhood" in the sampled domain, the four "directional differences" sum to a fixed scalar multiple of the density function; see Fig. 3.9.) Given a set of boundary conditions, this discretization is conveniently viewed as a set of linear equations to which a variety of solution methods may be applied. Taking $\mathbf{x}_P$ to be the vector of spatial samples of the function ϕ, $\mathbf{y}_P$ to be the vector of spatial samples of the density function ρ, and $\mathbf{A}_P$ to be the appropriate matrix of integer coefficients, we have

$$\mathbf{A}_P \mathbf{x}_P = \mathbf{y}_P \tag{3.92}$$

The matrix $\mathbf{A}_P$ has only five nonzero entries in each of its rows. Exploiting the sparse structure of $\mathbf{A}_P$ is crucial for solving this problem efficiently. Iterative solutions

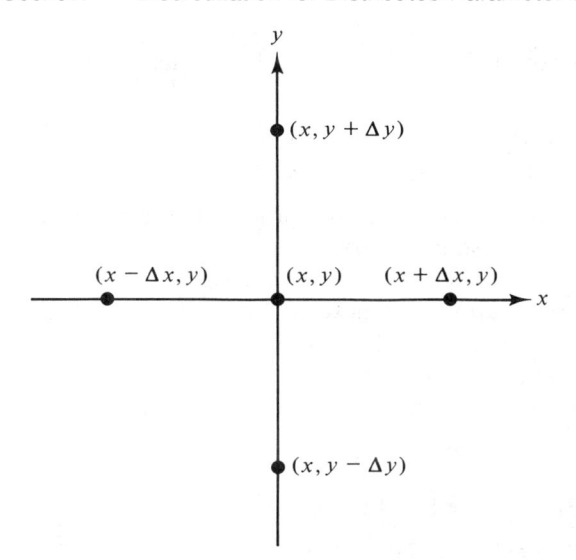

Sec. 3.4 Discretization for Distributed Parameter Systems

Figure 3.9 Five points used in spatial discretization of Poisson's equation.

methods such as Jacobi and Gauss-Seidel (recall the discussion in Section 1.2) may be used. Many specialized algorithms and parallel computing architectures have been developed for this problem and similar problems.

A third example to be considered involves the "heat equation," named for one of its important applications; it also describes other similar processes in fluid flow, electromagnetic theory, and so on. In a simple instance, the ends of a unit length bar with uniform thermal conductivity are placed in contact with constant-temperature heat sources; the bar neither loses nor gains any heat except through its ends (Fig. 3.10). It is assumed that the cross section of the bar is small so that the temperature variation in the bar is only a function of the position x along the bar and time t. Using $T(x,t)$ to denote the temperature leads to the equation

$$\frac{\partial T(x,t)}{\partial t} = \frac{\partial^2 T(x,t)}{\partial x^2} \qquad (3.93)$$

It is assumed that the bar has an initial temperature distribution at $t=0$.

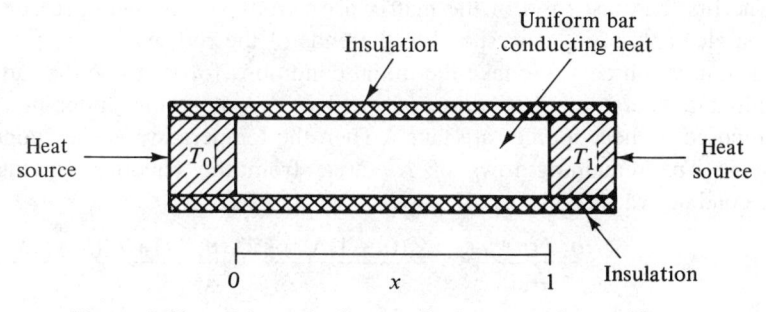

Figure 3.10 Schematic diagram for heat equation problem.

Should the conductivity of the bar vary along its length, the equation is modified to become

$$\frac{\partial T(x,t)}{\partial t} = \frac{\partial}{\partial x}\left(c(x)\frac{\partial T(x,t)}{\partial x}\right) \tag{3.94}$$

where $c(x)$ is the conductivity profile. For purposes of this brief discussion, the simpler form of the model will suffice to illustrate our points.

This third example provides a setting to mention a variation on the finite difference idea known as *semi-discretization*. Basically, a finite difference discretization in the spatial variable only can be used to obtain an ordinary differential equation model for the spatially sampled values of the function. Using the central difference approximation for the second derivative term, as in the Poisson equation example earlier, gives

$$\dot{\mathbf{x}}_N(t) = \mathbf{A}_N \mathbf{x}_N(t) \tag{3.95}$$

where $\mathbf{x}_N(t)$ is the vector of spatial samples of the temperature at time t,

$$\mathbf{x}_N^T(t) = \begin{bmatrix} T(0,t) & T(\Delta,t) & T(2\Delta,t) & \cdots & T((N-1)\Delta,t) \end{bmatrix} \tag{3.96}$$

where Δ is the spatial sampling length, $\Delta = 1/(N-1)$, and where $\mathbf{A}_N$ is a matrix arising from the discretization of the spatial second derivative (Fig. 3.11):

$$\mathbf{A}_N = \frac{1}{\Delta^2}\begin{bmatrix} 0 & 0 & \cdots & & 0 & 0 \\ 1 & -2 & 1 & & & \cdot \\ 0 & 1 & -2 & & & \cdot \\ \cdot & & 0 & & & \cdot \\ \cdot & & & & 1 & 0 \\ \cdot & & & & -2 & 1 \\ 0 & \cdot & \cdots & & 0 & 0 \end{bmatrix} \tag{3.97}$$

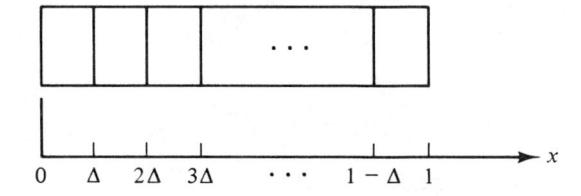

Figure 3.11 Semidiscretization of spatial variable; $\Delta = 1/(N-1)$.

The first and last rows of the matrix arise from the following reasoning. The first and last elements of $\mathbf{x}_N$ correspond to the ends of the rod, which are placed in contact with ideal heat sources. We take the initial conditions for $\mathbf{x}_N$ to be the sampled values of the initial temperature distribution in the bar except at the endpoints, where the values imposed by heat sources are taken. Then the temperature at the endpoints remains constant. The remaining rows of $\mathbf{A}_N$ arise from the equations giving the approximate second derivative:

$$\frac{\partial^2 T(n\Delta,t)}{\partial x^2} \approx \frac{T((n-1)\Delta,t) - 2T(n\Delta,t) + T((n+1)\Delta,t)}{\Delta^2} \tag{3.98}$$

for $1 \leq n \leq N-2$.

Sec. 3.4 Discretization for Distributed Parameter Systems

The matrix $\mathbf{A}_N$ clearly has (at least!) two zero eigenvalues, with corresponding eigenvectors that may be obtained by inspection: the vector of all 1's, and the vector whose elements form an arithmetic progression. The physical interpretation of this observation is natural and appealing. Functions of the form $a + bx$ are (obviously) the steady-state solutions to the heat equation; sampled values of such a function may be uniquely expressed as a linear combination of the two identified eigenvectors of $\mathbf{A}_N$ corresponding to its zero eigenvalue. On the other hand, any linear combination of these eigenvectors is an equilibrium solution of the semi-discretized state equations; can you argue that $\mathbf{A}_N$ has no other linearly independent eigenvectors corresponding to its zero eigenvalue? What is perhaps even more interesting is the following question: Can you show that for every initial condition with nonnegative components, the solution remains nonnegative for all time?

Should a full discretization of the heat equation be required, the methods described in the early part of this chapter may be applied to the semi-discretized differential equation model. Notice that the eigenvalues of $\mathbf{A}_N$ are proportional to N^2 (the largest is about $4/\Delta^2$); thus we expect there to be a very large spread in the eigenvalues when small spatial sampling is used: The semi-discretized model is "stiff." Using the simplest time-discretization method, the Euler method, is usually impractical in such cases because of the extremely small discretization steps required for stability; $h < \Delta^2/2$. In numerical methods books, the common choice of discretization method is called the *Crank-Nicolson method*, which is exactly the trapezoidal discretization method introduced earlier; it imposes no restrictions on h in terms of Δ while assuring stability. Thus the choices of the discretization steps can be made on the basis of desired accuracy only.

Finite Element Methods

The second general class of discretization techniques mentioned above, the finite element methods, is a collection of methods that exploit the idea of superimposing easily computed approximations to solutions on disjoint small regions. Typically, simple low-order polynomial functions are used to give approximate solutions on the small regions (usually triangular or rectangular regions for problems involving two independent variables), and the approximate solutions in neighboring regions are constrained to coincide along their common boundary. The approximate solutions are computed using discretization points located on the boundaries of the regions.

For example, a first-degree polynomial in two variables, x_1 and x_2, takes the form

$$p_1(x_1, x_2) = c_0 + c_1 x_1 + c_2 x_2 \tag{3.99}$$

and since there are three undetermined coefficients, its value over any triangular region is determined by its values at the vertices of the triangle. Furthermore, along any side of the triangle $p(x_1, x_2)$ varies linearly as determined by the values at the two connected vertices. Thus if two triangles share a common side and two first-degree polynomials are determined by choosing values at the vertices, the polynomials agree along the common side. There are many other possibilities for the choices of approximating function and regions. Second-degree polynomials in two variables, which have six coefficients, can be determined by the values taken at the vertices and midpoints of the sides of a triangle. Using values at the corners of a rectangle will give the four coefficients of the general bilinear function, $c_0 + c_1 x_1 + c_2 x_2 + c_{1,2} x_1 x_2$. For problems involving three variables, tetrahedrons can be used to define regions, as can rectangular parallelepipeds.

In order to illustrate the general approach, we will discuss the discretization of Poisson's equation,

$$\frac{\partial^2 \phi(x,y)}{\partial x^2} + \frac{\partial^2 \phi(x,y)}{\partial y^2} = -\rho(x,y) \tag{3.100}$$

assuming zero boundary conditions on a polygon enclosing a finite region W in the plane, and using triangular regions with piecewise linear approximations. Thus, suppose that the region W has been subdivided into triangles as a first step (Fig. 3.12), and suppose that the interior vertices are numbered, $1 \leq k \leq N$. A convenient way to represent an approximation that is piecewise linear on the selected triangular subdivisions is to use the set of pyramidal test functions $\{p_i(x,y); 1 \leq i \leq N\}$, where $p_k(x,y)$ is a continuous, piecewise linear function taking the value 1 at (interior) vertex k and the value 0 at all other vertices (Fig. 3.13). Graphically, $p_k(x,y)$ is the zero function, except that on the region formed by the triangles containing vertex k, it looks like a pyramid with a peak value of 1. Because it is determined by its values at the vertices of the triangles, an arbitrary continuous piecewise linear function on the triangular subdivisions may then be expressed as the linear combination

$$\Phi_{PL}(x,y) = \sum_{i=1}^{N} \Phi_i p_i(x,y) \tag{3.101}$$

where the coefficients are just the values of the function at the vertices.

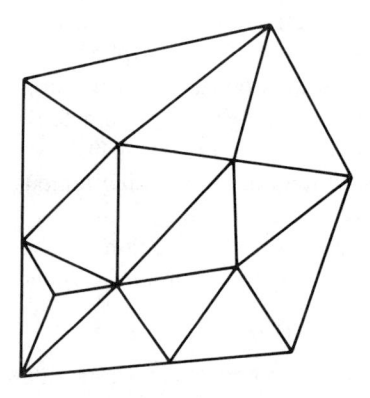

Figure 3.12 Region W subdivided into triangles.

The finite element method is simply a procedure for obtaining an approximating function of this form for the solution of Poisson's equation. Since an analytical form of the solution is not available to allow this approximation to be found, the equation itself must be employed. If the approximation were a solution, we could multiply by an arbitrary function $p(x,y)$ and integrate over the entire region W to obtain

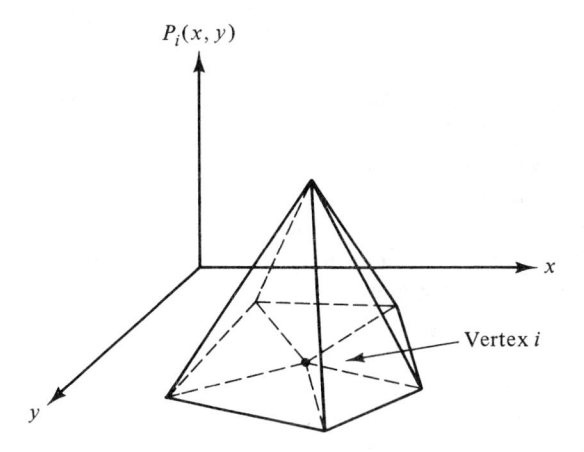

Sec. 3.4 Discretization for Distributed Parameter Systems

Figure 3.13 Pyramidal test function.

$$\iint_W \left(\frac{\partial^2 \Phi_{PL}(x,y)}{\partial x^2} + \frac{\partial^2 \Phi_{PL}(x,y)}{\partial y^2} \right) p(x,y)\, dx\, dy = -\iint_W \rho(x,y) p(x,y)\, dx\, dy \qquad (3.102)$$

Since the approximation is determined by N coefficients, a natural thing to try is to choose N different choices for $p(x,y)$ and find, if possible, solutions for the linear equations. And what more natural choice could there be than to use the same pyramidal test functions that go into making up the approximation?

Choosing these functions leads to two simplifications. First, since all of the $p_i(x,y)$ functions satisfy the boundary conditions for the problem, we may simplify the integral on the left side by integrating by parts, giving

$$\iint_W \left(\frac{\partial \Phi_{PL}(x,y)}{\partial x} \frac{\partial p_k(x,y)}{\partial x} + \frac{\partial \Phi_{PL}(x,y)}{\partial y} \frac{\partial p_k(x,y)}{\partial y} \right) dx\, dy$$

$$= \iint_W \rho(x,y) p_k(x,y)\, dx\, dy \qquad (3.103)$$

Now the form of the linear equations for the unknown coefficients can easily be determined. Using the expression for $\Phi_{PL}(x,y)$ gives

$$\iint_W \left(\sum_{i=1}^N \Phi_i \frac{\partial p_i(x,y)}{\partial x} \frac{\partial p_k(x,y)}{\partial x} + \sum_{i=1}^N \Phi_i \frac{\partial p_i(x,y)}{\partial y} \frac{\partial p_k(x,y)}{\partial y} \right) dx\, dy$$

$$= \iint_W \rho(x,y) p_k(x,y)\, dx\, dy \qquad (3.104)$$

which may be written compactly as

$$\mathbf{A}_F \mathbf{x}_F = \mathbf{y}_F \qquad (3.105)$$

where $\mathbf{x}_F$ is the vector whose components are the unknown coefficients, $\Phi_1, \ldots, \Phi_N$, $\mathbf{y}_F$ is the vector of results obtained from the integrals involving the products of ρ and p_k, and $\mathbf{A}_F$ is the $N \times N$ symmetric matrix whose elements are given by

$$a_{ik} = a_{ki} = \iint_W \left(\frac{\partial p_i(x,y)}{\partial x} \frac{\partial p_k(x,y)}{\partial x} + \frac{\partial p_i(x,y)}{\partial y} \frac{\partial p_k(x,y)}{\partial y} \right) dx\, dy \qquad (3.106)$$

The fact that $p_k(x,y)$ is nonzero only on the region formed by triangles containing vertex k means that the matrix $\mathbf{A}_F$ is sparse. Also notice that the integrals for computing the elements of the matrix $\mathbf{A}_F$ are simplified even further by the piecewise linear form of the pyramidal test functions.

In applications, software routines for automatic generation of the matrix perform the tedious calculations. These software routines typically allow the user to construct triangular subdivisions optimized for accuracy (which requires the choice of "nearly equiangular" triangles to the greatest extent possible, while assuring that no triangular region gets too large to lead to accuracy problems).

3.5 NOTES AND REFERENCES

Computational techniques for solving systems of differential equations are discussed in many numerical analysis books; Strang's book [8] is strongly recommended for its insights, particularly in connection with partial differential equation models. In fluid dynamics, combustion, structural mechanics, and many other fields, simulation is an indispensable tool. Simulation of electromagnetic phenomena is another area of growing interest; see the book by Steele [7].

Circuit simulation is widely used in teaching electrical engineering courses; this topic is covered in the book by Chua and Lin [2] and the one by Mastascusa [4]. Tools such as SPICE play an important role in the design of large-scale electronic systems, and as integrated circuit technology has advanced, applications of simulation in semiconductor device modeling and in IC fabrication process modeling have gained in importance.

Discretization is important in fields such as signal processing and control, where digital computers are required to deal with systems and signals evolving continuously in time. Control applications are covered in Franklin, Powell, and Workman [3] and in Åström and Wittenmark [1]. Digital signal processing applications are discussed in Peled and Liu [6] and in Oppenheim and Schafer [5].

BIBLIOGRAPHY

[1] K.J. Åström and B. Wittenmark, *Computer-Controlled Systems*, 2nd ed., Prentice-Hall, Englewood Cliffs, NJ, 1990.

[2] L.O. Chua and P.-M. Lin, *Computer-Aided Analysis of Electronic Circuits: Algorithms and Computational Techniques*, Prentice-Hall, Englewood Cliffs, NJ, 1975.

[3] G.F. Franklin, J.D. Powell, and M.L. Workman, *Digital Control of Dynamic Systems*, 2nd ed., Addison-Wesley, Reading, MA, 1990.

[4] E.J. Mastascusa, *Computer-Assisted Network and System Analysis*, John Wiley & Sons, New York, 1988.

[5] A.V. Oppenheim and R.W. Schafer, *Discrete-Time Signal Processing*, Prentice-Hall, Englewood Cliffs, NJ, 1989.

[6] A. Peled and B. Liu, *Digital Signal Processing: Theory, Design, and Implementation*, John Wiley & Sons, New York, 1976.

[7] C.W. Steele, *Numerical Computation of Electric and Magnetic Fields*, Van Nostrand Reinhold, New York, 1987.

[8] G. Strang, *Introduction to Applied Mathematics*, Wellesley-Cambridge Press, Wellesley, MA, 1986.

PROBLEMS

1. Find the z-domain transfer functions of the sample-and-hold discretizations of the continuous-time systems described by the s-domain transfer functions

$$H_1(s) = \frac{1}{s+a}, \quad a > 0$$

and

$$H_2(s) = \frac{1}{s^2 + 2\zeta s + 1}, \quad 0 < \zeta < 1$$

2. The system

$$\dot{x}_1(t) = -2x_1(t) + x_2(t)$$
$$\dot{x}_2(t) = -x_2(t) + u(t)$$
$$y(t) = x_1(t)$$

is to be discretized using the sample-and-hold method. Find the resulting discretized system when the sampling interval is $h = 0.05$.

3. The system

$$\dot{x}_1(t) = -2x_1(t) + x_2(t)$$
$$\dot{x}_2(t) = -x_2(t) + u(t)$$
$$y(t) = x_1(t)$$

is to be discretized using Euler's method (forward difference method). Find the resulting discretized system if the sampling interval is $h = 0.05$. What is the relationship between the transfer function of the original system and that of its discretized version?

4. The system

$$\dot{x}_1(t) = -x_1(t) + x_2(t)$$
$$\dot{x}_2(t) = -2x_2(t) + u(t)$$
$$y(t) = x_1(t)$$

is to be discretized using the backward difference method. Find the resulting discretized system if the sampling interval is $h = 0.05$. What is the relationship between the transfer function of the original system and that of its discretized version?

5. Show that when the Runge-Kutta method given in equations (3.24–3.28) is applied to the linear system

$$\dot{\mathbf{x}}(t) = \mathbf{A}\mathbf{x}(t); \quad \mathbf{x}(0) = \mathbf{x}_0$$

the value of $\mathbf{x}_d(h)$ is accurate to fourth order in h.

6. Find the Euler discretization of the system

$$\dot{\mathbf{x}}(t) = \begin{bmatrix} -2 & 1 \\ -2 & 0 \end{bmatrix} \mathbf{x}(t) + \begin{bmatrix} 1 \\ 1 \end{bmatrix} \mathbf{u}(t)$$

$$\mathbf{y}(t) = [\, 1 \quad 0 \,] \, \mathbf{x}(t)$$

using step size h. For what range of step sizes is the resulting discrete-time system stable?

7. The first-order system

$$\dot{x}(t) = -2x(t) + u(t)$$

is to be discretized using sample-and-hold with a sampling rate of 12 samples per second. Find the resulting discretized state equations, and find the resulting z-domain transfer function. Compare the latter to the transfer function obtained from using Tustin discretization. Find all frequencies at which the frequency responses of the two discretizations are equal.

8. Study the discretization error for the Euler, backward difference, and Tustin discretizations of the unforced linear oscillator described by the equations

$$\dot{x}_1(t) = x_2(t)$$

$$\dot{x}_2(t) = -x_1(t)$$

In particular, show that for the Euler method, if $x_1^2(0) + x_2^2(0) = r^2$, then after k time steps of size h, $x_1^2(0) + x_2^2(0) = r^2(1+h^2)^k$. Examine the other two methods to see if the "energy conservation" of the continuous-time system is preserved after discretization.

9. For the filters designed in Section 3.3, compare the Bode plots and the step responses.

10. Design a digital bandpass filter based on the prototype transfer function

$$H_p(s) = \frac{1}{s^2 + 2\zeta s + 1}$$

The center frequency should be 1.21 MHz, the sampling frequency 6 MHz, and ζ should be chosen so that the half-power bandwidth is approximately 10 kHz.

11. Use the discretized form of the telegraph equations, (3.83–3.84), to compute the unit step response (voltage $v(0,t)$ arising from input $i(0,t)$) of a line having $L_n = 1$ and $C_n = 1$. Assume that the line has no stored energy at $t = 0$. Use three successively smaller choices of the discretization step sizes for both x and t to investigate convergence.

12. Compute the temperature distribution in a uniform bar of unit length, initially at environmental equilibrium at temperature T_0, when heat sources of temperature T_1 are brought into contact with the two ends for a period of time and then removed. Apply the semidiscretization method with $N = 10$ and use trapezoidal discretization of the resulting linear system.

13. For the linear equations $\mathbf{A}\mathbf{x} = \mathbf{y}$, the condition that $\mathbf{A}^{-1}$ has nonnegative elements is necessary and sufficient for every nonnegative $\mathbf{y}$ to produce a nonnegative $\mathbf{x}$. Prove this. Find a condition on $\mathbf{A}$ that is sufficient to ensure that $\mathbf{A}^{-1}$ has nonnegative elements. An analogous question arises in connection with the semi-discretization of the heat equation. What condition on $\mathbf{A}_N$ will assure that the entries of $\exp(\mathbf{A}_N t)$ are all nonnegative for all $t > 0$?

14. Verify analytically, and confirm experimentally, that the eigenvalues of the matrix $\mathbf{A}_N$ arising in the semi-discretization of the heat equation are spread out over the interval $[0, 4/\Delta^2]$. Also, prove that 0 is an eigenvalue of $\mathbf{A}_N$ having multiplicity 2.

15. Use the discretized form of the Poisson equation, (3.91), to compute the potential in an infinite slab with "L-shaped" cross section when the density ρ is constant and the boundary conditions are zero. Repeat the computation using finite element discretization.

CHAPTER 4

Nonlinear Systems

We now turn to a discussion of nonlinear systems, which we take as described by the set of equations

$$\dot{\mathbf{x}}(t) = \mathbf{f}(\mathbf{x}(t), \mathbf{u}(t)) ; \quad \mathbf{x}(0) = \mathbf{x}_0 \tag{4.1}$$

$$\mathbf{y}(t) = \mathbf{h}(\mathbf{x}(t)) \tag{4.2}$$

where as before $\mathbf{x}(t)$ is an n-dimensional state vector, $\mathbf{u}(t)$ is the system input vector, assumed m-dimensional, and $\mathbf{y}(t)$ is the p-dimensional system output vector. Thus $\mathbf{h}(\mathbf{x}(t))$ is p-vector of real-valued functions and $\mathbf{f}(\mathbf{x}(t), \mathbf{u}(t))$ is an n-vector. Strictly speaking, the equations describe a time-invariant finite-dimensional nonlinear system, since neither $\mathbf{f}(\mathbf{x}(t), \mathbf{u}(t))$ nor $\mathbf{h}(\mathbf{x}(t))$ depend explicitly on t and $n < \infty$.

For linear systems we have a special case of this model, with

$$\mathbf{f}(\mathbf{x}(t), \mathbf{u}(t)) = \mathbf{A}\mathbf{x}(t) + \mathbf{B}\mathbf{u}(t) \tag{4.3}$$

$$\mathbf{h}(\mathbf{x}(t)) = \mathbf{C}\mathbf{x}(t) \tag{4.4}$$

By *nonlinear*, we simply mean "not necessarily linear" so that **f** and **h** may assume much more general functional forms, subject to some weak restrictions necessary to assure existence and uniqueness of solutions to the differential equation. We will not concern ourselves with precise technical considerations in this book. However, the reader should be warned of the need to proceed with caution when using nonlinear systems as mathematical models for real-world phenomena because the existence and uniqueness of solutions is necessary for the use of such models in most applications.

A detailed study of nonlinear systems would require more mathematical sophistication than is assumed for readers of this book. And of course there remains the difficulty of finding general results about nonlinear systems, that is, results that are valid for an "interestingly large" subclass of the set of all possible nonlinear systems. What we hope to convey by our coverage of nonlinear systems is that there are some simple, but general ideas that provide a great deal of insight into the behavior of many nonlinear systems of interest.

Previous experience with mathematics and engineering courses may suggest some of the ideas that have proven useful in studying nonlinear systems. Keeping in mind the kinds of physical systems that are commonly described by linear systems, e.g., RLC electrical circuits and mechanical systems of masses and springs, it should be clear that the linear systems are only idealizations that are valid over certain limited ranges of the physical variables involved. For example, no resistor provides a perfectly linear relationship between terminal voltage and current; for high enough powers, nonlinear effects, including eventual burnout, must be included in a valid physical model. However, for low power levels, the linear Ohm's law model suffices as a model for a resistor's behavior. Similarly, the linear force-displacement relationship assumed for an ideal spring does not remain valid over all operating conditions due to the stress-strain characteristics of materials used to fabricate real springs such as those found in automobile suspensions or retractable ball-point pens; no real spring can be extended indefinitely without breaking.

An analogous situation in calculus is the use of a linear function as an approximation to a (possibly) nonlinear one over some range of validity. This is the idea behind representing a function as a power series and then approximating it with the zeroth- and first-order terms only. You may recall an example of this kind from earlier in this book, namely the approximation of the matrix exponential function implicit in the Euler method of discretization:

$$e^{\mathbf{A}h} = \mathbf{I} + \mathbf{A}h + \frac{\mathbf{A}^2 h^2}{2} + \cdots$$

$$\approx \mathbf{I} + \mathbf{A}h \qquad (4.5)$$

which is a linear approximation of the nonlinear function (of h) $e^{\mathbf{A}h}$ that is quite accurate in the vicinity of $h=0$ (i.e., it is valid and thus useful for "small" h).

We might expect that linear approximation is useful for "smooth" nonlinear systems, where smooth means that there are no abrupt changes, at least over the operating range of the variables involved. This is the kind of situation arising in RLC circuits, for example; under normal operating conditions, variations of components' behaviors from the linear, "ideal" equations are small. Intuitively, we would thus expect that their effects on circuit behavior would be small also, and this is in fact justified by mathematical analysis of such models. This idea of *local approximation*, in which derivatives play a fundamental role, will be central to the discussions in this chapter.

Other topics to be covered in this chapter include an overview of the qualitative behavior of nonlinear systems, a description of methods for modeling input-output behavior for nonlinear systems, and a brief discussion of the role of piecewise linear models as a methodology for *global approximation*.

Nonlinear Systems   Chap. 4   **161**

Just as we did in Chapter 2, we will conclude our introduction with a collection of examples to indicate some of the tremendous variety of applications for nonlinear systems.

Examples

Electronic circuits provide a vast collection of examples where nonlinear behavior is crucial for achieving desired functions. We will give two simple examples arising from different kinds of applications. The first is a simple diode circuit that could be used for demodulation of an amplitude-modulated (AM) waveform. The second is a simple digital transistor circuit.

A diode is a two-terminal device with a nonlinear functional relationship between current and terminal voltage; let $i = N_D(v)$ denote this function (Fig. 4.1). Demodulation of an AM waveform may be accomplished by a circuit consisting of a diode in series with a simple parallel RC circuit (Fig. 4.2). The diode performs half-wave rectification of the voltage difference across its terminals, and the RC current divider performs low-pass filtering to produce a demodulated message.

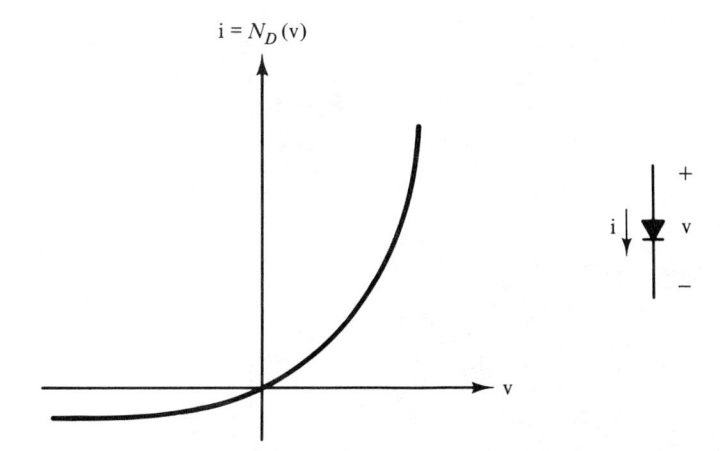

Figure 4.1   Nonlinear current-voltage relationship for diode.

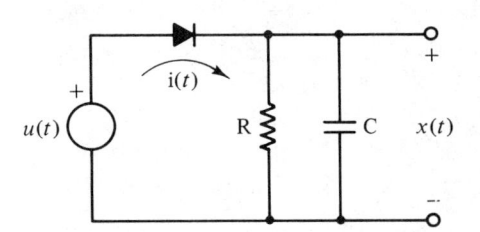

Figure 4.2   AM demodulation circuit.

We will adopt the following notation. The input, $u(t)$, is the voltage waveform to be demodulated; typically,

$$u(t) = (A + m(t)) \cos \omega_c t$$

where $m(t)$ is the modulating (message) signal, which has a bandwidth much smaller than the carrier frequency ω_c, and where $A \gg |m(t)|$. The RC filter bandwidth is matched to the message bandwidth. We take i(t) to be the current in the diode. The mathematical model for the circuit is obtained by choosing the state $x(t)$ to be the terminal voltage across the parallel RC current divider; then

$$i(t) = \frac{x(t)}{R} + C\dot{x}(t)$$

The terminal voltage across the diode is $u(t) - x(t)$, so the current-voltage relationship of the diode may be used to eliminate i(t), giving

$$N_D(u(t) - x(t)) = \frac{x(t)}{R} + C\dot{x}(t)$$

which may be rewritten in the standard form for a nonlinear system as

$$\dot{x}(t) = f(x(t), u(t)) = -\frac{x(t)}{RC} + \frac{N_D(u(t) - x(t))}{C}$$

In contrast to the small-signal or incremental model of a bipolar junction transistor introduced in Chapter 2, a nonlinear model that incorporates saturation effects is the *Ebers-Moll* bipolar junction transistor model (Fig. 4.3). This model consists of two diodes and two current-controlled current sources, and it produces families of current-voltage relationships for this three-terminal device that are suitable for models of dc and low-frequency circuit applications. Additional capacitances may be added to the model so that high-frequency and switching circuit behavior is more accurately captured.

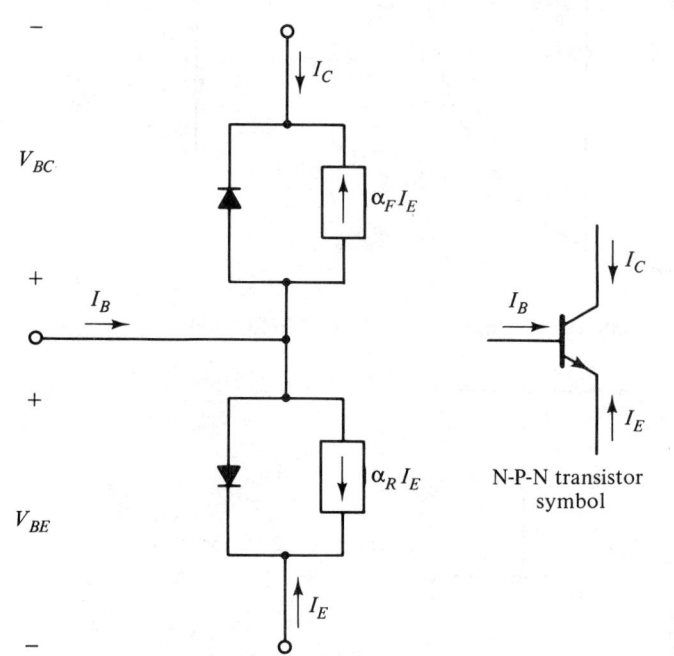

Figure 4.3 Ebers-Moll transistor model.

A two-transistor square-wave generator (i.e., a clock signal generator) is a concrete example of a system that can be expressed in state equation form by employing the Ebers-Moll transistor model (Fig. 4.4). It is worth mentioning that this example demonstrates a

Nonlinear Systems   Chap. 4

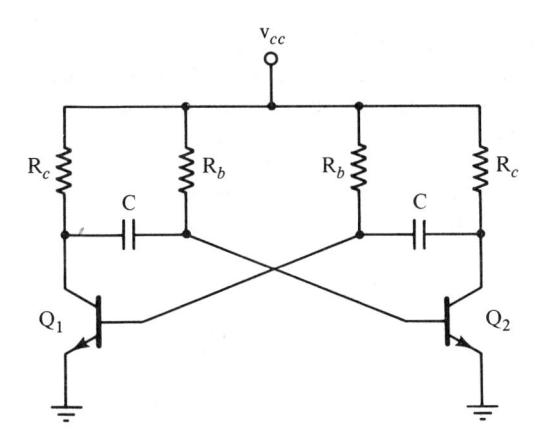

Figure 4.4 Transistor clock signal generator.

type of periodic behavior known as a *limit cycle solution*. Limit cycles have certain stability properties that are important for clocks. This will be discussed in a later section.

Turning to mechanical systems for another group of examples, we will briefly mention two simple systems studied in a beginning physics course and describe a third, more complicated system that serves as a prototype for nonlinear control system design.

First consider the classic example of a frictionless mechanical pendulum acted upon by gravitational forces only (Fig. 4.5). Assuming a unit mass suspended on an ideal (massless, rigid) pendulum of length L, the equation of motion is obtained by applying Newton's laws, giving

$$\ddot{\theta}(t) = -\frac{g}{L} \sin \theta(t)$$

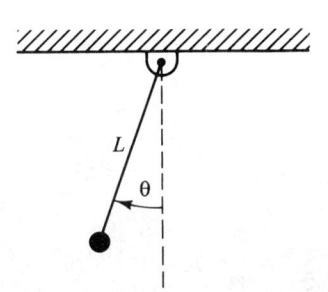

Figure 4.5 Ideal pendulum.

where θ measures the angle of the pendulum with respect to its vertical rest position; g is the acceleration due to gravity. Choosing state variables in the obvious way: $x_1(t) = \theta(t)$, $x_2(t) = \dot{\theta}(t)$, we obtain the state equations

$$\dot{\mathbf{x}}(t) = \begin{bmatrix} \dot{x}_1(t) \\ \dot{x}_2(t) \end{bmatrix} = \begin{bmatrix} x_2(t) \\ -g \sin x_1(t)/L \end{bmatrix}$$

Since there is no input to this system, $\mathbf{u}(t) = 0$ and

$$\mathbf{f}(\mathbf{x}(t), \mathbf{u}(t)) = \begin{bmatrix} x_2(t) \\ -g \sin x_1(t)/L \end{bmatrix} = \mathbf{f}(\mathbf{x})$$

A second elementary example concerns a satellite (idealized as a unit point mass) in an orbit in a central, inverse-square-law force field. We suppose that the satellite is equipped with thrusters capable of exerting radial forces (Fig. 4.6). We choose a polar coordinate system in the plane of motion of the satellite, whose origin is the center of the force field. The equation describing the motion of the satellite can be obtained from Newton's laws and written in state space form by taking $x_1 = r$, the radial distance to the satellite, $x_2 = \dot{r}$, the radial velocity, $x_3 = \theta$, the angular displacement of the satellite, and $x_4 = \dot{\theta}$, the angular velocity. We let $\mathbf{u}(t)$ be the radial thrust applied at the satellite. Then the equations of motion are

$$\ddot{r}(t) = r(t)\,\dot{\theta}^2(t) - \frac{\alpha}{r^2(t)} + \mathbf{u}(t)$$

$$\ddot{\theta}(t) = -\frac{2\,\dot{\theta}(t)\,\dot{r}(t)}{r(t)}$$

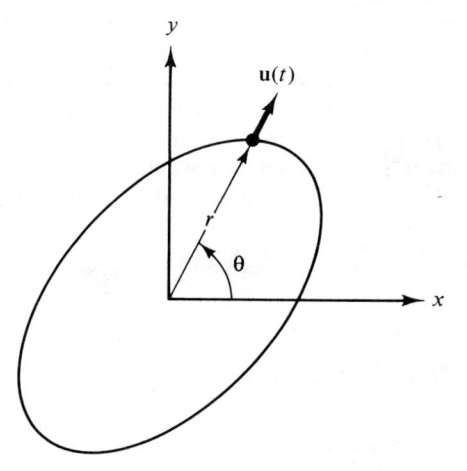

Figure 4.6 Orbiting satellite.

where α is a constant determining the strength of the force field. In terms of the state variables

$$\dot{\mathbf{x}}(t) = \begin{bmatrix} x_2(t) \\ x_1(t)\,x_4^2(t) - \dfrac{\alpha}{x_1^2(t)} + \mathbf{u}(t) \\ x_4(t) \\ \dfrac{-2x_4(t)\,x_2(t)}{x_1(t)} \end{bmatrix}$$

For a final example mechanical system, we describe the "stick balancer" system, where a cart of mass M can move horizontally along a line without friction, driven by an applied force $\mathbf{u}(t)$ (Fig. 4.7). Attached to the cart by a frictionless pivot is an ideal inverted pendulum of length l, topped by a mass of size m. This model provides a natural control design problem: Find a control input $\mathbf{u}(t)$ that will move the cart so as to return the pendulum to a full upright position from initially perturbed configurations; think of the similar problem of balancing a broomstick on the palm of your hand. The motion of the cart and

Sec. 4.1 Derivatives and Applications

Figure 4.7 Stick balancer.

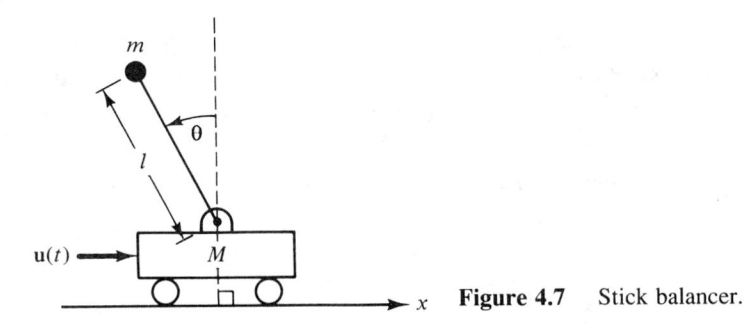

pendulum can be described with four state variables. We take x_1 to be the horizontal displacement of the cart from some reference position, x_2 to be the cart velocity, $x_3 = \theta$, the angular displacement of the pendulum away from its desired vertical orientation, and $x_4 = \dot\theta$, the angular velocity. Newton's laws again determine the system equations:

$$(M+m)\ddot{x}_1(t) + ml\cos\theta(t)\,\ddot\theta(t) - ml\,\dot\theta^2(t)\sin\theta(t) = \mathbf{u}(t)$$

$$ml\cos\theta(t)\,\ddot{x}_1(t) - ml\sin\theta(t)\,\dot{x}_1(t)\,\dot\theta(t) + ml^2\,\ddot\theta(t) - mgl\sin\theta(t) = 0$$

where g is the acceleration due to gravity. The reader is invited to write these equations in state space form using the variables mentioned above.

Discrete-time nonlinear systems arise in a variety of applications. For example, when a nonlinear system described by equation (4.1) is discretized using the second-order Ralston-Runge-Kutta method, a system of the following form is obtained:

$$\mathbf{x}_{k+1} = \mathbf{F}(\mathbf{x}_k, \mathbf{u}_k)$$

where the nonlinear function $\mathbf{F}$ is determined by equations (3.21–3.23) describing the discretization method.

Many other examples could also be given, and the range of applications where nonlinear systems models are employed seems to be virtually unlimited.

4.1 DERIVATIVES AND APPLICATIONS

From your knowledge of calculus, you will recall that the derivative of a real-valued function $\phi(x)$ gives the slope of the graph of $\phi(x)$ as a function of x. So, for any particular point x_0, $\phi'(x_0)$ gives the slope of the tangent line to the graph of $\phi(x)$ at the point x_0. Thus, near x_0, the function $\phi(x)$ is well approximated by a linear function with slope $\phi'(x_0)$; to make the linear function agree with the value of $\phi(x)$ at x_0, we must choose the approximating linear function to be $\psi(x) = \phi(x_0) + \phi'(x_0)(x - x_0)$. The (Extended) Mean Value Theorem of calculus spells out the notion of "well approximated" that applies when $\phi(x)$ is twice continuously differentiable:

$$|\phi(x) - \psi(x)| \le \frac{M|x - x_0|^2}{2} \quad (4.6)$$

where M is the least upper bound of the magnitude of the second derivative of ϕ over the interval (a,b) where the approximation is to be made (i.e., $a \le x, x_0 \le b$).

This notion of a derivative as providing a "good" linear approximation may be easily extended to cover much more general situations, including ones that are of great importance for understanding nonlinear systems. For a function $\mathbf{F}$ mapping a normed vector space $\mathbf{X}$ into a normed vector space $\mathbf{Y}$, the derivative at $\mathbf{x}_0$, denoted $\mathbf{F}'(\mathbf{x}_0)$ or $\dfrac{d\mathbf{F}}{d\mathbf{x}}(\mathbf{x}_0)$, is the *linear mapping* of $\mathbf{X}$ into $\mathbf{Y}$ that makes the following limit exist (if possible):

$$\lim_{\|\Delta \mathbf{x}\| \to 0} \frac{\|\mathbf{F}(\mathbf{x}_0 + \Delta \mathbf{x}) - \mathbf{F}(\mathbf{x}_0) - \mathbf{F}'(\mathbf{x}_0)\Delta \mathbf{x}\|}{\|\Delta \mathbf{x}\|} = 0 \tag{4.7}$$

Here we have used the usual symbol $\|\cdot\|$ for the *norm* (length) of a vector.

The intuitive content of this formulation can be easily understood: With $\mathbf{x} = \mathbf{x}_0 + \Delta \mathbf{x}$,

$$\mathbf{F}(\mathbf{x}) \approx \mathbf{F}(\mathbf{x}_0) + \mathbf{F}'(\mathbf{x}_0)(\mathbf{x} - \mathbf{x}_0) \tag{4.8}$$

for those $\mathbf{x}$ "close" to $\mathbf{x}_0$.

For ordinary vector spaces of real numbers, say $\mathbf{X} = \mathbb{R}^p$ and $\mathbf{Y} = \mathbb{R}^q$, the norms in equation (4.7) are the usual Euclidean lengths for p- and q-dimensional vectors, and $\mathbf{F}'(\mathbf{x}_0)\Delta \mathbf{x}$ is the multiplication of the $q \times p$ matrix $\mathbf{F}'(\mathbf{x}_0)$ times the p-vector $\Delta \mathbf{x}$. Furthermore, the Mean Value Theorem for vector-valued functions of several variables may be used to show that

$$\mathbf{F}'(\mathbf{x}_0) = \begin{bmatrix} \dfrac{\partial F_1}{\partial x_1} & & \dfrac{\partial F_1}{\partial x_p} \\ \cdot & & \cdot \\ \cdot & \cdots & \cdot \\ \cdot & & \cdot \\ \dfrac{\partial F_q}{\partial x_1} & & \dfrac{\partial F_q}{\partial x_p} \end{bmatrix} \tag{4.9}$$

where the subscripts denote the components of $\mathbf{F}$ and of $\mathbf{x}$ and the partial derivatives are evaluated at $\mathbf{x} = \mathbf{x}_0$. In words, $\mathbf{F}'(\mathbf{x}_0)$ is the matrix of partial derivatives of the q components of $\mathbf{F}$ with respect to the p components of $\mathbf{x}$. When $p = q = 1$ (so $\mathbf{F}$ is a scalar function of one variable), we get the usual derivative familiar from elementary calculus. If $q = 1$ and $p > 1$ (so $\mathbf{F}$ is a scalar function of several variables), we get a row vector that is often called the *gradient* of $\mathbf{F}$ and denoted $\nabla \mathbf{F}$; this case is particularly important in optimization theory, as will be seen in Chapter 5. The case of $p = q = n$ arises in many applications concerning nonlinear systems (as will be seen shortly) and optimization.

Derivatives and System Linearization

As a first application of derivatives, we consider system linearization, and we introduce the basic ideas by using the example of a mechanical pendulum acted upon by gravitational forces only. As described earlier, the equation of motion is

$$\ddot{\theta}(t) = -\frac{g}{L} \sin \theta(t) \tag{4.10}$$

Sec. 4.1 Derivatives and Applications

where θ measures the angle of the pendulum with respect to its vertical rest position. Choosing state variables in the obvious way: $x_1(t) = \theta(t)$, $x_2(t) = \dot{\theta}(t)$, we obtain the state equations

$$\dot{\mathbf{x}}(t) = \begin{bmatrix} \dot{x}_1(t) \\ \dot{x}_2(t) \end{bmatrix} = \begin{bmatrix} x_2(t) \\ -g \sin x_1(t)/L \end{bmatrix} = \mathbf{f}(\mathbf{x}(t)) \tag{4.11}$$

If we consider a solution to this differential equation, $\mathbf{x}_0(t)$, with

$$\dot{\mathbf{x}}_0(t) = \mathbf{f}(\mathbf{x}_0(t)), \quad \mathbf{x}_0(0) = \mathbf{x}^* \tag{4.12}$$

we can use linearization to get an approximation for the solution $\mathbf{x}(t)$ with perturbed initial condition $\mathbf{x}(0) = \mathbf{x}^* + \Delta\mathbf{x}^*$. We assume that $\Delta\mathbf{x}(t) = \mathbf{x}(t) - \mathbf{x}_0(t)$ is small (in norm) over the interval of interest. Then

$$\frac{d}{dt} \Delta\mathbf{x}(t) = \dot{\mathbf{x}}(t) - \dot{\mathbf{x}}_0(t) = \mathbf{f}(\mathbf{x}(t)) - \mathbf{f}(\mathbf{x}_0(t)) \tag{4.13}$$

and linearization provides an approximation for $\mathbf{f}(\mathbf{x}(t))$:

$$\mathbf{f}(\mathbf{x}(t)) \approx \mathbf{f}(\mathbf{x}_0(t)) + \mathbf{f}'(\mathbf{x}_0(t)) \Delta\mathbf{x}(t) \tag{4.14}$$

Using this approximation in equation (4.13) gives a linear differential equation describing the behavior of the perturbation:

$$\frac{d}{dt} \Delta\mathbf{x}(t) = \mathbf{f}'(\mathbf{x}_0(t)) \Delta\mathbf{x}(t) \tag{4.15}$$

This is the so-called linearized system with respect to the nominal solution $\mathbf{x}_0(t)$.

For many problems the only kind of solution that can be easily found in order to apply linearization in this way is an *equilibrium solution* (i.e., a solution that is constant as a function of time). If $\mathbf{x}_0(t) = \mathbf{x}^*$ for all t, then $\dot{\mathbf{x}}(t) = \mathbf{0}$, so equilibrium solutions are found by solving the equation

$$\mathbf{f}(\mathbf{x}^*) = \mathbf{0} \tag{4.16}$$

Furthermore, the linearized system about an equilibrium solution is described by a linear, time-invariant differential equation, so its solutions are described in terms of the corresponding matrix exponential function.

Example

For the pendulum problem, the "rest" position of the system is an equilibrium solution, namely

$$\mathbf{x}^* = \begin{bmatrix} 0 \\ 0 \end{bmatrix} = \mathbf{0}$$

and we have

$$\mathbf{f}'(\mathbf{0}) = \begin{bmatrix} \frac{\partial f_1}{\partial x_1} & \frac{\partial f_1}{\partial x_2} \\ \frac{\partial f_2}{\partial x_1} & \frac{\partial f_2}{\partial x_2} \end{bmatrix}(\mathbf{0}) = \begin{bmatrix} 0 & 1 \\ -g/L & 0 \end{bmatrix}$$

so that

$$\frac{d}{dt}\Delta\mathbf{x}(t) = \begin{bmatrix} 0 & 1 \\ -g/L & 0 \end{bmatrix} \Delta\mathbf{x}(t)$$

This may be seen to be the usual "small-angle" approximation: $\Delta x_1(t) = \Delta\theta(t)$ and

$$\frac{d^2}{dt^2}\Delta\theta(t) = -\frac{g}{L}\Delta\theta(t)$$

This linear differential equation has sinusoids of frequency $\sqrt{g/L}$ as its solutions. These solutions may be obtained by using the matrix exponential:

$$\Delta\mathbf{x}(t) = \exp(\mathbf{f}'(\mathbf{0})t)\,\Delta\mathbf{x}(0)$$

where

$$\exp(\mathbf{f}'(\mathbf{0})t) = \begin{bmatrix} \cos\omega_0 t & \omega_0^{-1}\sin\omega_0 t \\ -\omega_0 \sin\omega_0 t & \cos\omega_0 t \end{bmatrix}$$

and $\omega_0 = \sqrt{g/L}$.

The crucial assumption in the development of the linearized system equations is that $\Delta\mathbf{x}(t)$ stays "small"; this is obviously a kind of stability condition on solutions. We will see more about this later.

Example

Nonlinear discrete-time systems are linearized using a similar approach. Suppose that a system is described by

$$\mathbf{x}_{i+1} = \mathbf{f}(\mathbf{x}_i)$$

and that $\mathbf{x}_e$ is an equilibrium solution, that is,

$$\mathbf{x}_e = \mathbf{f}(\mathbf{x}_e)$$

Linearization is useful in studying solutions that start out and remain in the vicinity of a known solution, and the case of an equilibrium solution arises in many applications. If we take the initial condition of the system to be

$$\mathbf{x}_0 = \mathbf{x}_e + \Delta\mathbf{x}_0$$

and ask for an approximate expression for the resulting solution, we find that

$$\mathbf{x}_1 = \mathbf{f}(\mathbf{x}_0) \approx \mathbf{f}(\mathbf{x}_e) + \mathbf{f}'(\mathbf{x}_e)(\mathbf{x}_0 - \mathbf{x}_e)$$

Letting $\Delta\mathbf{x}_1 = \mathbf{x}_1 - \mathbf{x}_e$, rearranging, and ignoring higher-order terms in the approximation gives the linearized equation, accurate to first order:

$$\Delta\mathbf{x}_1 = \mathbf{f}'(\mathbf{x}_e)\,\Delta\mathbf{x}_0$$

Assuming that $\Delta\mathbf{x}_1$ is small when $\Delta\mathbf{x}_0$ is, we may repeat this process indefinitely, giving the state equations for the linearized discrete-time system, with state $\Delta\mathbf{x}_i = \mathbf{x}_i - \mathbf{x}_e$, for $i \geq 0$,

$$\Delta\mathbf{x}_{i+1} = \mathbf{f}'(\mathbf{x}_e)\,\Delta\mathbf{x}_i$$

The following expression provides its solution:

$$\Delta\mathbf{x}_i = (\mathbf{f}'(\mathbf{x}_e))^i\,\Delta\mathbf{x}_0$$

Sec. 4.1 Derivatives and Applications

For systems with inputs, linearization can be achieved by following the same general approach as above, except that input variations are taken into account too. For the sake of the following discussion, we treat the case of a single input only, but the extension to multiple inputs is straightforward. For the differential equation

$$\dot{\mathbf{x}}(t) = \mathbf{f}(\mathbf{x}(t), \mathbf{u}(t)), \quad \mathbf{x}(0) = \mathbf{x}^* \tag{4.17}$$

we suppose that $\mathbf{x}_0(t)$ is a solution (that now depends on both $\mathbf{x}^*$ and $\mathbf{u}(t)$) and consider what new solution is obtained if both the initial condition and the input function are perturbed. We assume that the new input is $\mathbf{u}(t) + \Delta\mathbf{u}(t)$, where the perturbation $\Delta\mathbf{u}(t)$ is small (in norm) over the interval of interest, and we assume that the resulting solution $\mathbf{x}_0(t) + \Delta\mathbf{x}(t)$ is close (in norm) to $\mathbf{x}_0(t)$ over the same time interval. Then, using linearization as before leads to the following linear differential equation for the perturbation $\Delta\mathbf{x}(t)$:

$$\frac{d}{dt}\Delta\mathbf{x}(t) = \mathbf{f}'(\mathbf{x}_0(t), \mathbf{u}(t)) \begin{bmatrix} \Delta\mathbf{x}(t) \\ \Delta\mathbf{u}(t) \end{bmatrix} \tag{4.18}$$

where the derivative is

$$\mathbf{f}'(\mathbf{x}_0(t), \mathbf{u}(t)) = \begin{bmatrix} \frac{\partial f_1}{\partial x_1} & & \frac{\partial f_1}{\partial x_n} & \frac{\partial f_1}{\partial u} \\ \cdot & & \cdot & \cdot \\ \cdot & \cdots & \cdot & \cdot \\ \cdot & & \cdot & \cdot \\ \frac{\partial f_n}{\partial x_1} & & \frac{\partial f_n}{\partial x_n} & \frac{\partial f_n}{\partial u} \end{bmatrix} \tag{4.19}$$

and the partial derivatives are evaluated at $\mathbf{x}_0(t)$ and $\mathbf{u}(t)$ as suggested by the notation on the left side of the equation. A more common way to write the linearized equation is to split up the derivative into two submatrices (in the natural way), giving

$$\mathbf{f}' = \begin{bmatrix} \mathbf{f_x} & \mathbf{f_u} \end{bmatrix} \tag{4.20}$$

so that the equation can be written in the usual form for a linear system:

$$\frac{d}{dt}\Delta\mathbf{x}(t) = \mathbf{f_x}\,\Delta\mathbf{x}(t) + \mathbf{f_u}\,\Delta\mathbf{u}(t) \tag{4.21}$$

You should note that this is generally going to be a *time-varying* linear system because the partial derivative matrices appearing in the system equations depend on the solution $\mathbf{x}_0(t)$ and the nominal input $\mathbf{u}(t)$. If both of these quantities are constant over time, the linearized system will be time invariant; this is a special case of some importance.

Examples

We will give two more examples to illustrate system linearization. For the first, recall the model for an orbiting satellite in a central, inverse-square-law force field, where we assume that the satellite is a unit point mass equipped with thrusters capable of exerting radial forces. The equation describing the motion of the satellite can be written in state space form by taking $x_1 = r$, the radial distance to the satellite, $x_2 = \dot{r}$, the radial velocity, $x_3 = \theta$, the angular displacement of the satellite, and $x_4 = \dot{\theta}$, the angular velocity. We let $\mathbf{u}(t)$ be the radial thrust applied at the satellite. Then the equations of motion are

$$\ddot{r}(t) = r(t)\,\dot{\theta}^2(t) - \frac{\alpha}{r^2(t)} + \mathbf{u}(t)$$

$$\ddot{\theta}(t) = -\frac{2\,\dot{\theta}(t)\,\dot{r}(t)}{r(t)}$$

or in terms of the state variables

$$\dot{\mathbf{x}}(t) = \begin{bmatrix} x_2(t) \\ x_1(t)\,x_4^2(t) - \dfrac{\alpha}{x_1^2(t)} + \mathbf{u}(t) \\ x_4(t) \\ -\dfrac{2x_4(t)\,x_2(t)}{x_1(t)} \end{bmatrix}$$

The satellite moves in an elliptical orbit when no thrust is being applied, and we may choose such a solution to the equations as the nominal one about which linearization is to be carried out. The simplest case is for a circular orbit: $x_1 = r(t) = r_0$; $x_4 = \dot{\theta}(t) = \omega_0$; and therefore $x_2 = \dot{r}(t) = 0$ and $x_3 = \omega_0 t$. The angular velocity and orbital radius are not independent quantities, since for this choice of $\mathbf{x}(t)$ to be a solution we must have $r_0^3 \omega_0^2 = \alpha$. The calculation of the linearized system using this nominal circular orbit is easily carried out, with the key step being the evaluation of the derivative:

$$\mathbf{f}' = \begin{bmatrix} \mathbf{f}_\mathbf{x} & \mathbf{f}_\mathbf{u} \end{bmatrix} = \begin{bmatrix} 0 & 1 & 0 & 0 & 0 \\ 3\omega_0^2 & 0 & 0 & 2r_0\omega_0 & 1 \\ 0 & 0 & 0 & 1 & 0 \\ 0 & -2\omega_0/r_0 & 0 & 0 & 0 \end{bmatrix}$$

As a second example, we return to the "stick balancing" system introduced earlier. Recall that the cart of mass M can move horizontally without friction, driven by an applied force $\mathbf{u}(t)$. Attached to the cart is an inverted pendulum of length l, topped by a mass of size m. The motion of the cart and pendulum can be described with four state variables: x_1, the horizontal displacement of the cart; $x_2 = \dot{x}_1$, the cart velocity; $x_3 = \theta$, the angular displacement of the pendulum; and $x_4 = \dot{\theta}$, the angular velocity. Newton's laws lead to the equations

$$(M+m)\ddot{x}_1(t) + ml\cos\theta(t)\,\ddot{\theta}(t) - ml\,\dot{\theta}^2(t)\sin\theta(t) = \mathbf{u}(t)$$

$$ml\cos\theta(t)\,\ddot{x}_1(t) - ml\sin\theta(t)\,\dot{x}_1(t)\,\dot{\theta}(t) + ml^2\,\ddot{\theta}(t) - mgl\sin\theta(t) = 0$$

from which a set of state equations is easily obtained.

This system has an (unstable) equilibrium solution $\mathbf{x}_e = \mathbf{0}$ for zero applied force, $\mathbf{u}(t) = 0$. This is one point about which the system may be linearized. The reader is invited to carry out the calculation. The system also admits some "quasi-static" solutions that are easily obtained; for a constant input force, $\mathbf{u}(t) = \mathbf{u}^*$, there is a solution with constant x_3, meaning that there is a solution where the angular displacement of the pendulum is constant (i.e., $x_4 = 0$ and the pendulum doesn't fall over), although the cart is moving in the direction of the applied force at constant acceleration. In particular, $\ddot{x}_1^* = \mathbf{u}^*/(M+m)$ and $x_3^*(t) = \theta^*(t) = \tan^{-1}(\ddot{x}_1^*/g)$. Linearization can be carried out using this nominal solution. It turns out that thinking of the system in terms of this family of linearized versions,

Sec. 4.1   Derivatives and Applications   171

parametrized by the value of the applied force, provides a useful framework for designing feedback control laws for keeping the pendulum in balance with advantages over using a single linearized version of the system corresponding to the equilibrium solution.

Solution of Nonlinear Equations

Derivatives are important for both theoretical and practical aspects of the problem of solving nonlinear equations. Our main concerns will be practical ones, but one important theoretical result will be described first. Suppose that a nonlinear functional relationship of the form

$$\mathbf{y} = \mathbf{F}(\mathbf{x}) \tag{4.22}$$

relates two n-vectors. If it is known that the choice of $\mathbf{y} = \mathbf{y}_0$ can be "solved" by taking $\mathbf{x} = \mathbf{x}_0$, it is of some interest to know when there is a function, say $\mathbf{H}$, that serves as an inverse for $\mathbf{F}$ in a small enough neighborhood of $\mathbf{y}_0$. In other words, does there exist a function $\mathbf{H}$ so that taking

$$\mathbf{x} = \mathbf{H}(\mathbf{y}) \tag{4.23}$$

gives

$$\mathbf{y} = \mathbf{F}(\mathbf{H}(\mathbf{y})) \tag{4.24}$$

at least when $\|\mathbf{y} - \mathbf{y}_0\|$ is sufficiently small? An answer to this question is provided by the *Inverse Function Theorem* from advanced calculus: Assuming that $\mathbf{F}$ has a continuous derivative for all $\mathbf{x}$ with $\|\mathbf{x} - \mathbf{x}_0\|$ sufficiently small, $\mathbf{H}$, the desired inverse function, exists and is continuously differentiable for all $\mathbf{y}$ with $\|\mathbf{y} - \mathbf{y}_0\|$ sufficiently small provided that $\mathbf{F}'(\mathbf{x}_0)$ is a nonsingular (i.e., an invertible) matrix.

Example

As a simple application, consider the problem of determining a point of intersection between a line with slope 1 and a circle of radius r. Using x_1 and x_2 to denote Cartesian coordinates, the two equations to be satisfied are

$$x_1^2 + x_2^2 = r^2$$
$$-x_1 + x_2 = b$$

where b is the x_2 intercept of the line (Fig. 4.8). These equations are written compactly in the form $\mathbf{F}(\mathbf{x}) = \mathbf{y}$ by taking

$$\mathbf{F}(\mathbf{x}) = \begin{bmatrix} x_1^2 + x_2^2 \\ -x_1 + x_2 \end{bmatrix} \quad \text{and} \quad \mathbf{y} = \begin{bmatrix} r^2 \\ b \end{bmatrix}$$

A direct calculation of the derivative gives

$$\mathbf{F}' = \begin{bmatrix} 2x_1 & 2x_2 \\ -1 & 1 \end{bmatrix}$$

which is continuous at every $\mathbf{x}$ and is a nonsingular matrix provided that $x_1 \neq -x_2$. This result has a nice geometrical interpretation: When $x_1 \neq -x_2$, the line is not tangent to the circle at the point of intersection and there is a continuously differentiable inverse function valid for all small changes in the parameters r and b. However, when the intersection

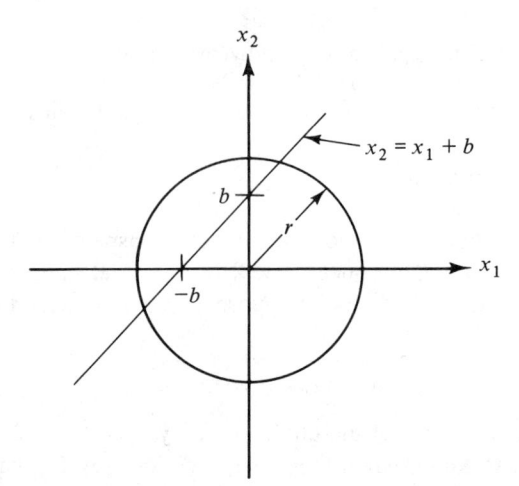

Figure 4.8 Cartesian plane with circle of radius r and line of slope 1.

occurs at a point with $x_1 = -x_2$, the line is tangent to the circle, and for circles with a smaller radius r there will be no point of intersection (i.e., the equations are inconsistent for some choices of **y** arbitrarily close to $\mathbf{y}_0$).

Newton's Method

We now turn to a more practical aspect of the application of derivatives to solving nonlinear equations. We will consider *Newton's method*, which has many important uses. The basic problem involves solving a nonlinear (vector-valued) equation of the form $\mathbf{G}(\mathbf{x}) = \mathbf{0}$. We have already encountered this problem in two places: In implicit discretization methods for nonlinear differential equations, such as the trapezoidal or backward difference methods, and in the determination of equilibrium solutions of nonlinear differential equations. In general, for solving equations of the form $\mathbf{F}(\mathbf{x}) = \mathbf{y}$, we transform to the equivalent form $\mathbf{G}(\mathbf{x}) = \mathbf{F}(\mathbf{x}) - \mathbf{y} = \mathbf{0}$ to obtain a problem of the desired form. Since finding any closed-form solution to most sets of nonlinear equations is impossible, an alternative, more computational approach will be adopted. *Newton's method* is an iterative method for generating a convergent sequence of vectors whose limit solves $\mathbf{G}(\mathbf{x}) = \mathbf{0}$.

Newton's method arises from the use of derivatives to provide a sequence of linear approximations to the nonlinear equation of interest; since linear equations are (relatively) easy to solve, this is often an effective way of solving the nonlinear equation. Writing the linear approximation to $\mathbf{G}(\mathbf{x})$ at the point $\mathbf{x}_0$, we have

$$\mathbf{G}(\mathbf{x}) \approx \mathbf{G}(\mathbf{x}_0) + \mathbf{G}'(\mathbf{x}_0)(\mathbf{x} - \mathbf{x}_0) \tag{4.25}$$

Taking $\mathbf{x}_0$ as a "guess" for a zero of the function $\mathbf{G}$, this suggests solving the equation

$$\mathbf{0} = \mathbf{G}(\mathbf{x}_0) + \mathbf{G}'(\mathbf{x}_0)(\mathbf{x} - \mathbf{x}_0) \tag{4.26}$$

for **x** to obtain a (hopefully!) better guess. Writing this in the familiar form for linear equations,

$$\mathbf{G}'(\mathbf{x}_0)\mathbf{x} = -\mathbf{G}(\mathbf{x}_0) + \mathbf{G}'(\mathbf{x}_0)\mathbf{x}_0 \tag{4.27}$$

must be solved for **x**. Assuming that $\mathbf{G}'(\mathbf{x}_0)$ is a square, nonsingular matrix (i.e., det $\mathbf{G}'(\mathbf{x}_0) \neq 0$) we obtain

$$\mathbf{x} = \mathbf{x}_0 - (\mathbf{G}'(\mathbf{x}_0))^{-1} \mathbf{G}(\mathbf{x}_0) \qquad (4.28)$$

If $\mathbf{G}'(\mathbf{x}_0)$ is not a square, nonsingular matrix, then the pseudo-inverse may be used in place of $(\mathbf{G}'(\mathbf{x}_0))^{-1}$ to obtain the approximation **x**; the details of this generalization are left for the reader.

Newton's method consists of successive repetitions of this linearizing approximation, giving the sequence

$$\mathbf{x}_{k+1} = \mathbf{x}_k - (\mathbf{G}'(\mathbf{x}_k))^{-1} \mathbf{G}(\mathbf{x}_k), \qquad k \geq 0 \qquad (4.29)$$

(Fig. 4.9). From our derivation, it is to be expected that the method works best for "nearly linear" functions, in particular for determining a zero within a neighborhood where the function **G** has no singular points (i.e., a neighborhood where the derivative is always nonsingular). Without requiring such regularity conditions, convergence may not even occur; with suitable regularity conditions Newton's method converges *quadratically*. This means that for a sufficiently good initial guess, each iteration provides about double the number of additional decimal places of accuracy. Even better performance can be achieved in certain cases; for example, the solution to a linear equation is achieved in one step from any initial guess!

Examples

Consider the use of Newton's method for computing square roots. To find $\sqrt{2}$, we look for a positive solution to the equation $x^2 - 2 = G(x) = 0$. We have $G'(x) = 2x$, so Newton's method takes the explicit form

$$x_{k+1} = x_k - \frac{x_k^2 - 2}{2 x_k}$$

Choosing $x_0 = 1$ we find that $x_1 = 3/2 = 1.5$, $x_2 = 17/12 = 1.4666...$, $x_3 = 577/408 = 1.41421569...$.

Similar calculations can be carried out to determine π, using the property $\sin \pi = 0$ to suggest solving the equation $\sin x = 0$. Try starting with $x_0 = 22/7$.

In a similar way, e, the base of natural logarithms, can be determined from the property $\ln(e) = 1$. Try solving the equation $\ln(x) - 1 = 0$ with the starting point $x_0 = 2$.

Finally, consider the use of Newton's method in carrying out the (implicit) backward difference discretization of the nonlinear system

$$\dot{\mathbf{x}}(t) = \mathbf{f}(\mathbf{x}(t), \mathbf{u}(t))$$

Using equation (3.8), the backward difference discretization is given by

$$\mathbf{x}((k+1)h) = \mathbf{x}(kh) + h\mathbf{f}(\mathbf{x}((k+1)h), \mathbf{u}((k+1)h))$$

Simplifying notation in the obvious way and rearranging leads to the following equation to be solved for $\mathbf{x}((k+1)h)$:

$$\mathbf{x}_{k+1} - \mathbf{x}_k - h\mathbf{f}(\mathbf{x}_{k+1}, \mathbf{u}_{k+1}) = \mathbf{G}(\mathbf{x}_{k+1}) = \mathbf{0}$$

Newton's method provides a means of determining the solution, $\mathbf{x}_{k+1}$, and carrying out the discretization. For small discretization step size, h, it is reasonable to suppose that $\mathbf{x}_{k+1} \approx \mathbf{x}_k$, and this provides an initial condition for starting the Newton method iteration.

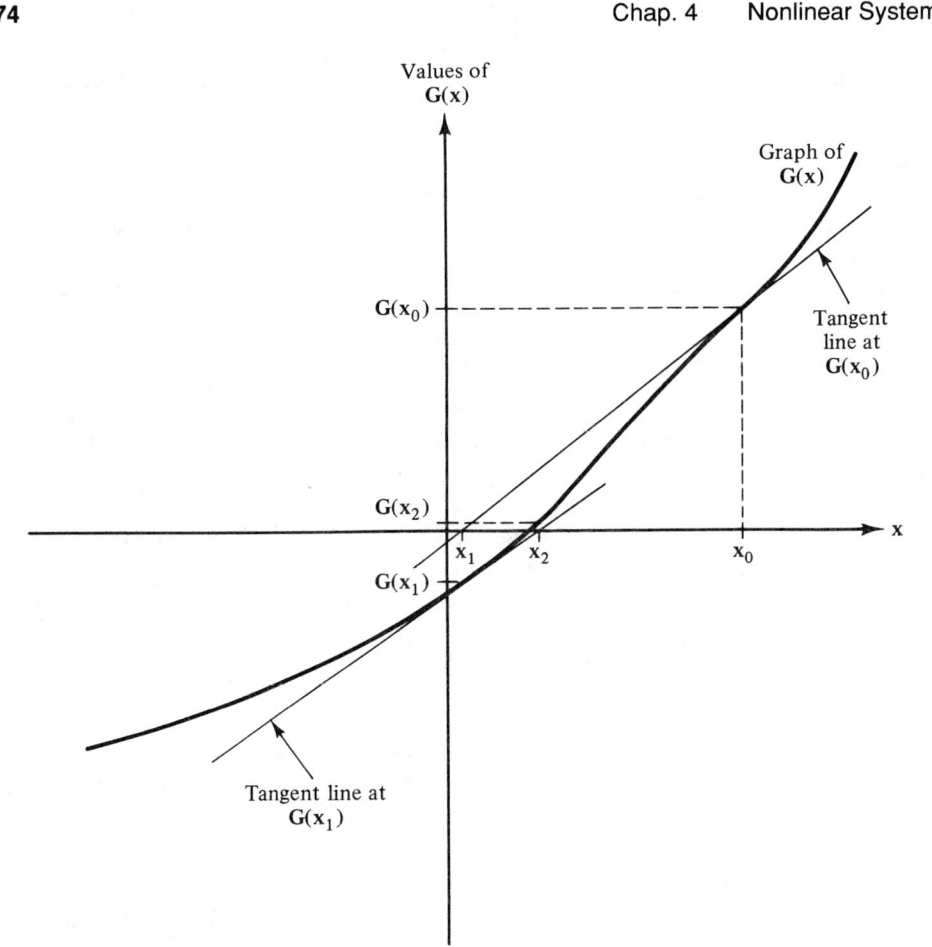

Figure 4.9 Steps taken by Newton's method for solving $\mathbf{G}(\mathbf{x}) = \mathbf{0}$.

Continuing with this example, it is interesting to point out a connection with system linearization. For sufficiently small discretization step size, h, not only will $\mathbf{x}_k$ be a good approximation to $\mathbf{x}_{k+1}$, but the iteration should converge very quickly. Recalling that the backward difference discretization is accurate only to first order in h, this suggests taking only a single Newton step and using the result as a (hopefully) sufficiently accurate solution for $\mathbf{x}_{k+1}$. The resulting approximate backward difference discretization is an explicit discretization method given by the equation

$$\mathbf{x}((k+1)h) = \mathbf{x}(kh) + (\mathbf{I} - h\mathbf{f}'(\mathbf{x}(kh), \mathbf{u}((k+1)h)))^{-1} f(\mathbf{x}(kh), \mathbf{u}((k+1)h))$$

where $\mathbf{f}'(\mathbf{x}(kh), \mathbf{u}((k+1)h))$ denotes the derivative of $\mathbf{f}(\mathbf{x}, \mathbf{u})$ with respect to $\mathbf{x}$ (with $\mathbf{u}$ fixed at the value $\mathbf{u}((k+1)h)$), evaluated at $\mathbf{x} = \mathbf{x}(kh)$. This can be recognized as a system obtained by a combination of linearization and backward difference discretization.

One important variation on Newton's method is the *modified* Newton iteration

$$\mathbf{x}_{k+1} = \mathbf{x}_k - \delta_k (\mathbf{G}'(\mathbf{x}_k))^{-1} \mathbf{G}(\mathbf{x}_k) \tag{4.30}$$

Sec. 4.1 Derivatives and Applications

where the parameter δ_k, $0 < \delta_k \leq 1$ is chosen so that $\|\mathbf{G}(\mathbf{x}_{k+1})\| < \|\mathbf{G}(\mathbf{x}_k)\|$. This modification is designed to avoid problems caused by making large changes in $\mathbf{x}_k$ in Newton's method. The choice of the parameter δ_k may be viewed as a minimization problem in one variable and is usually solved by a search technique or by a curve-fitting technique. We will return to a discussion of this idea in Chapter 5.

We will close this discussion of Newton's method with an illustration of the generality of the notion of derivative as a linear operator on normed spaces; the example is another one related to systems. For $\mathbf{x}(t)$ an n-vector of continuous functions of time t on the interval $[0,T]$, with $\mathbf{x}(0) = \mathbf{x}^*$, the mapping

$$\mathbf{z}(t) = \mathbf{G}(\mathbf{x}(\tau)\,;\, 0 \leq \tau \leq T) = \mathbf{x}(t) - \mathbf{x}^* - \int_0^t \mathbf{f}(\mathbf{x}(\tau), \mathbf{u}(\tau))\,d\tau, \qquad 0 \leq t \leq T \qquad (4.31)$$

produces another n-vector of continuous functions $\mathbf{z}(t)$ with $\mathbf{z}(0) = \mathbf{0}$. If $\mathbf{z}(t) \equiv \mathbf{0}$ over the interval $[0,T]$, then $\mathbf{x}(t)$ is the solution of the differential equation

$$\dot{\mathbf{x}}(t) = \mathbf{f}(\mathbf{x}(t), \mathbf{u}(t)), \qquad \mathbf{x}(0) = \mathbf{x}^* \qquad (4.32)$$

A norm for the vector $\mathbf{x}(t)$ (and similarly $\mathbf{z}(t)$), whose components are continuous functions, is defined by

$$\|\mathbf{x}(t)\|_\infty = \max_{1 \leq i \leq n}\; \max_{0 \leq t \leq T} |x_i(t)| \qquad (4.33)$$

In words, the norm is the largest of the uniform norms of the components of $\mathbf{x}(t)$ (see equation (A.41) in the Chapter 1 Appendix).

The derivative of the mapping $\mathbf{G}$ at the "point" $\mathbf{x}_0$, which is an n-vector of continuous functions with $\mathbf{x}_0(0) = \mathbf{x}^*$, is the linear operator $\mathbf{G}'(\mathbf{x}_0)$ that acts on n-vectors of continuous functions, say $\Delta\mathbf{x}(t)$, according to the formula

$$\mathbf{G}'(\mathbf{x}_0)\,\Delta\mathbf{x}(t) = \Delta\mathbf{x}(t) - \int_0^t \mathbf{f}'(\mathbf{x}_0(\tau), \mathbf{u}(\tau))\,\Delta\mathbf{x}(\tau)\,d\tau \qquad (4.34)$$

where $\mathbf{f}'(\mathbf{x}_0(\tau), \mathbf{u}(\tau))$ denotes the derivative of $\mathbf{f}(\mathbf{x}, \mathbf{u})$ with respect to $\mathbf{x}$ (with $\mathbf{u}$ fixed at the value $\mathbf{u}(\tau)$), evaluated at $\mathbf{x} = \mathbf{x}_0(\tau)$.

Using this formulation, we may generate a sequence of approximate solutions to the differential equation by applying Newton's method. Let $\mathbf{x}^{[k]}(t)$ denote the kth approximate solution. Then the linear equations for Newton's method can be written formally as

$$\mathbf{G}'(\mathbf{x}^{[k]})\,\mathbf{x}^{[k+1]} = -\mathbf{G}(\mathbf{x}^{[k]}) + \mathbf{G}'(\mathbf{x}^{[k]})\,\mathbf{x}^{[k]} \qquad (4.35)$$

This may be written out explicitly for the problem at hand, giving

$$\mathbf{x}^{[k+1]}(t) - \int_0^t \mathbf{f}'(\mathbf{x}^{[k]}(\tau), \mathbf{u}(\tau))\,\mathbf{x}^{[k+1]}(\tau)\,d\tau = -\mathbf{x}^{[k]}(t) + \mathbf{x}^*$$
$$+ \int_0^t \mathbf{f}(\mathbf{x}^{[k]}(\tau), \mathbf{u}(\tau))\,d\tau + \mathbf{x}^{[k]}(t) - \int_0^t \mathbf{f}'(\mathbf{x}^{[k]}(\tau), \mathbf{u}(\tau))\,\mathbf{x}^{[k]}(\tau)\,d\tau \qquad (4.36)$$

This can be recognized as the integrated form of the (linear, time varying) differential equation

$$\frac{d}{dt}\mathbf{x}^{[k+1]}(t) = \mathbf{A}_k(t)\,\mathbf{x}^{[k+1]}(t) + \mathbf{B}_k(t) \qquad (4.37)$$

where

$$\mathbf{A}_k(t) = \mathbf{f}'(\mathbf{x}^{[k]}(t), \mathbf{u}(t)) \tag{4.38}$$

$$\mathbf{B}_k(t) = \mathbf{f}(\mathbf{x}^{[k]}(t), \mathbf{u}(t)) - \mathbf{A}_k(t)\mathbf{x}^{[k]}(t) \tag{4.39}$$

Example

Consider how this approach is applied in the case of the nonlinear pendulum problem. Then

$$\mathbf{f} = \begin{bmatrix} x_2 \\ -(g \sin x_1)/L \end{bmatrix}$$

and

$$\mathbf{f}' = \begin{bmatrix} 0 & 1 \\ -g/L & 0 \end{bmatrix}$$

A typical choice of initial guess for the solution would be the solution to the linearized problem with the correct initial conditions. Suppose that the pendulum starts at angle θ_0 at time $t=0$ with zero initial angular velocity. Then from a previous example, we obtain

$$\mathbf{x}^{[0]}(t) = \begin{bmatrix} \theta_0 \cos \omega_0 t \\ -\theta_0 \omega_0 \sin \omega_0 t \end{bmatrix}$$

where $\omega_0 = \sqrt{g/L}$. The differential equation for $\mathbf{x}^{[k+1]}$ for this problem takes the explicit form

$$\frac{d}{dt}\mathbf{x}^{[k+1]}(t) = \begin{bmatrix} 0 & 1 \\ -g/L & 0 \end{bmatrix} \mathbf{x}^{[k+1]}(t) + \begin{bmatrix} 0 \\ g(x_1^{[k]} - \sin x_1^{[k]})/L \end{bmatrix}$$

The solution of this equation, even for $\mathbf{x}^{[1]}$, must be carried out numerically.

One application where this approach has been exploited is in electronic circuit simulation, where the terminology *waveform relaxation* is used. It has been found to offer some advantages over discretization methods of solving the associated large systems of differential equations, thanks to the underlying network structure and the kinds of nonlinear functions encountered in common MOS transistor models.

Successive Approximation

Having demonstrated a wide range of equation-solving problems to which Newton's method may be applied, it is worthwhile describing a somewhat simpler approach that is also widely used. Motivated by the rather complicated nature of the Newton-based approach to solving differential equations, we will introduce another method, called *successive approximation*, that provides an alternative way of using the integral representation

$$\mathbf{x}(t) = \mathbf{x}^* + \int_0^t \mathbf{f}(\mathbf{x}(\tau), \mathbf{u}(\tau)) \, d\tau \tag{4.40}$$

for the solution of the equation

$$\dot{\mathbf{x}}(t) = \mathbf{f}(\mathbf{x}(t), \mathbf{u}(t)), \quad \mathbf{x}(0) = \mathbf{x}^* \tag{4.41}$$

Sec. 4.1 Derivatives and Applications

Instead of rewriting the problem to be solved as $\mathbf{G}(\mathbf{x}) = \mathbf{0}$, we will preserve the form of the integral equation by writing it as

$$\mathbf{x} = \mathbf{K}(\mathbf{x}) \tag{4.42}$$

This suggests a simple iterative method of solution, known as successive approximation (or Picard iteration): For any trial solution $\mathbf{x}^{[k]}(t)$, let

$$\mathbf{x}^{[k+1]} = \mathbf{K}(\mathbf{x}^{[k]}) \tag{4.43}$$

In words, the iteration involves substituting an old approximation into the right-hand side of the equation to provide a new approximation and repeating this procedure until (hopefully) it converges. For the problem of interest, we have

$$\mathbf{x}^{[k+1]}(t) = \mathbf{x}^* + \int_0^t \mathbf{f}(\mathbf{x}^{[k]}(\tau), \mathbf{u}(\tau))\, d\tau \tag{4.44}$$

usually taking the constant function, $\mathbf{x}^{[0]}(t) = \mathbf{x}^*$, as the initial approximation. Other choices of the initial approximation are possible, and for nonzero input $\mathbf{u}(t)$ it makes good sense to choose the initial guess in a way that depends on the input (e.g., $\mathbf{x}^{[0]}(t) = \mathbf{x}^* \mathbf{u}(t)/\mathbf{u}(0)$ when $\mathbf{u}(0) \neq 0$).

The method of successive approximation has one clear advantage over Newton's method in that each new approximation is obtained by a simple integration, rather than by solving a differential equation.

Examples

To illustrate the method of successive approximation in a little more detail, two examples will be considered. First we choose to approximate the solution to the linear equation

$$\dot{\mathbf{x}}(t) = \mathbf{A}\mathbf{x}(t), \quad \mathbf{x}(0) = \mathbf{x}^*$$

even though we know that the solution is $\mathbf{x}(t) = e^{\mathbf{A}t}\mathbf{x}^*$, because we can carry out the steps of successive approximation explicitly and compare the results with the known solution. Starting with the approximation $\mathbf{x}^{[0]}(t) = \mathbf{x}^*$ we find that

$$\mathbf{x}^{[1]}(t) = (\mathbf{I} + \mathbf{A}t)\mathbf{x}^*$$

$$\mathbf{x}^{[2]}(t) = \left(\mathbf{I} + \mathbf{A}t + \frac{\mathbf{A}^2 t^2}{2}\right)\mathbf{x}^*$$

and for general k,

$$\mathbf{x}^{[k]}(t) = \left[\sum_{i=0}^{k} \frac{\mathbf{A}^i t^i}{i!}\right]\mathbf{x}^*$$

so the sequence of approximations is simply the sequence of partial sums of the infinite series expression for the known analytical solution. (The reader is invited to investigate the sequence of approximations obtained from applying Newton's method.)

For our second example, the pendulum system, we propose to use the periodic solution available from the linearized system to provide a better first approximation to the solution than a constant. For simplicity, we again treat the case of

$$\mathbf{x}^* = \begin{bmatrix} \theta_0 \\ 0 \end{bmatrix}$$

corresponding to starting the pendulum with initial angular displacement θ_0 and zero initial angular velocity. Then the linearized solution is $\theta(t) = \theta_0 \cos(\sqrt{g/L}\, t)$, from which we can compute $\dot{\theta}(t)$ and thus the initial approximation

$$\mathbf{x}^{[0]}(t) = \begin{bmatrix} \theta(t) \\ \dot{\theta}(t) \end{bmatrix}$$

Then we obtain the next approximation,

$$\mathbf{x}^{[1]}(t) = \mathbf{x}^* + \int_0^t \begin{bmatrix} \dot{\theta}(\tau) \\ -\dfrac{g}{L} \sin \theta(\tau) \end{bmatrix} d\tau = \begin{bmatrix} \theta_0 \cos(\sqrt{g/L}\, t) \\ -\dfrac{g}{L} \int_0^t \sin(\theta_0 \cos(\sqrt{g/L}\, \tau))\, d\tau \end{bmatrix}$$

Notice that this expression involves an integral that must be computed numerically. The reader is invited to ponder whether or not the periodic nature of the solutions to the nonlinear pendulum problem can be deduced from this approach.

Having introduced successive approximation in the context of solving differential equations, we hasten to point out that it is a method applicable to other kinds of *fixed-point problems*, where solutions to an equation of the form $\mathbf{x} = \mathbf{K}(\mathbf{x})$ are sought (Fig. 4.10). When such an equation involves a quantity $\mathbf{x}$ having an associated norm, $\|\mathbf{x}\|$, a convenient sufficient condition for convergence of the successive approximation iteration is that the function $\mathbf{K}$ is a *contraction mapping*, which means that it satisfies the property

$$\|\mathbf{K}(\mathbf{x})\| < \gamma \|\mathbf{x}\| \qquad (4.45)$$

for all $\mathbf{x}$ and for some number $\gamma < 1$.

Examples

An iterative method for solution of linear equations, sometimes known as the Richardson method, is obtained by a suitable rewriting of the equations, say

$$\mathbf{A}\mathbf{x} = \mathbf{y}$$

in the form

$$\mathbf{x} = \mathbf{x} - \mu(\mathbf{A}\mathbf{x} - \mathbf{y})$$

to which the method of successive approximation may be applied. The Richardson method is given by the equation

$$\mathbf{x}_{k+1} = (\mathbf{I} - \mu \mathbf{A})\mathbf{x}_k + \mu \mathbf{y}$$

The analysis of the Richardson method is particularly simple because this equation is just a linear discrete-time state equation. As shown in Chapter 2, the solution takes the form

$$\mathbf{x}_i = (\mathbf{I} - \mu \mathbf{A})^i \mathbf{x}_0 + \mu \sum_{k=0}^{i-1} (\mathbf{I} - \mu \mathbf{A})^k \mathbf{y}$$

With μ chosen so that the eigenvalues of $(\mathbf{I} - \mu \mathbf{A})$ all have magnitude less than 1, the system is asymptotically stable and $\mathbf{x}_i$ approaches a limiting value, $\mathbf{x}_\infty$, satisfying

$$\mathbf{x}_\infty = \mathbf{x}_\infty - \mu \mathbf{A} \mathbf{x}_\infty + \mu \mathbf{y}$$

Sec. 4.1   Derivatives and Applications   179

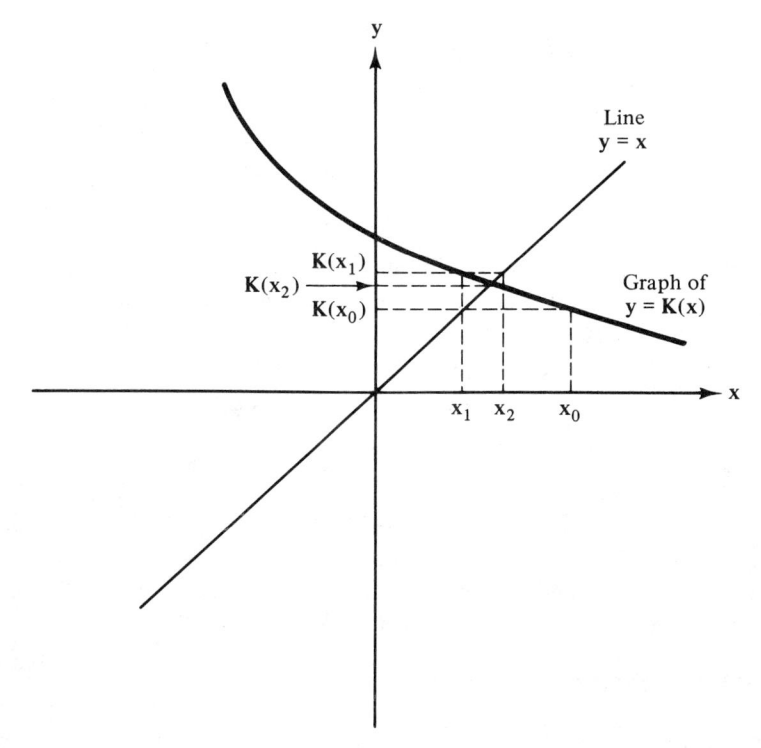

Figure 4.10   Steps taken by the method of successive approximation for solving $x = K(x)$.

Solving for $\mathbf{x}_\infty$, we find

$$\mathbf{x}_\infty = \mathbf{A}^{-1}\mathbf{y}$$

as expected. As for the choice of μ, the stability condition requires that all eigenvalues of $\mathbf{A}$ lie inside of a circle of radius $1/\mu$ centered at the point $1/\mu$.

As another application of successive approximation, consider the problem of computing the *midpoint discretization* of the nonlinear system

$$\dot{\mathbf{x}}(t) = \mathbf{f}(\mathbf{x}(t))$$

which is defined by the implicit equation

$$\mathbf{x}((k+1)h) = \mathbf{x}(kh) + h\mathbf{f}\left(\frac{\mathbf{x}((k+1)h) + \mathbf{x}(kh)}{2}\right)$$

(For linear systems, the midpoint method coincides with trapezoidal discretization, but the two methods differ when applied to nonlinear systems.) This equation is already in the form to which successive approximation may be applied; with $\mathbf{x} = \mathbf{x}((k+1)h)$,

$$\mathbf{x} = \mathbf{K}(\mathbf{x}) = \mathbf{x}(kh) + h\mathbf{f}\left(\frac{\mathbf{x} + \mathbf{x}(kh)}{2}\right)$$

Various initial guesses for $\mathbf{x}((k+1)h)$ are reasonable, such as $\mathbf{x}(kh)$ or $\mathbf{x}(kh) + h\mathbf{f}(\mathbf{x}(kh))$.

4.2 LINEARIZATION AND STABILITY

Linearization plays an important role in the study of qualitative behavior of nonlinear systems thanks to a mathematical result due to Lyapunov. The setup is as follows: A system with no input is described by the equations

$$\dot{\mathbf{x}}(t) = \mathbf{f}(\mathbf{x}(t)) \tag{4.46}$$

and the n-vector $\mathbf{x}_e$ is an equilibrium solution of this equation, so that $\mathbf{f}(\mathbf{x}_e) = \mathbf{0}$. We want to determine, if possible, the qualitative behavior of solutions to the equation starting from initial conditions in the vicinity of $\mathbf{x}_e$. By a simple translation of coordinates, taking $\mathbf{w}(t) = \mathbf{x}(t) - \mathbf{x}_e$, we may rewrite the differential equation in terms of $\mathbf{w}(t)$ as

$$\dot{\mathbf{w}}(t) = \mathbf{f}(\mathbf{w}(t) + \mathbf{x}_e) = \overline{\mathbf{f}}(\mathbf{w}(t)) \tag{4.47}$$

where $\overline{\mathbf{f}}(\cdot)$ is simply a different nonlinear function obtained from $\mathbf{f}(\cdot)$ and the constant vector $\mathbf{x}_e$. Furthermore, $\mathbf{w}_e = \mathbf{0}$ is an equilibrium solution of this rewritten equation corresponding to the equilibrium solution $\mathbf{x}_e$ of the original equation. Hence it is without loss of generality that we may limit our attention to the case of a zero equilibrium solution. So hereafter we assume that the original system has $\mathbf{x}_e = \mathbf{0}$ as an equilibrium solution that is to be investigated for stability.

Stability Definitions

In order to describe some of the possible kinds of behavior that might occur for initial conditions in the vicinity of $\mathbf{x}_e = \mathbf{0}$, we introduce a few definitions.

1. The equilibrium solution $\mathbf{x}_e = \mathbf{0}$ is *stable* if for every choice of a bound M, there is a sufficiently small neighborhood surrounding the equilibrium, $\{\mathbf{x}_i : \|\mathbf{x}_i\| < \delta\}$, such that every initial condition in this region produces a solution bounded in norm by M for all times $t \geq 0$.
2. The equilibrium solution $\mathbf{x}_e = \mathbf{0}$ is *asymptotically stable* if it is stable, in the sense of Definition 1, and in addition if for every initial condition of sufficiently small norm, the corresponding solution tends to $\mathbf{x}_e = \mathbf{0}$ as $t \to \infty$ (Fig. 4.11).
3. The equilibrium solution $\mathbf{x}_e = \mathbf{0}$ is *unstable* if it is not stable.

It is helpful to understand these definitions when applied to a linear system

$$\dot{\mathbf{x}}(t) = \mathbf{A}\mathbf{x}(t), \qquad \mathbf{x}(0) = \mathbf{x}_0 \tag{4.48}$$

whose solutions we can express analytically in terms of a matrix exponential function. If $\mathbf{A}$ is a nonsingular matrix, then the only equilibrium solution is $\mathbf{x}_e = \mathbf{0}$ since

$$\mathbf{0} = \mathbf{A}\mathbf{x}_e \quad \text{if and only if} \quad \mathbf{x}_e = \mathbf{0} \tag{4.49}$$

For simplicity, we'll also assume that $\mathbf{A}$ has distinct eigenvalues and is therefore diagonalizable. Since the solution of the linear differential equation is given by

$$\mathbf{x}(t) = e^{\mathbf{A}t}\mathbf{x}_0 \tag{4.50}$$

Sec. 4.2 Linearization and Stability

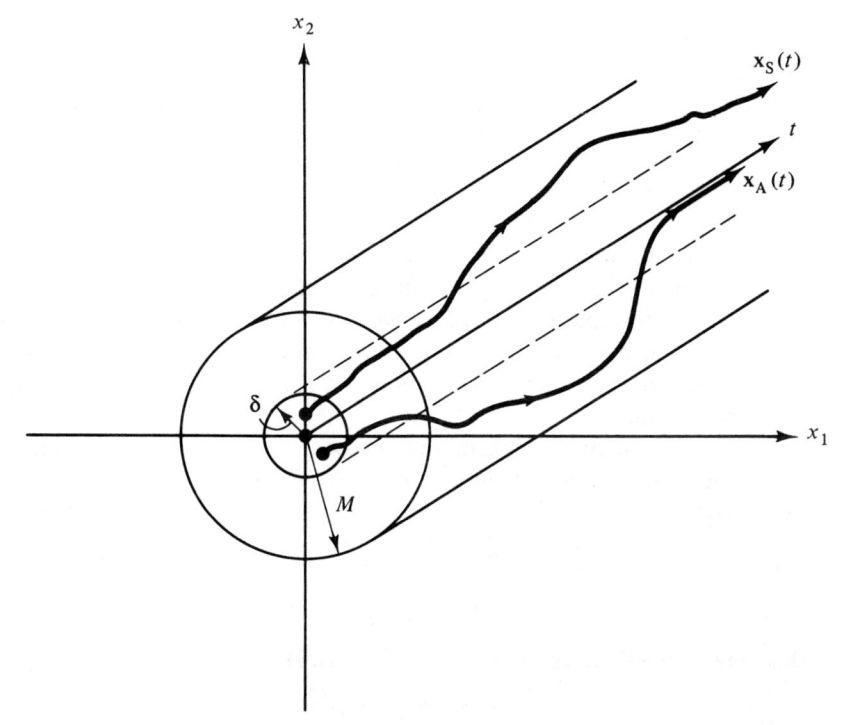

Figure 4.11 A stable equilibrium (trajectory $\mathbf{x}_S(\mathbf{t})$) and an asymptotically stable equilibrium (trajectory $\mathbf{x}_A(\mathbf{t})$).

we know that $\mathbf{x}(t)$ takes the form

$$\mathbf{x}(t) = \sum_{i=1}^{n} \mathbf{v}_i e^{\lambda_i t} \quad (4.51)$$

where the λ_i are the eigenvalues of $\mathbf{A}$ and the $\mathbf{v}_i$ are appropriately chosen n-vectors. Thus we can say that the equilibrium solution $\mathbf{x}_e = \mathbf{0}$ of this linear system is stable if the real part of every eigenvalue of $\mathbf{A}$ is less than or equal to 0; it is asymptotically stable if the real part of every eigenvalue of $\mathbf{A}$ is strictly less than 0; and it is unstable if $\mathbf{A}$ has one or more eigenvalues with positive real part.

Lyapunov's Theorem

We are now ready to describe an important result, the Lyapunov theorem, that indicates how linearization may be used to study the stability of an equilibrium solution of a nonlinear system. Suppose that $\mathbf{x}_e = \mathbf{0}$ is an equilibrium solution of the system

$$\dot{\mathbf{x}}(t) = \mathbf{f}(\mathbf{x}(t)) \quad (4.52)$$

where $\mathbf{f}$ is continuously differentiable in a neighborhood of $\mathbf{0}$ with derivative $\mathbf{f}'(\mathbf{0})$. Then if all of the eigenvalues of $\mathbf{f}'(\mathbf{0})$ have negative real part, $\mathbf{x}_e = \mathbf{0}$ is an asymptotically stable equilibrium solution, and if one or more eigenvalues of $\mathbf{f}'(\mathbf{0})$ has positive real part, $\mathbf{x}_e = \mathbf{0}$ is an unstable equilibrium solution.

Notice that the result means that asymptotic stability or instability of the equilibrium solution $\mathbf{x}_e = \mathbf{0}$ for the linearized system

$$\frac{d}{dt} \Delta \mathbf{x}(t) = \mathbf{f}'(\mathbf{0}) \Delta \mathbf{x}(t) \tag{4.53}$$

implies that the same property holds for the associated nonlinear system. Also notice that the theorem does not say anything about the case when the linearized system is stable but not asymptotically stable. Said another way, if all eigenvalues of $\mathbf{f}'(\mathbf{0})$ have nonpositive real part, and if the real part of one or more eigenvalues of $\mathbf{f}'(\mathbf{0})$ has zero real part, then nothing can be said about the stability of the equilibrium solution of the nonlinear system.

We will now give an example to show that the Lyapunov theorem is the best possible result of this kind that can be expected. Consider the following second-order nonlinear system:

$$\dot{x}_1(t) = \beta x_2(t) + \mu x_1(t)(x_1^2(t) + x_2^2(t)) \tag{4.54}$$

$$\dot{x}_2(t) = -\beta x_1(t) + \mu x_2(t)(x_1^2(t) + x_2^2(t)) \tag{4.55}$$

This may be rewritten in polar coordinates (ρ, θ) as

$$\dot{\rho}(t) = \mu \rho^3(t) \tag{4.56}$$

$$\dot{\theta}(t) = \beta \tag{4.57}$$

from which it may be seen that the equilibrium solution $\mathbf{x}_e = \mathbf{0}$ of the original system is asymptotically stable for negative μ, stable for $\mu = 0$, and unstable for positive μ (Fig. 4.12). The linearized system at the origin is described by the equations

$$\Delta \dot{x}_1(t) = \beta \Delta x_2(t) \tag{4.58}$$

$$\Delta \dot{x}_2(t) = -\beta \Delta x_1(t) \tag{4.59}$$

which has purely imaginary eigenvalues. Clearly, it is impossible to say anything at all about stability or instability of the nonlinear system based on its linearization. The crucial parameter, μ, does not even appear in the linearized system!

Example

The connections between linearization and stability are valid for discrete-time systems too. Suppose that the system

$$\mathbf{x}_{k+1} = \mathbf{f}(\mathbf{x}_k)$$

has $\mathbf{x}_e = \mathbf{0}$ as an equilibrium solution. The linearized system is

$$\Delta \mathbf{x}_{k+1} = \mathbf{f}'(\mathbf{0}) \Delta \mathbf{x}_k$$

If all eigenvalues of $\mathbf{f}'(\mathbf{0})$ have magnitude less than 1, this linear discrete-time system is asymptotically stable, and $\mathbf{x}_e = \mathbf{0}$ is an asymptotically stable equilibrium of the original nonlinear discrete-time system.

Sec. 4.2 Linearization and Stability 183

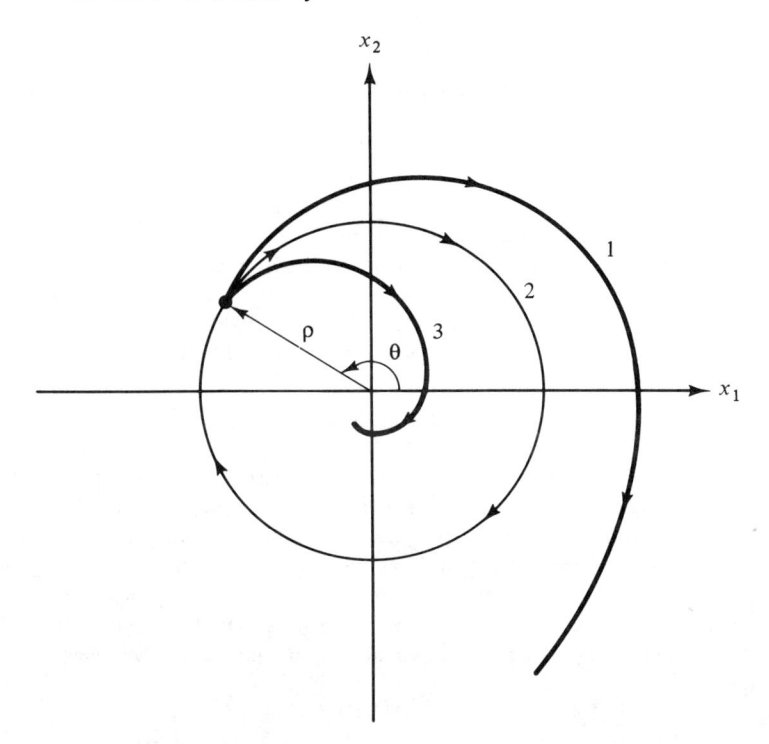

Figure 4.12 Phase plane plot for system given by equations (4.54, 4.55). Arrows show direction of motion for increasing t when β is negative: case 1, $\mu > 0$; case 2, $\mu = 0$; case 3, $\mu < 0$.

It is worth noting that the results on stability analysis via linearization, for both continuous-time and discrete-time systems, can be combined to extend the stability analysis of the Euler discretization method, given in the preceding chapter, to nonlinear systems.

Lyapunov Functions for Stability

To conclude our discussion of stability, we want to mention that Lyapunov's name is also attached to a second, more general approach to the study of stability of nonlinear systems. Briefly described, this alternative method involves finding a scalar function of the state, $V(\mathbf{x})$, that plays a role analogous to the total energy function in a "lossy" electrical circuit or mechanical system. $V(\mathbf{x})$ must have certain properties: $V(\mathbf{x}_e) = 0$; $V(\mathbf{x}) > 0$ for $\mathbf{x} \neq \mathbf{x}_e$; also the time derivative of $V(\mathbf{x})$, $V'(\mathbf{x}(t))\mathbf{f}(\mathbf{x}(t))$, must be zero at $\mathbf{x}(t) = \mathbf{x}_e$, must be nonpositive for $\mathbf{x}(t) \neq \mathbf{x}_e$, and must not be identically zero along any nonequilibrium solution $\mathbf{x}(t)$. If such a function $V(\mathbf{x})$ can be found, it is called a "Lyapunov function," and the equilibrium solution is asymptotically stable. The intuitive idea is that the "energy" of the system decreases along solution trajectories and so solutions move toward some point where minimum energy is achieved. Since the minimum value of energy is 0, which occurs at $\mathbf{x}(t) = \mathbf{x}_e$, solutions $\mathbf{x}(t)$ tend to $\mathbf{x}_e$.

Example

The case of a damped pendulum system will be used to illustrate this idea. The system is described by the second-order differential equation

$$\ddot{\theta}(t) + d(\dot{\theta}(t)) + \frac{g}{L} \sin \theta(t) = 0$$

where the middle term on the left side arises from a damping force depending on angular velocity. We assume that the function $d(\cdot)$ is continuous and satisfies a "generalized odd-symmetry" property: $\omega d(\omega) \geq 0$; this is a weak assumption, requiring only that the damping force be in the direction opposite to the angular velocity.

A candidate Lyapunov function is obtained by using the total energy function for the pendulum:

$$E = \frac{1}{2} (L \dot{\theta}(t))^2 + gL (1 - \cos \theta(t))$$

The first term on the right side is the kinetic energy of the system (we are assuming unit mass), and the second term is the potential energy. When no damping is present (i.e., $d(\cdot) = 0$) physical arguments show that E is constant; this will also be clear from the mathematical analysis that follows. In any case, E is clearly a nonnegative quantity.

To use these facts, we translate them into the standard notation for state equations. The state vector $\mathbf{x}(t)$ has two components, $x_1(t) = \theta(t)$ and $x_2(t) = \dot{\theta}(t)$. $V(\mathbf{x}) = E$ is the candidate Lyapunov function. Computing its time derivative gives

$$\frac{d}{dt} V(\mathbf{x}) = V'(\mathbf{x}(t)) \mathbf{f}(\mathbf{x}(t))$$

$$= \begin{bmatrix} gL \sin x_1 & L^2 x_2 \end{bmatrix} \begin{bmatrix} x_2 \\ -\frac{g}{L} \sin x_1 - d(x_2) \end{bmatrix}$$

$$= -L^2 x_2 \, d(x_2)$$

which is nonpositive, so $V(\mathbf{x})$ is decreasing (as expected!).

Finding a suitable choice of $V(\mathbf{x})$ for any particular nonlinear system is not always straightforward. A typical function that might be tried as a candidate Lyapunov function is $V(\mathbf{x}) = \|\mathbf{x}(t)\|^2$, the squared norm of the state vector. In fact, for any asymptotically stable linear system a suitable quadratic Lyapunov function, $V(\mathbf{x}) = \mathbf{x}^T \mathbf{P} \mathbf{x}$, may be found. When applied to a system obtained by linearizing a nonlinear system around an equilibrium solution, the quadratic Lyapunov function associated with the linearized system may be used to prove the Lyapunov theorem about stability and linearization.

4.3 QUALITATIVE BEHAVIOR OF NONLINEAR SYSTEMS

The study of low-order nonlinear systems has an importance that goes well beyond that due to their applications in Newtonian mechanics of particle motion. In particular, second-order systems may be studied by graphical methods that are unavailable for higher-order systems (due to the difficulty of portraying the time evolution of a point in three- or higher-dimensional space). Such graphical methods have led to the develop-

Sec. 4.3 Qualitative Behavior of Nonlinear Systems

ment of sophisticated geometric methods that abstract many important features from the easily visualized two-dimensional case.

An additional motivation for giving special attention to two-dimensional systems arises from the study of families of systems depending on a (small) parameter. (Imagine circuit equations where the parameter represents a parasitic capacitance.) For the study of how system response varies as a function of the parameter, it turns out that it is often appropriate to think of a system as composed of a parallel connection of first- and second-order systems.

Second-order nonlinear systems offer a rich variety of possible kinds of behavior, including the simplest kind of "global" nonlinear behavior: *limit cycles*. (These are never present in the response of a linear system; this is the justification for identifying them as a truly nonlinear phenomenon.) Another important kind of truly nonlinear behavior, *chaotic* solutions, can arise in the forced response of certain second-order systems.

Numerical methods for studying nonlinear system behavior consist mainly of simulation techniques. Analytical expressions, in the form of certain kinds of series expansions, can be obtained for limited classes of systems, and these have possible application more generally as a means of providing approximations that improve on linearization. For the case of second-order systems, computers offer greatly improved capabilities for using graphical analysis methods. This provides the final motivation for the remainder of this section.

Phase Plane Analysis

The notation for two-dimensional systems follows that already established. We'll write the vector differential equation for $\mathbf{x}(t)$ in terms of its components as

$$\dot{x}_1(t) = f_1(x_1(t), x_2(t)) \tag{4.60}$$

$$\dot{x}_2(t) = f_2(x_1(t), x_2(t)) \tag{4.61}$$

As customary, we think of the solution at time t, $\mathbf{x}(t)$, as a point in the Cartesian plane (with the components $x_1(t)$ and $x_2(t)$). Then solution trajectories correspond to curves, parametrized by time t. Some properties of the collection of all such curves can be stated, based on the properties of solutions of differential equations. For example, at most one curve can pass though any point, due to uniqueness of solutions, unless the point is a *singular point*, where both f_1 and f_2 vanish. Thus curves cannot intersect, except at a singular point or unless they form simple closed curves; the latter are periodic solutions of the differential equation.

Because there is no explicit dependence of the function $\mathbf{f}$ on time, t, the curves in the $\mathbf{x}$-plane, commonly called the *phase plane*, can also be described, at least implicitly, only in terms of x_1 and x_2 by eliminating the parameter t between the two differential equations. For all x_1 and x_2 such that $f_1(x_1, x_2) \neq 0$, we have

$$\frac{dx_2}{dx_1} = \frac{f_2(x_1, x_2)}{f_1(x_1, x_2)} \tag{4.62}$$

by applying the chain rule and inverse function theorem. Thus at points in the plane satisfying the stated condition on f_1, x_2 can be obtained as a function of x_1, at least in principle, by solving a single first-order differential equation. At points where $f_2(x_1,x_2) \neq 0$, a similar argument can be given with the roles of x_1 and x_2 interchanged. If both f_1 and f_2 are zero, the corresponding x_1 and x_2 values are equilibrium solutions of the differential equation (i.e., solutions that are constant for all time). Such points are precisely the *singular points* of the differential equation, which we sometimes also call *equilibrium points*.

Examples

As a first example, we choose a linear system that can be solved in analytical form. We suppose that

$$\dot{x}_1 = x_2$$
$$\dot{x}_2 = -\omega^2 x_1$$

The solutions to these equations have the form

$$x_1(t) = A \cos \omega t + B \sin \omega t$$
$$x_2(t) = \omega(-A \sin \omega t + B \cos \omega t)$$

which may be seen to be the parametric equation for a family of ellipses centered at the origin of the (x_1, x_2)-plane (Fig. 4.13). The sense of movement around these curves, for increasing time t, is most easily obtained by considering the original differential equations: For positive x_1 and x_2, x_1 is increasing while x_2 is decreasing. Thus the sense of motion around the trajectories is in the clockwise direction.

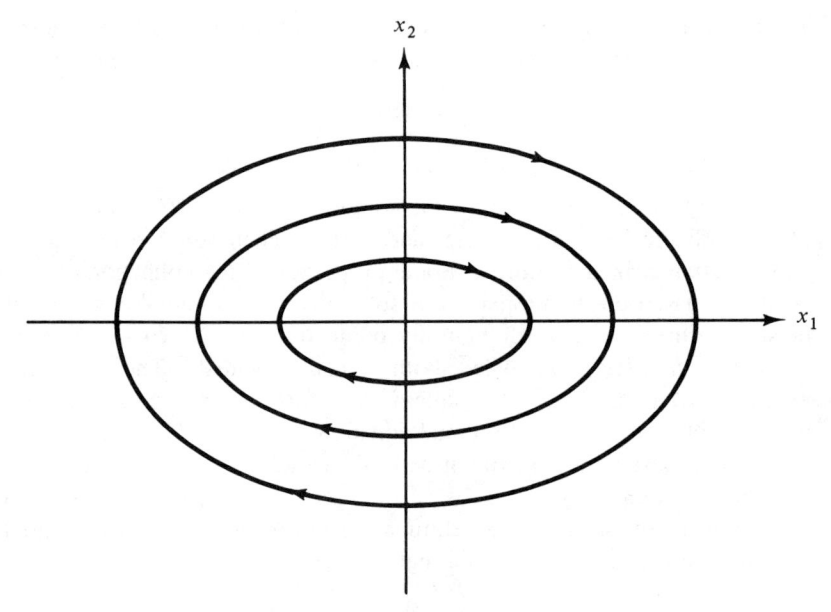

Figure 4.13 Phase plane plot for first example (linear oscillator) ($\omega_2 < 1$ when scales for x_1 and x_2 are the same).

Sec. 4.3 Qualitative Behavior of Nonlinear Systems

To determine the shape of the phase plane curves for this example another way, we consider the differential equation obtained by eliminating t:

$$\frac{dx_2}{dx_1} = -\frac{\omega^2 x_1}{x_2}$$

which holds except when $x_2 = 0$. This equation can easily be solved by separating variables, giving

$$x_2 \, dx_2 = -\omega^2 x_1 \, dx_1$$

which may be integrated to give

$$x_2^2 + \omega^2 x_1^2 = C$$

for some constant of integration, C. Once again we see that the phase plane curves are ellipses centered at the origin.

For a second example, we return to the nonlinear pendulum problem discussed earlier. Newton's laws provide a description in the form of the second-order differential equation

$$\ddot{\theta}(t) + \frac{g}{L} \sin \theta(t) = 0$$

and state equations are obtained by taking $x_1 = \theta$ and $x_2 = \dot{\theta}$. The resulting equations are

$$\dot{x}_1(t) = x_2(t)$$

$$\dot{x}_2(t) = -\frac{g}{L} \sin x_1(t)$$

The equation describing the solution trajectories in the phase plane is

$$\frac{dx_2}{dx_1} = -\frac{g \sin x_1}{L x_2}$$

Solving this equation gives

$$\frac{1}{2} x_2^2 - \frac{g}{L} \cos x_1 + \overline{C} = 0$$

for some constant of integration, $\overline{C}$. Rewriting this slightly brings it to a more familiar form:

$$\frac{1}{2} x_2^2 + \frac{g}{L} (1 - \cos x_2) = C$$

where $C = 1 - \overline{C}$ is another constant. We recognize the first term on the left side as the L^{-2} times the kinetic energy of the system (assuming unit mass), and the second term is L^{-2} times the potential energy. Contours of constant C are the phase plane trajectories. For small enough C, namely $C < 2g/L$, these contours are closed curves enclosing the origin; this reflects the fact that the pendulum admits a periodic solution with an angular amplitude of less than π radians. The trajectories with larger total energy correspond to periodic complete revolutions of the pendulum in either the clockwise or counterclockwise direction (Fig. 4.14).

The equation for the phase plane trajectories (i.e., the conservation of energy equation) may be used to determine the amplitude-frequency relationship of the nonlinear pendulum. Assume that $C < 2g/L$. When the angular displacement reaches its largest value,

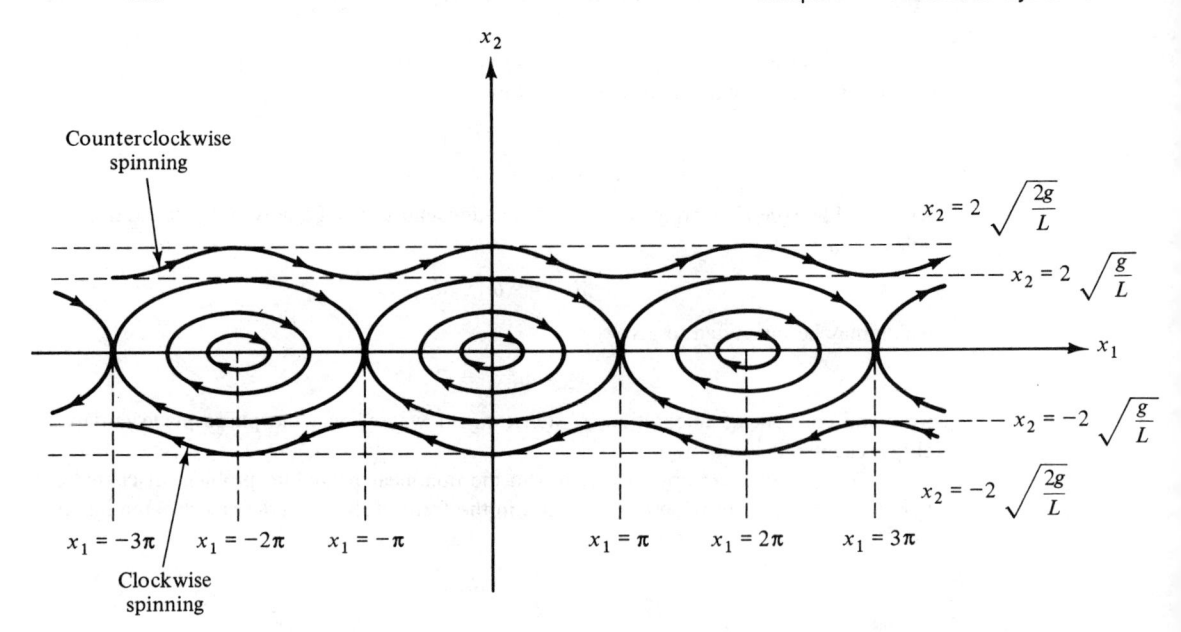

Figure 4.14 Phase plane plot for a nonlinear pendulum system.

θ_{max}, the angular velocity is zero and the total energy equals the potential energy, so

$$C = (1 - \cos \theta_{max})$$

Thus the equation for the phase plane trajectories may be written as

$$x_2^2 = \frac{2g}{L} (\cos x_1 - \cos \theta_{max})$$

Considering the quarter-period starting when the pendulum passes through $x_1 = \theta = 0$ with positive velocity and ending when the pendulum reaches maximum amplitude and zero velocity, we obtain

$$\frac{dx_1}{dt} = x_2 = \sqrt{(2g/L)(\cos x_1 - \cos \theta_{max})}$$

Inverting to obtain the derivative of t with respect to x_1 and integrating over the quarter-period of the motion gives

$$\frac{T}{4} = \left(\frac{L}{2g}\right)^{1/2} \int_0^{\theta_{max}} \frac{1}{\sqrt{\cos \theta - \cos \theta_{max}}} d\theta$$

Classification of Singular Points

For linear systems, it is always possible to obtain analytical expressions for the phase plane curves. We have previously seen that the solutions of a second-order linear system are described in terms of the system eigenvalues, and the location of these eigenvalues in the complex plane (not to be confused with the phase plane!) determines the

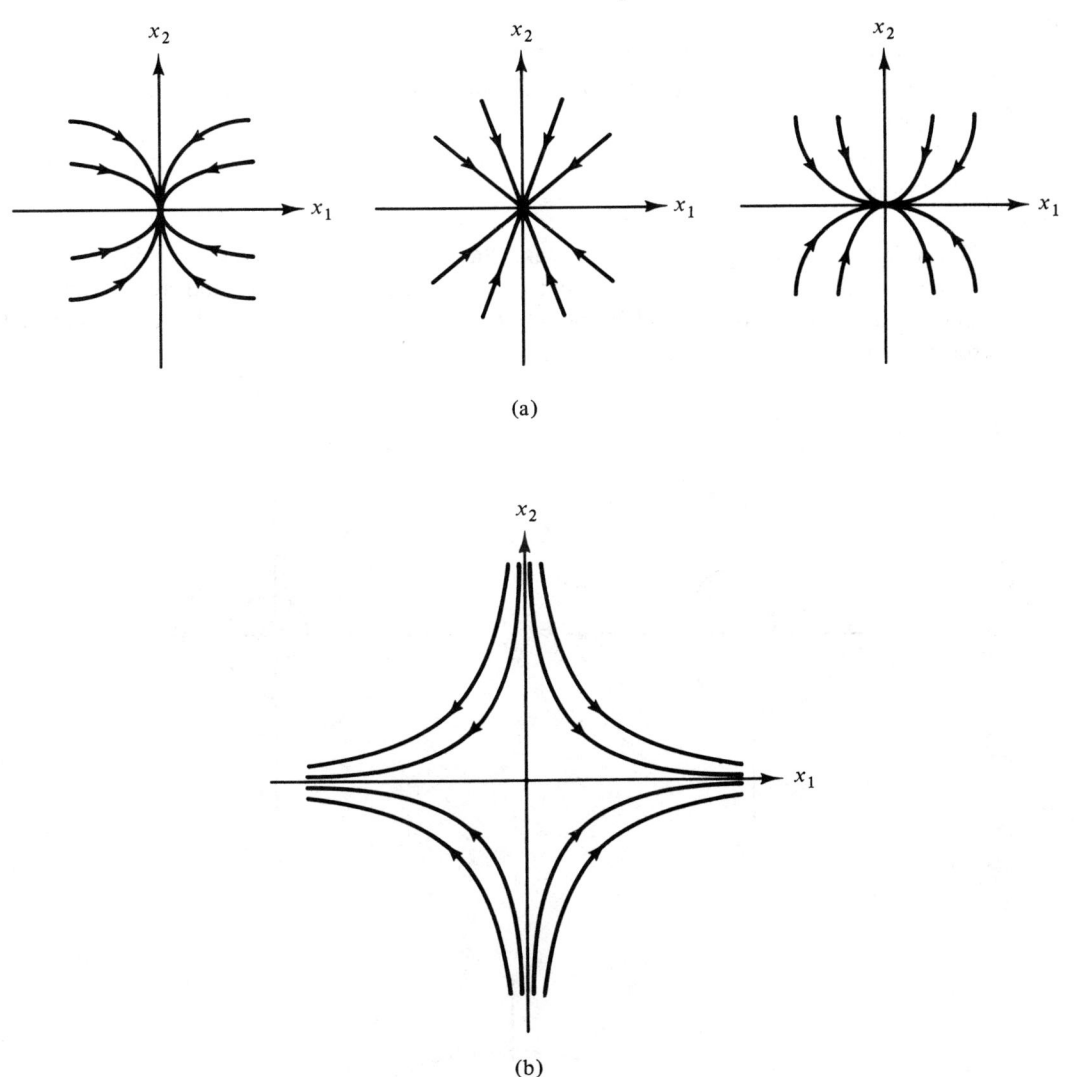

Sec. 4.3　Qualitative Behavior of Nonlinear Systems

Figure 4.15　Phase plane trajectories (real eigenvalues): (a) stable nodes; (b) saddle point.

qualitative form of the phase plane curves (Figs. 4.15, 4.16). The following terminology is fairly standard:

Stable focus: Corresponds to a complex conjugate pair of eigenvalues with negative real part.

Unstable focus: Corresponds to a complex conjugate pair of eigenvalues with positive real part.

Stable node: Corresponds to a pair of negative, real eigenvalues.

Unstable node: Corresponds to a pair of positive, real eigenvalues.

Saddle point: Corresponds to a pair of real eigenvalues, one positive and one negative.

Center (or vortex): Corresponds to a purely imaginary conjugate pair of eigenvalues.

The trajectories corresponding to foci take the form of spirals; those associated with a vortex are ellipses. Unstable nodes and foci give trajectories that tend away from the singular point at the origin and become unbounded; the same curves with the opposite sense of motion correspond to stable nodes or foci.

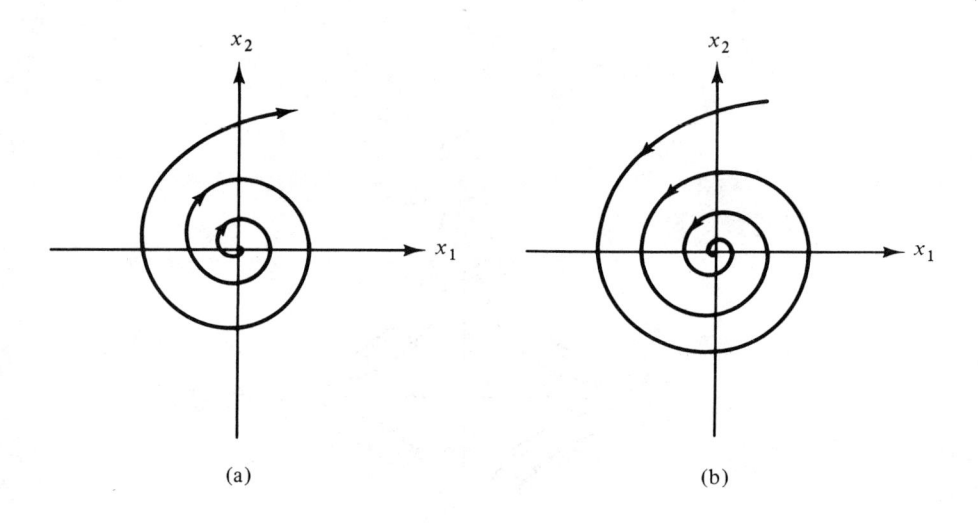

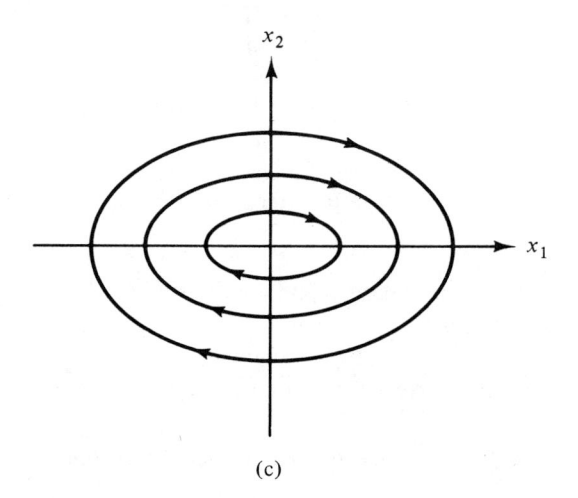

Figure 4.16 Phase plane trajectories (eigenvalues not purely real): (a) unstable focus; (b) stable focus; (c) center.

Sec. 4.3   Qualitative Behavior of Nonlinear Systems   191

For nonlinear systems, the same terms are used, justified by the Lyapunov theorem described in the preceding section which indicates that except for the case of a center, the behavior of a system in a sufficiently small neighborhood of a singular point is identical to the behavior of its linearized version. However, as mentioned earlier, nonlinear systems can produce responses that have no counterparts in the linear case; this means that we have to be careful not to put too much confidence in linearized systems operating "far" from a singular point.

Limit Cycles

A limit cycle is an isolated periodic solution trajectory. For second-order systems, a limit cycle shows up as an isolated closed curve in the phase plane (Fig. 4.17). By isolated, we mean that the curve can be surrounded by an open annulus with the closed curve contained in its interior, and the annulus contains no other closed curve in the phase plane. (When we speak of a curve in the phase plane, we mean those curves generated from solutions of the differential equations.) The limit cycle is thus a periodic solution surrounded by a region in the phase plane where other solution curves

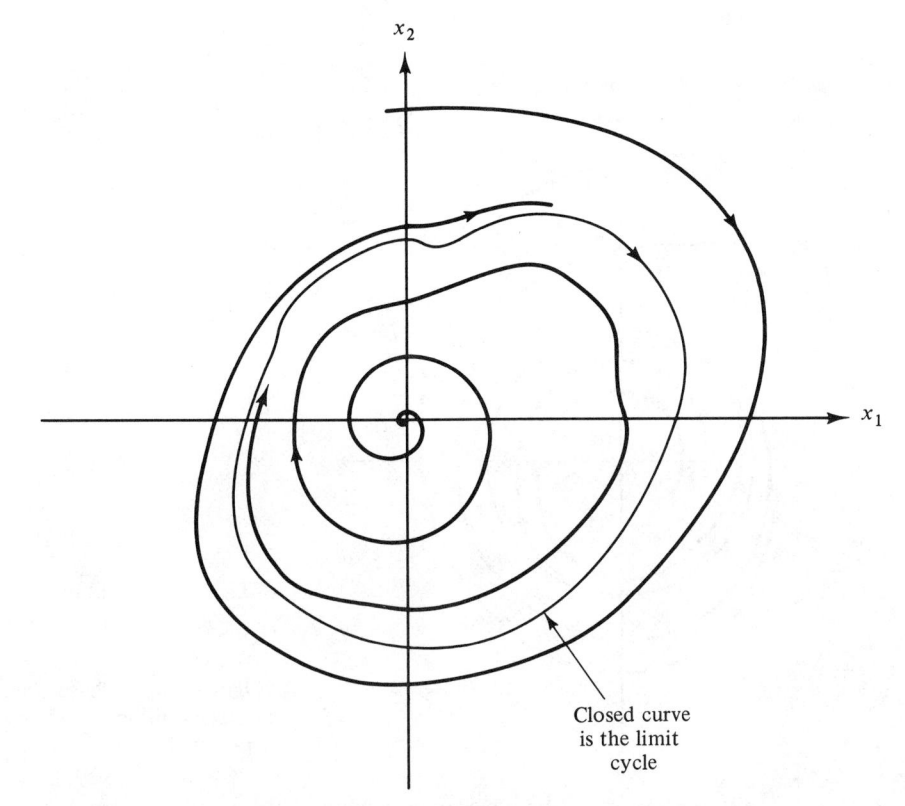

Figure 4.17   Phase plane plot showing a limit cycle with converging trajectories.

either spiral toward it or away from it. The closed elliptical trajectories found in linear systems with a vortex singular point are not limit cycles because they are not isolated; every neighborhood of each of the ellipses contains (infinitely many) other ellipses.

Example

As an example of a system exhibiting a limit cycle, consider

$$\dot{x}_1 = x_2 + x_1(1 - x_1^2 - x_2^2)$$
$$\dot{x}_2 = -x_1 + x_2(1 - x_1^2 - x_2^2)$$

Using polar coordinates (ρ, θ), the differential equation may be written as

$$\dot{\rho} = \rho(1 - \rho^2)$$
$$\dot{\theta} = -1$$

from which it may be seen that solution curves starting outside the unit circle in the phase plane spiral inward at a constant clockwise rate of rotation, approaching the unit circle as $t \to \infty$ (Fig. 4.18). Similarly, solution curves starting inside the unit circle (and not at the origin, which is the only singular point) spiral outward at a constant angular rate, approaching the unit circle as $t \to \infty$. Finally, the unit circle is a closed curve in the phase plane; thus it is a limit cycle.

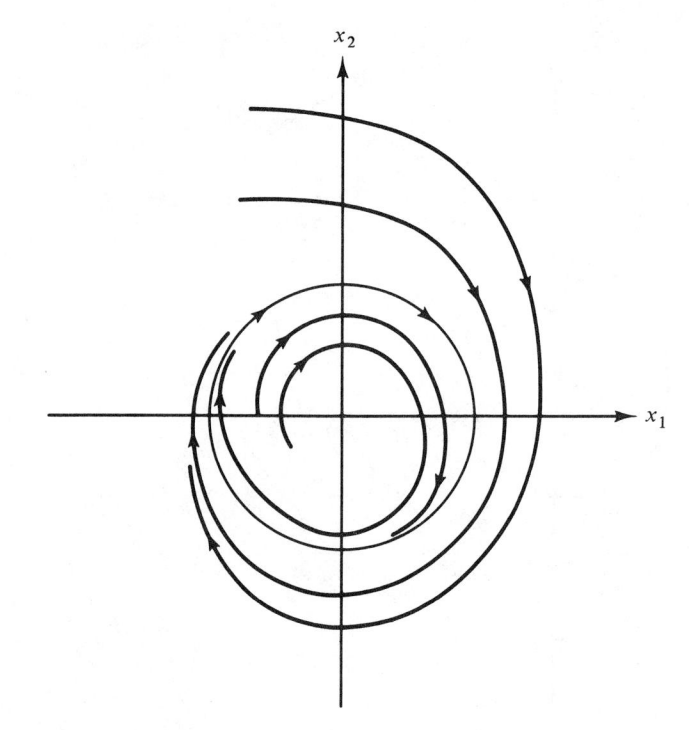

Figure 4.18 Phase plane plot for system with limit cycle given by $x_1^2 + x_2^2 - 1 = 0$.

Limit cycles are often desirable behavior; for instance, imagine a thermostat-controlled air conditioning system operating in a steady thermal environment and

Sec. 4.3 Qualitative Behavior of Nonlinear Systems

designed to keep temperatures within a desired range, or think of a clock generating circuit in a digital audio tape deck that is required to produce a repeatable, fixed-frequency square-wave signal every time it is switched on. In other cases limit cycles are undesirable; think of a photolithographic mask alignment control system for use in integrated circuit fabrication where the aligner must be globally asymptotically stable because persistent oscillations, even if very small in amplitude, would produce unacceptable exposures. Other motion control applications in robotics and automated machining also require that limit cycles be avoided. Thus results on existence and nonexistence of limit cycles are of great interest.

The easiest test to apply is based on the Poincaré index theorem, a strong necessary condition for existence that can sometimes be used to rule out the possibility of a limit cycle: For a closed curve in the phase plane to exist (and hence for a limit cycle to exist), the number of enclosed singular points which are nodes, centers, and foci, minus the number which are saddle points, must equal 1.

A more useful test to rule out the possibility of any closed curve in certain regions of the phase plane, therefore ruling out a limit cycle in particular, is the Bendixson test, which is described as follows. Let R be a simply connected region (i.e., a region containing no holes) in the phase plane where

$$\frac{\partial f_1}{\partial x_1} + \frac{\partial f_2}{\partial x_2} \neq 0 \qquad (4.63)$$

and where this quantity does not change sign. Then R contains no closed curves; in particular, it contains no limit cycles.

Examples

For the example given above, a simple calculation gives

$$\frac{\partial f_1}{\partial x_1} + \frac{\partial f_2}{\partial x_2} = 2 - 4x_1^2 - 4x_2^2$$

so the system has no limit cycle inside the circle of radius $1/\sqrt{2}$.

As a second example, we introduce the van der Pol equation,

$$\ddot{z} - \varepsilon(1 - z^2)\dot{z} + z = 0$$

Choosing $x_1 = z$ and $x_2 = \dot{z}$, this may be written in the form

$$\dot{x}_1 = x_2$$
$$\dot{x}_2 = \varepsilon(1 - x_1^2)x_2 - x_1$$

For this equation

$$\frac{\partial f_1}{\partial x_1} + \frac{\partial f_2}{\partial x_2} = \varepsilon(1 - x_1^2)$$

so there is no closed curve, and in particular no limit cycle, contained entirely in the strip $|x_1| < 1$.

The proof of the Bendixson criterion for nonexistence of limit cycles involves an interesting application of Green's theorem from vector analysis. From

$$\frac{dx_2}{dx_1} = \frac{f_2(x_1,x_2)}{f_1(x_1,x_2)} \tag{4.64}$$

the differential equation

$$f_1(x_1,x_2)\,dx_2 - f_2(x_1,x_2)\,dx_1 = 0 \tag{4.65}$$

is obtained. Choosing a simple closed contour consisting of a closed curve in the phase plane, we obtain a line integral that must be zero:

$$\oint f_1(x_1,x_2)\,dx_2 - f_2(x_1,x_2)\,dx_1 = 0 \tag{4.66}$$

But by Green's theorem, the value of the line integral equals the value of a certain integral over the region enclosed by the contour, giving for this case

$$\iint \left[\frac{\partial f_1}{\partial x_1} + \frac{\partial f_2}{\partial x_2} \right] dx_1\,dx_2 = 0 \tag{4.67}$$

For this equality to hold, the integrand must be identically zero or it must change sign somewhere inside the contour. Thus there can be no limit cycle inside a simply connected region where this quantity is not identically zero and does not change sign.

It is possible to give a complete characterization of limit cycles using the celebrated Poincaré-Bendixson theorem, although it is not always easy to apply. (In particular, the demonstration that the van der Pol equation has a limit cycle is rather involved.) We assume that f_1 and f_2 are continuously differentiable and there is a domain D in the phase plane bounded by two smooth closed curves C_1 and C_2 with no singular points in D or on C_1 or C_2. Then if solution curves enter D through every point on C_1 and C_2 there exists at least one limit cycle in D (Fig. 4.19). (The same conclusion follows if solution curves leave D through every point on its boundaries.) Conversely, if a limit cycle exists, then such a domain can be found.

The Volterra-Lotka Population Model

The nonlinear pendulum system provides one example to show that nonlinear systems can have periodic solutions without having limit cycles. We will give another example with an interesting history, the Volterra-Lotka model for populations of two coexisting, competing species in a closed ecosystem. Historically, the model arose in connection with a question about the fluctuations of two fish species observed by Italian fishermen. To give the punch line before the story, it turns out that the simple model used to describe the ecosystem produces periodic solutions for the two species living in a predator-prey relationship, a somewhat surprising conclusion. A variation on the simple model has a limit cycle solution.

To describe the mathematical model, continuous variables are used for the populations of two species. Let $x_1(t)$ denote the (biomass of the) population of the predatory species; $x_2(t)$ corresponds to the prey. A simple model for the population variation over time is that the logarithmic growth rate of the predator species is an increasing function of the prey population while the logarithmic growth rate of the prey species is a

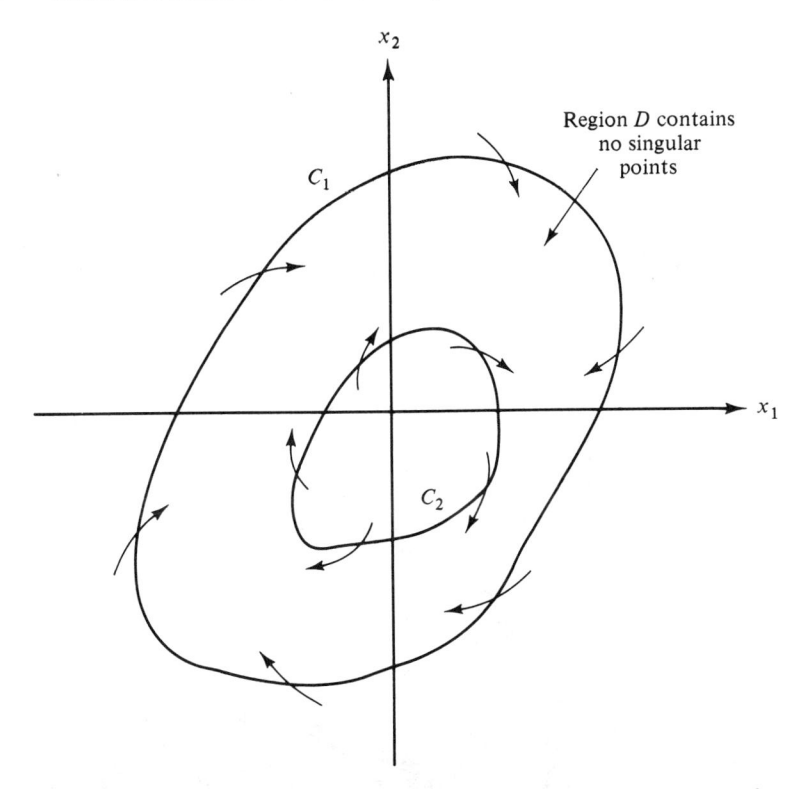

Sec. 4.3 Qualitative Behavior of Nonlinear Systems

Figure 4.19 Phase plane plot showing Poincaré-Bendixson conditions for a limit cycle.

decreasing function of the predator population. Lacking other information about the form of the functions, we will simply assume them to be affine (linear plus a constant) so that

$$\dot{x}_1(t) = (b_{12}x_2(t) - a_1)x_1(t) \tag{4.68}$$

$$\dot{x}_2(t) = (a_2 - b_{21}x_1(t))x_2(t) \tag{4.69}$$

All of the parameters in the model are positive quantities. The signs of the constant terms inside the parentheses were selected so that if there are no predators, the prey population will increase (exponentially fast), while if the prey population is zero, the predators will die out. (The model can be elaborated to include self-limiting effects on the population, and it can be extended to include more species and other kinds of competition between species.)

To lend some credence to our claim that the solutions of this system are periodic, notice that linearization of the system about the equilibrium solution $x_1(t) = a_2/b_{21}$ and $x_2 = a_1/b_{12}$ leads to a linear system with purely imaginary eigenvalues. Caution is required, however, since this is a case where the Lyapunov theorem does not apply, and we must resort to a phase plane analysis to show that the nonlinear system has periodic solutions. The phase plane trajectories satisfy

$$\frac{dx_2}{dx_1} = \frac{x_2}{x_1} \frac{a_2 - b_{21}x_1}{b_{12}x_2 - a_1} \tag{4.70}$$

which may be integrated to give the equation

$$b_{12}x_2 - a_1 \ln x_2 + C = -b_{21}x_1 + a_2 \ln x_1 \tag{4.71}$$

This defines a family of closed curves surrounding the nonzero equilibrium solution and lying in the first quadrant of the phase plane (Fig. 4.20). Hence the populations vary periodically.

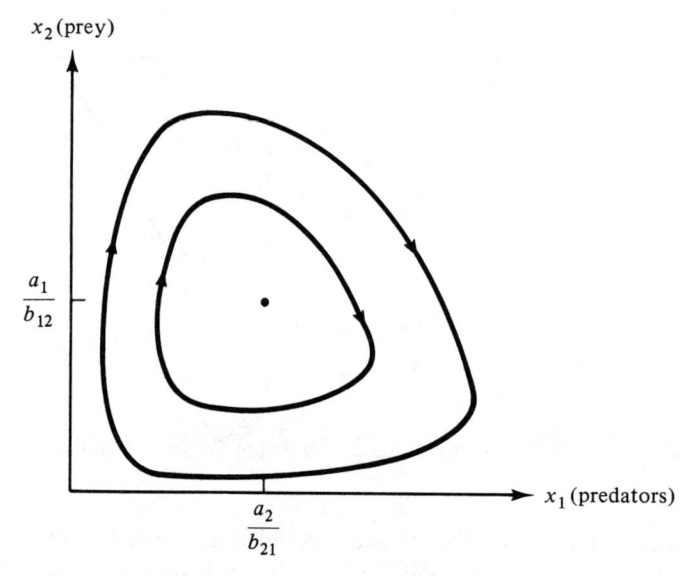

Figure 4.20 Phase plane plot for a predator-prey system.

We will show that some (reasonable!) variations on this model produce a system with a limit cycle, where any nonequilibrium initial populations asymptotically produce a unique periodic solution. We suppose that the prey species is self-limiting in population (reasonable because its food supply is ultimately limited); this can be accomplished by modifying the logarithmic grow rate of x_2 by adding a decreasing function of x_2. The modified differential equation is

$$\dot{x}_2(t) = (a_2 - b_{21}x_1(t) - c_{22}x_2(t))x_2(t) \tag{4.72}$$

Intuitively, this change should have a stabilizing effect on the growth of the prey population and, thereby, on the growth of the predator population also. Linearization confirms this intuition and shows that the new equilibrium solution is $x_1(t) = (a_2 - c_{22}a_1/b_{12})/b_{21}$ and $x_2(t) = a_1/b_{12}$. The only change is a decrease in the equilibrium population of predators; it must be assumed that c_{22} is a small enough positive number to keep the new predator equilibrium positive and hence meaningful in the context of this application. The linearized system is easily evaluated and, for sufficiently

Sec. 4.3 Qualitative Behavior of Nonlinear Systems

small c_{22}, its eigenvalues are complex conjugate quantities having negative real parts, $-c_{22}/2$, so the Lyapunov theorem may be applied to conclude that the nonlinear system is also asymptotically stable (Fig. 4.21).

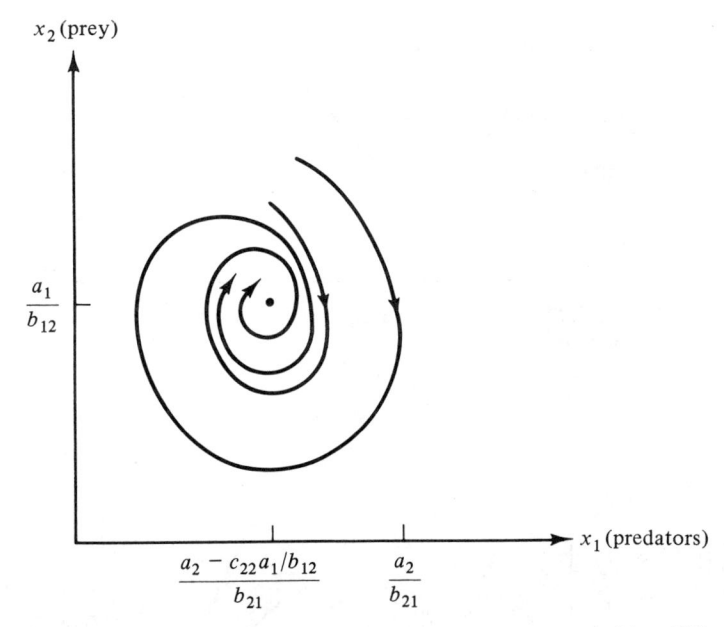

Figure 4.21 Phase plane plot for a predator-prey system with self-limited prey.

To obtain a limit cycle, another change is required in order to "destabilize" populations near equilibrium without upsetting stability for populations far from equilibrium. Close to equilibrium, the phase plane trajectories are spirals so that the population fluctuations are roughly $\pi/2$ radians out of phase with each other, and in fact $-\dot{x}_2$ is roughly in phase with x_1 (Fig. 4.22). This suggests that adding a small positive multiple of $-\dot{x}_2$ to the logarithmic growth rate of x_1 will be destabilizing. Furthermore, the differential equation for x_2 provides a convenient expression for $\dot{x}_2$ that may be used in the equation for $\dot{x}_1$, giving

$$\dot{x}_1(t) = (b_{12}x_2(t) - a_1)x_1(t) - b_{12}T(a_2 - b_{21}x_1(t) - c_{22}x_2(t))x_2(t)x_1(t) \quad (4.73)$$

We have written the equation to make it appear that we have replaced $x_2(t)$ in the original equation with the quantity $x_2(t) - T\dot{x}_2(t)$. The analysis of this new system by linearization shows that the equilibrium populations remain the same and for appropriate choices of c_{22} and T the equilibrium is unstable. This strongly suggests that a limit cycle exists (Fig. 4.23). Experimental verification is easily carried out, and an analytical verification using the Poincaré-Bendixson theorem is also possible.

Since the quantity $x_2(t) - T\dot{x}_2(t)$ is the first-order Taylor series approximation to $x_2(t-T)$, it may be argued that the modified model is a plausible one for interacting

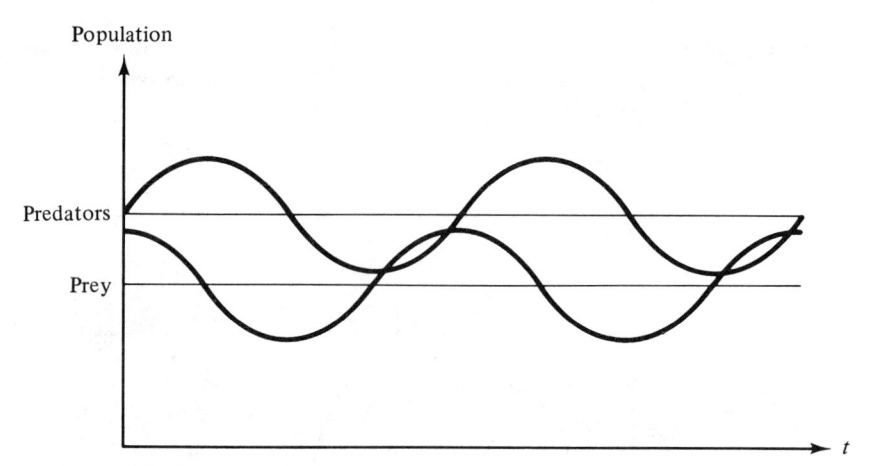

Figure 4.22 Near equilibrium, populations are varying roughly $\pi/2$ radians out of phase.

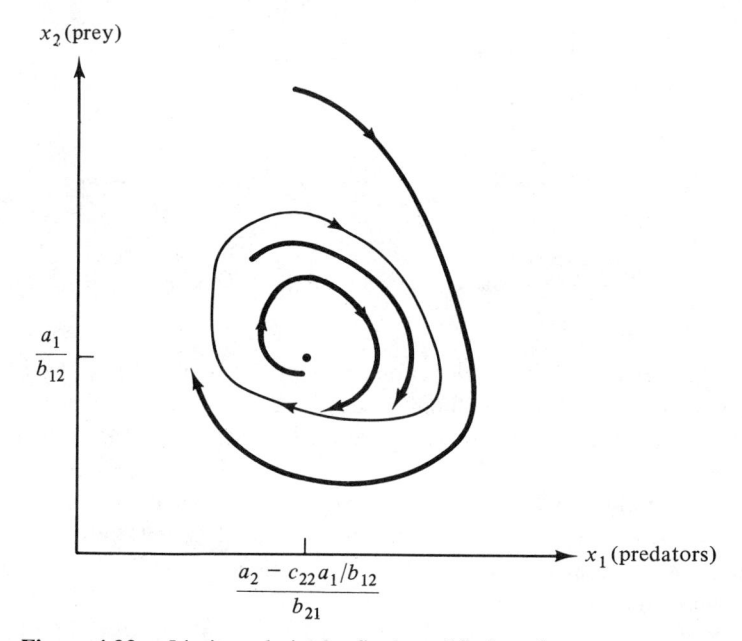

Figure 4.23 Limit cycle in the final modified predator-prey system.

species. Introducing a time delay in the response of the predator population to the prey population is a reasonable way to account for the time required for predator reproduction and maturation. By this reasoning, $x_2(t)$ should be replaced with the delayed quantity, $x_2(t-T)$, but this would completely change the character of the mathematical model. (It would lead to a so-called "delay-differential equation.") To obtain an ordinary differential model, a natural approximation is the one already mentioned, $x_2(t-T) \approx x_2(t) - T\dot{x}_2(t)$.

The Lorenz Attractor

To close this section, we will describe a well-studied example of a third-order nonlinear system whose solutions are much more complicated than the solutions arising in undriven second-order systems. These so-called "chaotic solutions" provide a kind of dynamic behavior similar to what might be expected from systems containing truly "random" phenomena. Solutions can remain in a bounded region of state space without approaching an equilibrium point or a limit cycle. Such "wild" behavior arising from a seemingly benign system has attracted considerable attention to possible applications of chaotic systems for modeling a variety of systems; for example, the drastic stock market collapse of October 1987 convinced many economists that chaotic models might improve upon some serious shortcomings of statistical models. More "plausible" applications to technological systems such as networks of coupled electric power generators, where sophisticated protective relaying systems are used to protect against massive power failures arising from spontaneous loss of synchronization of a single generator, have also received considerable attention. Similar models have also been applied by geophysicists to study reversals of the Earth's magnetic field, by physicists to study a dripping water faucet, and by physiologists to study the onset of cardiac fibrillation.

The particular example presented here goes by the name *Lorenz attractor* and arose first in a study of convective processes in fluid dynamics aimed at modeling atmospheric phenomena important for understanding weather and climate. More recently, the same system has been found to model some irregular behavior in lasers. The details of these applications will be left for the reader to investigate; here we focus simply on the mathematical model, which is governed by the three equations

$$\dot{x}_1(t) = \alpha(x_2(t) - x_1(t)) \tag{4.74}$$

$$\dot{x}_2(t) = (1 + \beta - x_3(t))x_1(t) - x_2(t) \tag{4.75}$$

$$\dot{x}_3(t) = x_1(t)x_2(t) - \gamma x_3(t) \tag{4.76}$$

All of the parameters are positive. The system has three equilibrium points, one at **0** and two others in symmetrically located positions: $x_1 = x_2 = \pm\sqrt{\gamma\beta}$ and $x_3 = \beta$. Linearizing about the equilibrium at the origin gives

$$\mathbf{f'(0)} = \begin{bmatrix} -\alpha & \alpha & 0 \\ \beta+1 & -1 & 0 \\ 0 & 0 & -\gamma \end{bmatrix} \tag{4.77}$$

The eigenvalues of this matrix may be found analytically, and one turns out to be positive for all positive choices of the parameters; thus the equilibrium **0** is unstable. For the two nonzero equilibria, the eigenvalues of the linearized system are not given by simple formulas. However, by using the Routh-Hurwitz test on the characteristic polynomial it is easily found that both nonzero equilibria are also unstable when the parameters satisfy the inequalities

$$\alpha > \gamma + 1 \tag{4.78}$$

$$\beta > \frac{\alpha^2 + \alpha\gamma + 2\alpha + \gamma + 1}{\alpha - \gamma - 1} \tag{4.79}$$

When such parameter values are chosen, say $\alpha = 10$, $\beta = 24$, and $\gamma = 2$, we obtain a system with three unstable equilibria. Yet a very surprising result is obtained by looking at state trajectories (using discretization to generate solutions numerically). Chaotic behavior is observed, with trajectories exhibiting pseudo-random behavior while remaining bounded (indefinitely); also the trajectories tend to "fill up" a complicated, highly intertwined region of the state space, which has the effect that solutions starting from initial conditions very close together can diverge widely after some time has elapsed. (The "filled" region characterizes the complex limiting behavior of trajectories and is the object described by the name "Lorenz attractor." The term "strange attractor" is used more generally to distinguish the profoundly different limiting behavior of a chaotic system from one of the ordinary attractors, a stable equilibrium or a limit cycle.) It is also a surprising fact that even for certain parameter choices that make the nonzero equilibria stable, chaotic solutions may exist in the Lorenz model (e.g., $\alpha = 10$, $\beta = 10$, and $\gamma = 2$; see Figs. 4.24 through 4.27). (Reducing β to 2 will eliminate the chaotic solutions.) A complete understanding of this system, particularly the variation of the qualitative form of the solution as the parameters are varied, is not yet available.

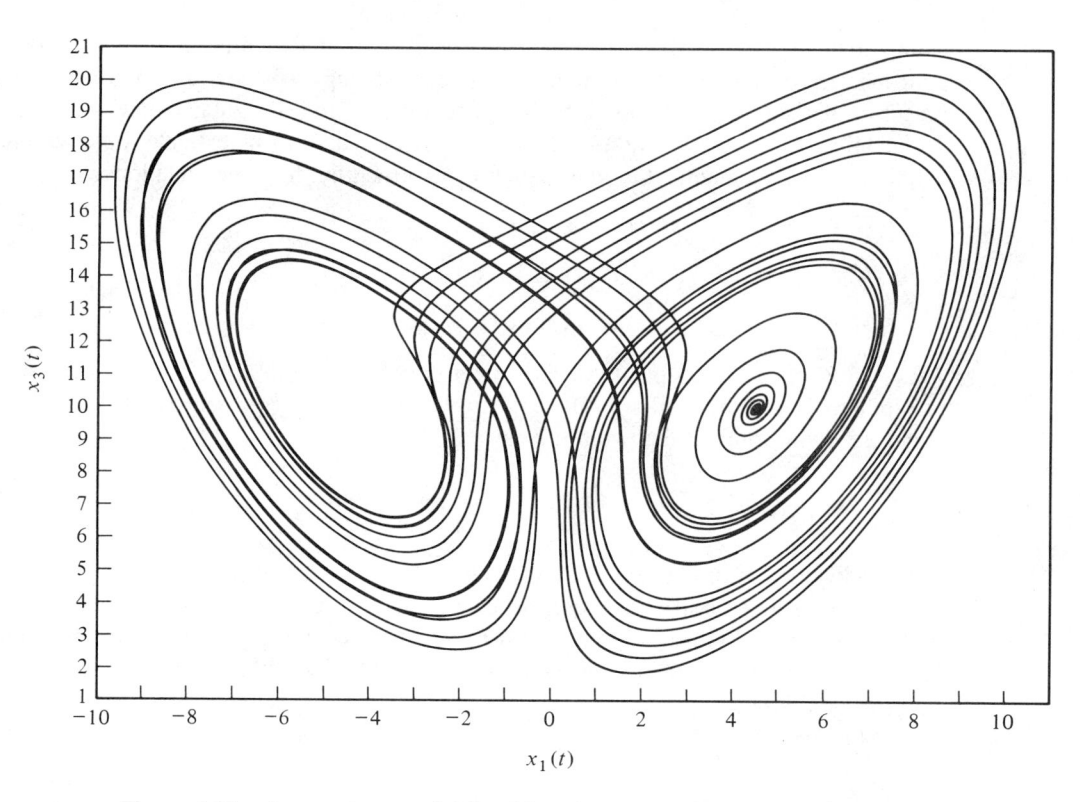

Figure 4.24 Lorenz attractor for $\beta = 10$; trajectory starting near (4,4,10), projected on (x_1, x_3)-plane.

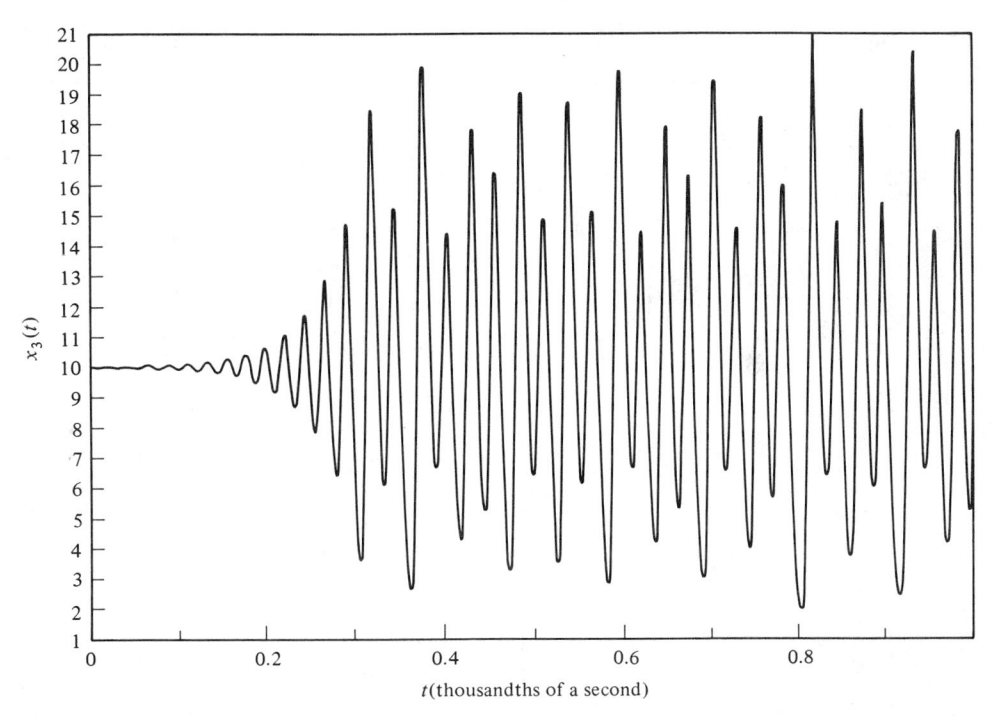

Figure 4.25 Lorenz attractor for $\beta = 10$; time response of state x_3.

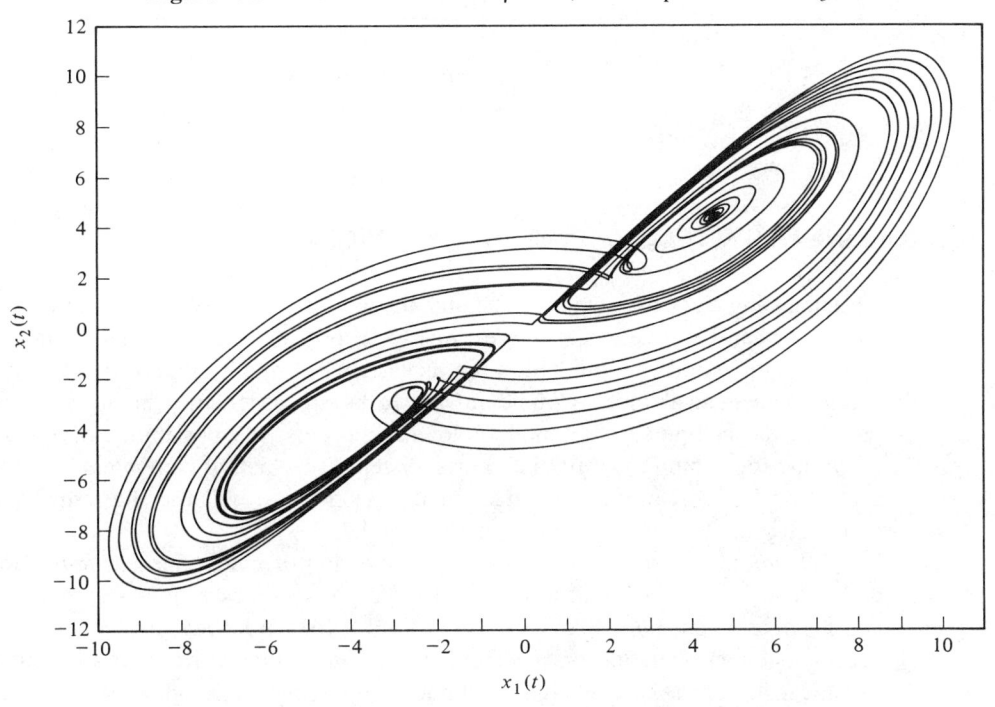

Figure 4.26 Lorenz attractor for $\beta = 10$; trajectory starting near (4,4,10), projected on (x_1, x_2)-plane.

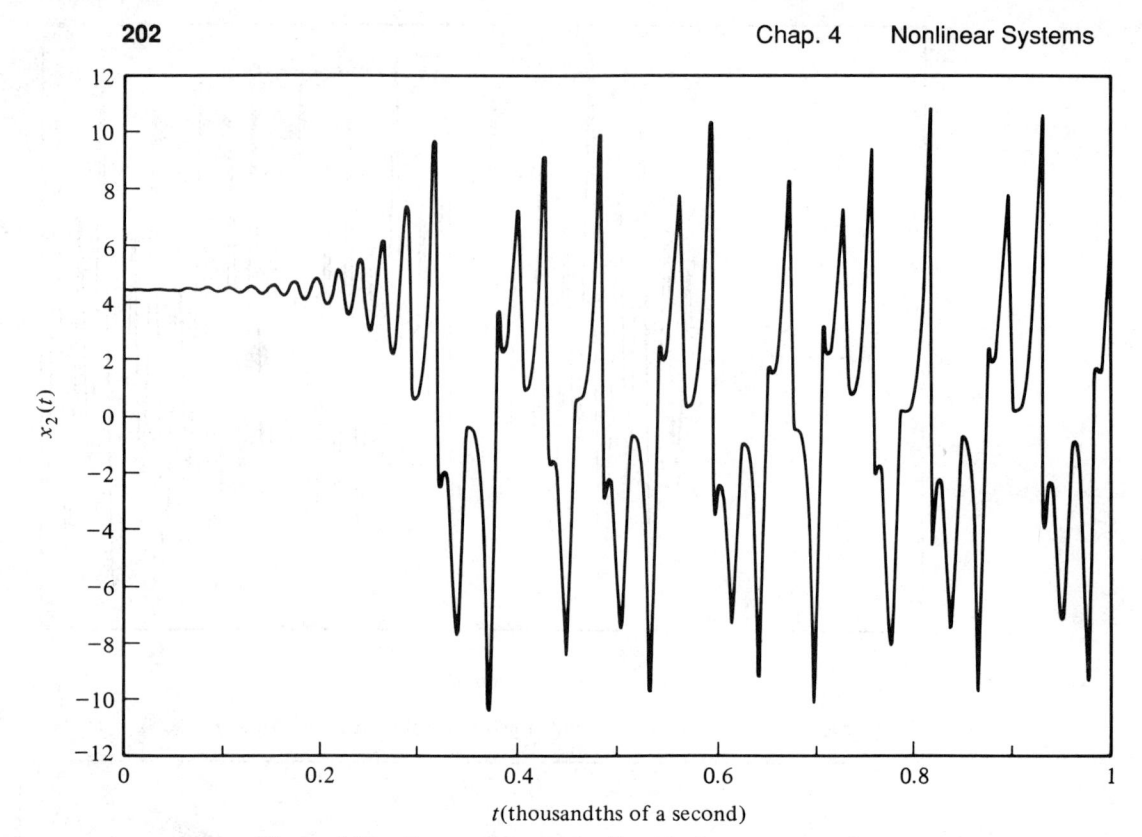

Figure 4.27 Lorenz attractor for β = 10; time response of state x_2.

4.4 NONLINEAR SYSTEMS AND NEURAL NETWORKS

An interesting class of nonlinear systems arises in the study of *neural networks*. This is an area of considerable recent interest due to its potential for providing insights into the kind of highly parallel computation that is carried out by physiological nervous systems. While mathematical models of neural networks are necessarily highly idealized compared to the complexities of real biological systems, the possibility of implementing electronic (or photonic) "artificial neural systems" or "neural computers" has motivated a great deal of research in this area. In this section we will use the term "neural network" only in the context of mathematical models.

Neural networks are composed of a collection of interconnected *units* (sometimes called "neurons"). Each unit has a set of input ports (representing synaptic connections) and a single output port. Each unit consists of a first-order low-pass filter (an approximate integrator) operating on a weighted sum of its inputs; the unit's output is a distorted version of the filtered signal obtained by passing it through a nonlinear *squashing* function (i.e., a monotonically increasing function with limits of −1 and 1 when its argument tends to −∞ and +∞, respectively). A scaled hyperbolic tangent function,

Sec. 4.4 Nonlinear Systems and Neural Networks

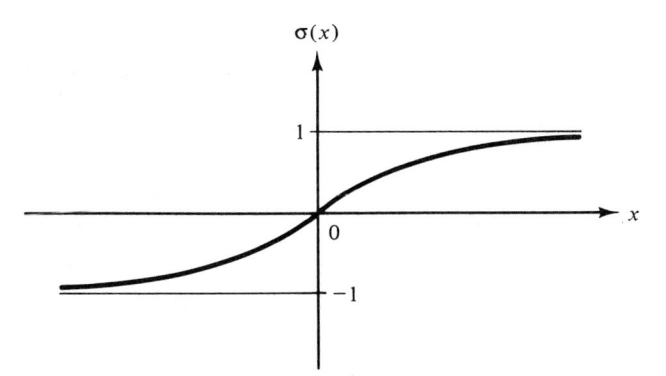

Figure 4.28 Squashing function.

$$\sigma(x) = \tanh \lambda x = \frac{e^{\lambda x} - e^{-\lambda x}}{e^{\lambda x} + e^{-\lambda x}} \tag{4.80}$$

where λ is a positive *gain* parameter (the slope of $\sigma(x)$ at $x=0$), is one commonly used squashing function (Fig. 4.28).

To give a mathematical description of the operation of each unit, take x_i to be the internal state of unit i; let $u_{i,k}$ denote its kth input and y_i its output value. Then the equations for unit i are

$$\dot{x}_i(t) = -\frac{x_i(t)}{\tau} + \sum_{k=1}^{N+1} w_{i,k} u_{i,k}(t) \tag{4.81}$$

$$y_i(t) = \sigma(x_i(t)) \tag{4.82}$$

Here τ is the time constant of the first-order filter (assumed to be the same for each unit), and the $w_{i,k}$ are the weights associated with the $N+1$ input ports (Fig. 4.29). A network is formed by taking N units and interconnecting outputs and inputs:

$$u_{i,k}(t) = y_k(t) = \sigma(x_k(t)), \quad 1 \leq k \leq N \tag{4.83}$$

(Fig. 4.30). The "extra" input to each unit, $u_{i,N+1}$, is viewed as an external input; we will assume that the external inputs are constants rather than varying in time, so their weighted values will be denoted simply u_i. Thus, the network equations are, for $1 \leq i \leq N$,

$$\dot{x}_i(t) = -\frac{x_i(t)}{\tau} + \sum_{k=1}^{N} w_{i,k} y_k(t) + u_i \; ; \quad y_i(t) = \sigma(x_i(t)) \tag{4.84}$$

Of course these equations could be collected into the usual state space form,

$$\dot{\mathbf{x}}(t) = \mathbf{f}(\mathbf{x}(t), \mathbf{u}(t)) \; ; \quad \mathbf{y}(t) = \mathbf{h}(\mathbf{x}(t)) \tag{4.85}$$

but we will have no need to do so in this section.

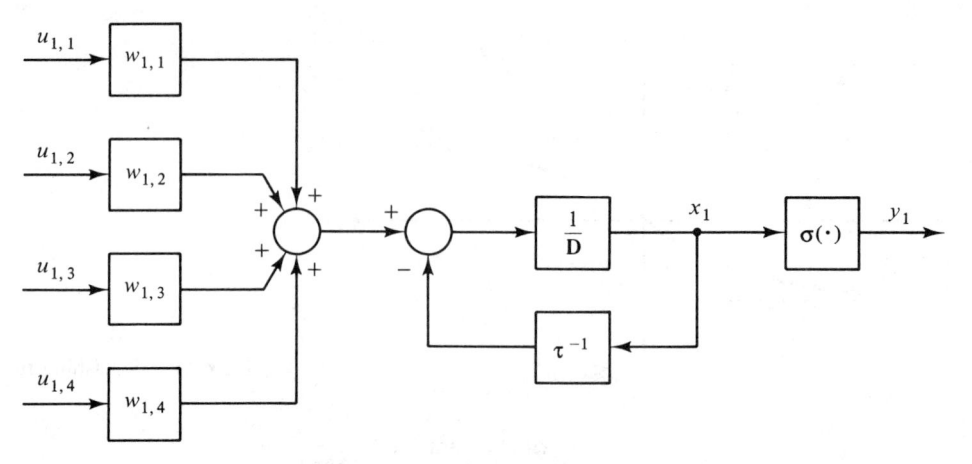

Figure 4.29 Neural network building block—unit 1 with four inputs shown: $\dot{x}_1 = (x_1/\tau) + \sum_{k=1}^{4} w_{1,k} u_{1,k}$, $y_1 = \sigma(x_1)$.

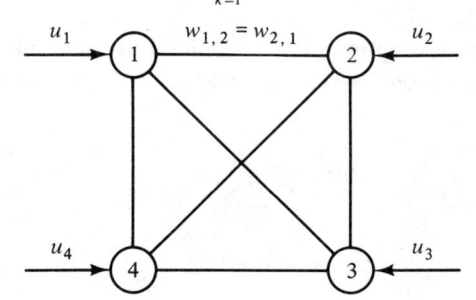

Figure 4.30 Schematic figure of a four-node neural network; branches represent symmetrically weighted connections.

We will assume that the time constant τ and squashing gain λ of the units are fixed. Then the weights and external inputs are the quantities available for designing particular kinds of behavior for the network. Our goal will be to choose these design parameters to achieve a "desirable" set of equilibrium solutions for the network. Then we will view the network as a computing device that transforms initial condition vectors into corresponding equilibrium solution vectors. When the squashing gain λ is large, we can expect that each equilibrium output value lies in the vicinity of ± 1, and thus we can view the network as carrying out computations that effectively transform initial condition vectors, whose components are real-valued, into binary-valued output vectors. In other words, each initial condition vector is classified as belonging to one of a finite set of "pattern classes," where each class is characterized by a binary "template" vector obtained by "squashing" an equilibrium solution of the network.

We will now describe one device that may be implemented with a neural network, a *content-addressable memory*, and then we will carry out some analysis to show how the device may be designed. For a content-addressable memory, the network weights are to be chosen so that a preselected set of binary N-vectors is obtained by squashing equilibrium solutions. (As noted above, when the squashing gain is large, the equilibrium output values will all be close to ± 1.) These binary vectors will be called *memory*

Sec. 4.4 Nonlinear Systems and Neural Networks

traces stored by the network. Any initial condition for the network state vector may be used to "probe" the network, and the resulting steady-state network output vector is called the *evoked response*. When one of the network's memory traces is used as a probe, it should also be the evoked response. More generally, each memory trace should be the evoked response of a (hopefully) large class of probe vectors that are "similar" to it. In this way, memory traces stored by the network may be recalled from probes that consist of imperfect (e.g., noisy) and partial information. Thus a memory trace is recalled on the basis of an incomplete specification of its content, hence the name "content-addressable memory."

A very important property that has been assumed above in the description of how a neural network would be used as a content-addressable memory is that all solutions to the associated differential equations tend to equilibrium solutions. We will now investigate this assumption to see how it can be guaranteed as part of the design process. For this purpose, a Lyapunov function (or energy function) will be employed. This approach may appear to be somewhat *ad hoc* compared to our analysis of a damped pendulum at the end of Section 4.2, but the use of Lyapunov functions for stability analysis of interconnected systems is a well-developed methodology.

Using the network state variables, define the function $V(x_1, \ldots, x_N)$ as

$$V = -\frac{1}{2} \sum_{i=1}^{N} \sum_{k=1}^{N} w_{i,k} y_i y_k - \sum_{i=1}^{N} y_i u_i + \tau^{-1} \int_0^{y_i} \sigma^{-1}(\xi) \, d\xi \qquad (4.86)$$

where we have written y_i rather than $\sigma(x_i)$ for notational simplicity. Because σ is the hyperbolic tangent function, the y_i variables take values between ± 1 and V is thus a bounded function. Computing the time derivative of V using the chain rule and the differential equations for the neural network, we find that if the symmetry condition $w_{i,k} = w_{k,i}$ holds, then

$$\frac{d}{dt} V = -\sum_{i=1}^{N} \dot{y}_i \left(\sum_{k=1}^{N} w_{i,k} y_k + u_i - \frac{x_i}{\tau} \right) = -\sum_{i=1}^{N} \dot{y}_i \dot{x}_i = -\sum_{i=1}^{N} \frac{d\sigma(x_i)}{dx} \dot{x}_i^2 \qquad (4.87)$$

where the dots indicate time derivatives. The derivative of σ is positive because σ is a monotonically increasing function. Thus the derivative of V is negative, except for states that are equilibrium states of the network, in which case the derivative is zero. We conclude that V is a decreasing function of time, that as $t \to \infty$, V tends to a limiting value (since V is bounded) and hence its derivative tends to 0, and finally that every solution of the neural network tends to an equilibrium solution.

Our next goal is to show how the network design parameters can be chosen to achieve desired equilibrium states. Two different approaches will be described. The equilibrium states of the network satisfy the equations

$$0 = -\frac{x_i}{\tau} + \sum_{k=1}^{N} w_{i,k} y_k + u_i \qquad (4.88)$$

By solving for x_i and "squashing" both sides of the equation we obtain

$$y_i = \sigma \left(\tau \sum_{k=1}^{N} w_{i,k} y_k + \tau u_i \right) \qquad (4.89)$$

In the limit of infinite gain, the function σ becomes the "signum" or "sign function," which we denote by sgn(·), that takes the values ±1 according to whether its argument is positive or negative. For both of the design approaches to be described, we proceed by setting the external inputs to zero and determining a symmetric weight matrix from the high-gain limiting form of the output equations, namely

$$y_i = \text{sgn}\left(\tau \sum_{k=1}^{N} w_{i,k} y_k\right) \tag{4.90}$$

It will be convenient to express these equations using matrices and vectors. Let **W** denote the $N \times N$ matrix of weights to be determined, and let $\mathbf{y}_m$ denote a column vector of the N elements of the mth memory trace that is to be stored by the network, for $1 \le m \le M$. Then the limiting output equations are written as

$$\mathbf{y}_m = \text{sgn}(\tau \mathbf{W} \mathbf{y}_m) \tag{4.91}$$

where we adopt the convention that the sgn function operates componentwise on vectors.

The first design approach follows from a simple observation: If the weight matrix **W** is chosen to make all of the components of $\tau \mathbf{W} \mathbf{y}_m$ equal to ±1, then the sgn function on the right side of equation (4.91) is superfluous. But the components of the vector $\mathbf{y}_m$ are all ±1, so the design equation for the weight matrix becomes linear:

$$\mathbf{y}_m = \tau \mathbf{W} \mathbf{y}_m \tag{4.92}$$

Let **Y** be the $N \times M$ matrix whose mth column is $\mathbf{y}_m$. Then the design equations for **W** based on the M memory trace vectors take the form of the following linear matrix equation:

$$\mathbf{Y} = \tau \mathbf{W} \mathbf{Y} \tag{4.93}$$

We assume that $M < N$ and use the pseudo-inverse (discussed in Section 1.2) to obtain a solution:

$$\mathbf{W} = \frac{1}{\tau} \mathbf{Y} \mathbf{Y}^\dagger \tag{4.94}$$

That this is a solution follows from one of the defining properties of the pseudo-inverse:

$$\mathbf{Y} \mathbf{Y}^\dagger \mathbf{Y} = \mathbf{Y} \tag{4.95}$$

Symmetry follows directly from another of the defining properties:

$$(\mathbf{Y} \mathbf{Y}^\dagger)^T = \mathbf{Y} \mathbf{Y}^\dagger \tag{4.96}$$

In the case when the columns of **Y** are linearly independent, we may write the solution in a more explicit form:

$$\mathbf{W} = \frac{1}{\tau} \mathbf{Y} (\mathbf{Y}^T \mathbf{Y})^{-1} \mathbf{Y}^T \tag{4.97}$$

Otherwise, **W** can be determined by using the Singular Value Decomposition (recall equation (1.79)).

Sec. 4.4   Nonlinear Systems and Neural Networks

The construction of **W** above was based on the equation for the network outputs at equilibrium and for an infinite-gain squashing function. For this case, the equilibrium outputs consist of certain corners of a cube in N-dimensional space, since the output values are ± 1. Using these weights when the squashing function gain is large, but not infinite, will usually produce acceptable results; the equilibrium outputs move from corners to the interior of the N-dimensional cube, but as long as λ is sufficiently large, they will be close to corners and the network will function as a mapping from initial states into binary vectors without ambiguities.

Example

Consider the design of a content-addressable memory that stores the following three vectors, expressed as the columns of the matrix **Y**:

$$\mathbf{Y} = \begin{bmatrix} 1 & -1 & 1 \\ 1 & 1 & 1 \\ -1 & 1 & 1 \\ -1 & 1 & 1 \\ 1 & -1 & 1 \end{bmatrix}$$

The columns are linearly independent, so the desired matrix of weights is given by

$$\mathbf{W} = \frac{1}{\tau}\mathbf{Y}(\mathbf{Y}^T\mathbf{Y})^{-1}\mathbf{Y}^T = \frac{1}{\tau}\begin{bmatrix} 0.5 & 0 & 0 & 0 & 0.5 \\ 0 & 1 & 0 & 0 & 0 \\ 0 & 0 & 0.5 & 0.5 & 0 \\ 0 & 0 & 0.5 & 0.5 & 0 \\ 0.5 & 0 & 0 & 0 & 0.5 \end{bmatrix}$$

Analytical and experimental work has shown that content-addressable memories designed with the pseudo-inverse method just described can provide relatively good performance. As long as the number of stored memory traces is smaller than $N/2$, accurate recall from noisy or incomplete probes is achieved.

The pseudo-inverse construction for design of content-addressable memories may be classified as a *batch learning* procedure. Given all of the memory traces to be stored (i.e., *learned*), the design is carried out with a one-shot (*batch*) procedure. A second approach to the design problem produces an *on-line* learning procedure, where memory traces may be presented one by one and the weight matrix is built up as a sum of individual contributions from each memory trace.

It is quite natural to view on-line learning as a discrete-time process with the network weights being determined by a sequence of training events (individual presentations of memory traces to be stored). In some cases, the implementation of a neural network (in hardware or in software) also involves a discretized model of the dynamics of the network. Thus before describing the second network design method, we will give a formulation for a discrete-time neural network.

Our discrete-time model will rely on an underlying discrete-time Lyapunov function so that it will again be true that all solutions tend to equilibrium solutions as time evolves. As a start, we first examine the Lyapunov function V in equation (4.86) in the limit of large squashing gain. The following argument shows that the contribution of the integral term in V becomes negligible as the gain gets large. Using notation that shows the λ dependence of σ explicitly gives the output relation

$$y_i(t) = \sigma(x_i(t)) = \tanh \lambda x_i(t) \tag{4.98}$$

from which we obtain

$$\int_0^{y_i} \sigma^{-1}(\xi)\, d\xi = \lambda^{-1} \int_0^{y_i} \tanh^{-1} \xi \, d\xi \tag{4.99}$$

which vanishes as $\lambda \to \infty$. Denoting the limiting form of V by V_∞ (recall that the external inputs are all zero), we have

$$V_\infty = -\frac{1}{2} \sum_{i=1}^{N} \sum_{k=1}^{N} w_{i,k} y_i y_k \tag{4.100}$$

We will now show that V_∞ may be naturally associated with a discrete-time neural network. The form and operation of this network follow from a consideration of the equilibrium output equations:

$$y_i = \text{sgn}\left(\tau \sum_{k=1}^{N} w_{i,k} y_k \right) \tag{4.101}$$

Since $\tau > 0$, its actual value is immaterial in solving this equation for the y_k; we take $\tau = 1$ and write equation (4.101) in vector notation as

$$\mathbf{y} = \text{sgn}(\mathbf{W}\mathbf{y}) \tag{4.102}$$

For a given $\mathbf{W}$, consider the following variation on the method of successive approximation for finding a solution $\mathbf{y}$ of this equation. Choose as an initial candidate solution some vector with components ± 1. If it is a solution, we're done. Otherwise, some of its components are not correct. Pick one of the incorrect components and change its sign. If the result is a solution vector, stop; otherwise, repeat this process until a solution is obtained. (The entire process is repeated with other initial guesses to obtain other equilibrium solutions.)

Suppose that the nth step of the algorithm involves changing the sign of element $y_{p(n)}$. Using the notation y_i^{n+1} to denote the value of the ith component after the nth step of the algorithm, we have

$$y_{p(n)}^{n+1} = \text{sgn}\left(\sum_{k=1}^{N} w_{p(n),k} y_k^n \right) \tag{4.103}$$

while

$$y_i^{n+1} = y_i^n, \qquad i \neq p(n) \tag{4.104}$$

With the following minor change, this algorithm provides the description of a discrete-time neural network mentioned above. Let $n_N = n \bmod N$. Then take $p(n) = n_N + 1$ so that p runs cyclically through the numbers 1 to N. The equations (4.103, 4.104) describing the updating steps are then valid for all $n \geq 1$, although some steps where equation (4.103) produces no change in $y_{p(n)}^n$ might now be included. (We adopt the convention that if the argument of the sgn function is exactly 0, then $y_{p(n)}^{n+1} = y_{p(n)}^n$.)

A discrete-time Lyapunov stability analysis employing the function V_∞ will be used to show that the successive approximation algorithm given above converges and that solutions of the discrete-time neural network model tend to equilibria. The function V_∞ is bounded, and we now will determine how it changes at each step of the algorithm. Clearly, there is no change except at those steps when a change in $y_{p(n)}$ is made. An easy calculation shows that when the weight matrix is symmetric,

$$V_\infty^{n+1} - V_\infty^n = y_{p(n)}^{n+1}(y_{p(n)}^n - y_{p(n)}^{n+1}) \qquad (4.105)$$

Thus V_∞ decreases by 2 at every "nontrivial" updating step of the discrete-time network equations. Since V_∞ is bounded and there are at most 2^N possibilities for the y vector, after a finite number of updating steps a solution to the nonlinear equations is attained. Thus the sequence of candidate solutions generated by the successive approximation algorithm produces a vector of outputs corresponding to an equilibrium solution as desired.

As already noted, V_∞ may be recognized as the high-gain limit of the Lyapunov function V introduced for studying stability of the original (continuous-time) neural network. The components y_i are always ± 1, so when the algorithm stops the resulting vector of discrete-time equilibrium outputs is not exactly equal to the continuous-time equilibrium outputs, which must have magnitude less than 1 because $\sigma(x) < 1$ for finite x and finite gain. It is to be hoped that design methods for choosing the weights in discrete-time networks will also produce acceptable results (equilibrium outputs close to the corners of the cube in N-space whose corners are the points with coordinate vectors having components ± 1) for large squashing gain. This turns out to be the case for the pseudo-inverse method described above and for the method to be described next.

The second content-addressable memory network design procedure to be described uses the *Hebb rule*,

$$\mathbf{W} = \sum_{m=1}^{M} \mathbf{W}_m \qquad (4.106)$$

where the component weight matrices are given in terms of previously introduced notation by

$$\mathbf{W}_m = \mathbf{y}_m \mathbf{y}_m^T - \mathbf{I} \qquad (4.107)$$

Each $\mathbf{W}_m$ has zero elements on its diagonal, so the same is true for $\mathbf{W}$.

Of course, we may express $\mathbf{W}$ as

$$\mathbf{W} = \mathbf{Y}\mathbf{Y}^T - M\mathbf{I} \qquad (4.108)$$

but this form does not display the additive decomposition in terms of individual memory traces, $\mathbf{y}_m$, which provides the on-line learning interpretation of this procedure. (Notice that the Hebb weight matrix must be multiplied by $1/\tau$ for use in a continuous-time network.)

Example

For the set of three memory traces given in the preceding example, the Hebb weight matrix is

$$\mathbf{W} = \mathbf{YY}^T - 3\mathbf{I} = \begin{bmatrix} 0 & 1 & -1 & -1 & 3 \\ 1 & 0 & 1 & 1 & 1 \\ -1 & 1 & 0 & 3 & -1 \\ -1 & 1 & 3 & 0 & -1 \\ 3 & 1 & -1 & -1 & 0 \end{bmatrix}$$

It is easily verified that each of the three vectors is an equilibrium solution of the discrete-time neural network equations; hence the Hebb rule successfully stores the three memory traces.

One motivation for the Hebb rule offers a bit of intuition about its form. If the components of all of the memory traces are chosen independently, at random, and are equally likely to be ±1, then for a particular memory, say $\mathbf{y}_{m_0}$,

$$\mathbf{Wy}_{m_0} = \sum_{m \neq m_0} \mathbf{W}_m \mathbf{y}_{m_0} + \mathbf{W}_{m_0} \mathbf{y}_{m_0} \qquad (4.109)$$

Notice that

$$\mathbf{W}_{m_0} \mathbf{y}_{m_0} = (N-1) \mathbf{y}_{m_0} \qquad (4.110)$$

whereas the terms in the sum are products involving independent, zero mean random variables and so will be zero "on the average." We omit the complicated details, but it turns out that the summation term has the effect of an additive "noise" that is unlikely to change the signs of the components of the "signal" term, $\mathbf{W}_{m_0}\mathbf{y}_{m_0}$, as long as the number of memory traces stored, M, does not get to be too large in comparison with N, the number of bits used to store each memory trace, say $M < N/(4 \log_2 N)$. Hence, the Hebb rule may be expected to work well in situations where a modest amount of random data is to be stored in a content-addressable memory.

Some words of caution are needed. For particular choices of the desired memory traces, $\{\mathbf{y}_m\}$, the Hebb rule may not store these traces due to "interference effects" between two or more of the vectors in the set. Also, additional equilibrium solutions of the network equations (i.e., "spurious memory traces") are often present in far greater numbers than the desired memory traces, and they reduce the capabilities for error correction.

The close interplay between the dynamics of a neural network and the minimization of an associated Lyapunov function has been extended to numerous other applications of the basic neural network structure given above. As we have seen, for a symmetric weight matrix, solutions of the network equations (4.84)

$$\dot{x}_i(t) = -\frac{x_i(t)}{\tau} + \sum_{k=1}^{N} w_{i,k} y_k(t) + u_i \, ; \quad y_i(t) = \sigma(x_i(t))$$

evolve along trajectories where the Lyapunov function in equation (4.86)

$$V = -\frac{1}{2} \sum_{i=1}^{N} \sum_{k=1}^{N} w_{i,k} y_i y_k - \sum_{i=1}^{N} y_i u_i + \tau^{-1} \int_0^{y_i} \sigma^{-1}(\xi) d\xi$$

Sec. 4.5   Input-Output Analysis of Nonlinear Systems   211

is decreasing. Since the contribution of the last term vanishes in the limit of large gain, this suggests trying to use neural networks to solve problems that can be cast as minimizing a quadratic function subject to the constraints that the solution variables are between −1 and 1. There is a large literature describing applications of this general idea. For illustrative purposes, a very simple example will serve to show how discrete problems can be "mapped" onto a neural network for solution.

Example

Suppose we would like to take a vector of ±1 elements and reverse the order of its elements; suppose the vectors have three components. Calling the original vector **p** and the desired output vector **q**, it is clear that the elements of **q** are to be chosen to minimize the function

$$\tfrac{1}{2}\sum_{i=1}^{3}(p_{4-i}-q_i)^2 = \tfrac{1}{2}\sum_{i=1}^{3}q_i^2 - \sum_{i=1}^{3}p_{4-i}q_i + \tfrac{1}{2}\sum_{i=1}^{3}p_k^2$$

Comparing with the form of equation (4.86) we see that the corresponding 3×3 weight matrix is −**I**, so the resulting neural network has no interconnections between units; the input variables are $u_i = -p_{4-i}$. The network equations are

$$\dot{x}_i(t) = -\frac{x_i(t)}{\tau} - q_i(t) - p_{4-i}; \qquad q_i(t) = \sigma(x_i(t))$$

4.5 INPUT-OUTPUT ANALYSIS OF NONLINEAR SYSTEMS

State space models, employing differential equations, provide one framework for studying the response of systems to inputs. This framework has the distinct advantage of generality: it may be used for both linear and nonlinear systems. On the other hand, because nonlinear differential equations rarely admit analytical solutions, state space methods often mask important qualitative (or semi-quantitative) information about the relations that exist between input signals and output signals behind a seemingly opaque screen of mathematical details. Linearization of a nonlinear system provides one way of obtaining approximations to input-output behavior for sufficiently well-behaved mathematical models, subject to "small variation" assumptions about the deviation of solutions from a nominal one. For models involving piecewise linear functions, which arise in a wide range of applications from semiconductor electronic circuits to mechanical positioning systems (e.g., robotic manipulators), the validity of linearization is questionable, and the ability to handle "global" rather that "local" nonlinear phenomena is of great importance. Furthermore, piecewise linear models can offer advantages in conceptual simplicity and in simulation of system behavior. Indeed, the use of piecewise linear models for the current-voltage response of a diode or for the voltage-torque response of a motor is often found to be sufficiently accurate for the end purposes of the models (such as determining delay characteristics of the circuit or determining the effectiveness of a position-control system), and the formulation of the models can be based on general qualitative features of the nonlinearities involved. This topic will be discussed briefly in Section 4.6.

Volterra Series Input-Output Models

To study the dynamic response of nonlinear systems to inputs, numerous methods have been proposed and explored. We will start with a description of one quite general approach, the Volterra series representation. This provides a kind of series expansion of the output of a nonlinear system in terms of its input. The mathematics required for a detailed study of Volterra series, especially their existence and convergence properties, is well beyond the scope of this book; however, the general idea is easily described in terms of the following example. Consider a system described by the following first-order differential equation:

$$\dot{x}(t) = a\,x(t) + b\,u(t) + n\,x(t)u(t) \qquad (4.111)$$

Notice that this equation differs from a linear equation through the addition of the *bilinear* term involving the product of $x(t)$ and $u(t)$. When $x(0)=0$, the equation can be integrated to give

$$x(t) = \int_0^t (a + n\,u(\tau))x(\tau)\,d\tau + \int_0^t b\,u(\tau)\,d\tau \qquad (4.112)$$

Now we proceed as in the method of successive approximation to get another expression for $x(t)$,

$$x(t) = \int_0^t (a + n\,u(\tau)) \left[\int_0^\tau (a + n\,u(\sigma))x(\sigma)\,d\sigma + \int_0^\tau bu(\sigma)\,d\sigma \right] d\tau + \int_0^t bu(\tau)\,d\tau \qquad (4.113)$$

We have simply substituted the expression for $x(t)$ into itself with the appropriate time argument. Clearly, we can repeat this process over and over, and each successive substitution creates higher-order multiple integrals. After N substitutions, the expression is

$$x(t) = \int_0^t bu(\tau)\,d\tau + \sum_{k=1}^{N} \left[\int_0^t \int_0^{\tau_0} \cdots \int_0^{\tau_{k-1}} \left[\prod_{j=0}^{k-1} (a + nu(\tau_j)) \right] (bu(\tau_k))\,d\tau_k \cdots d\tau_0 \right]$$

$$+ \int_0^t \int_0^{\tau_0} \cdots \int_0^{\tau_{N-1}} \left[\prod_{j=0}^{N-1} (a + nu(\tau_j)) \right] x(\tau_N)\,d\tau_N \cdots d\tau_0 \qquad (4.114)$$

Under certain conditions, say $(a + n\,u(t)) < 1$ for $0 < t \leq T$, the term with $x(\tau_N)$ in the integrand will tend to 0 as N grows large, giving an expression for the solution function $x(t)$ on the interval $[0,T]$ in terms of an infinite sum of multiple integral expressions involving the input $u(t)$.

This simple example suggests the form of the Volterra series representation of a nonlinear system's input-output relationship. The result is a "functional series" (analogous to a power series) expressing the output function $y(t)$ in the form

$$y(t) = \int_0^t h_1(t-\tau)u(\tau)\,d\tau \qquad (4.115)$$

$$+ \sum_{k=1}^{\infty} \int_0^t \int_0^{\tau_0} \cdots \int_0^{\tau_{k-1}} h_{k+1}(t-\tau_0, \ldots, t-\tau_k) u(\tau_0) \cdots u(\tau_k)\,d\tau_k \cdots d\tau_0 \qquad (4.116)$$

We have explicitly separated out the first term to show that it is the same form obtained for the input-output response of a time-invariant linear system.

Since the Volterra series provides one way of generalizing the kind of input-output representation that has proved so valuable for linear systems, there has been considerable study of the relationships between nonlinear state space representations and their corresponding Volterra series input-output representations. Bilinear state space systems, which take the form

$$\dot{\mathbf{x}}(t) = \mathbf{A}\mathbf{x}(t) + \mathbf{B}\mathbf{u}(t) + \mathbf{N}\mathbf{x}(t)\mathbf{u}(t) \tag{4.117}$$

$$\mathbf{y}(t) = \mathbf{C}\mathbf{x}(t) \tag{4.118}$$

can be used to model (essentially) any *finite* Volterra series. This might seem surprising, but in effect the products of several input variables that appear in higher terms of the Volterra series may be generated by "nested products," much as any polynomial function may be expressed using nested multiplications and additions:

$$\lambda^n + a_1\lambda^{n-1} + \cdots + a_{n-1}\lambda + a_n = a_n + (\lambda(a_{n-1} + \lambda(\cdots(a_1 + \lambda)\cdots))) \tag{4.119}$$

Speaking loosely, we can say that multiplication is a sufficiently strong nonlinearity to be combined with linear operations and allow approximation of very general nonlinear functions. For the case of functions, the famous Weierstrass approximation theorem states that on any finite interval, an arbitrary continuous function can be uniformly approximated by a polynomial. The analogous result for systems is that any sufficiently well-behaved Volterra series can be uniformly approximated by a finite Volterra series, and hence by a system whose state space description is bilinear. Of course, continuous functions can be uniformly approximated by piecewise linear functions too, with much greater simplicity in many cases; for the task of polynomial approximation of a piecewise linear function, it is clear that a high-degree polynomial will be required for a good fit. Such observations as this emphasize the necessity of proceeding cautiously when faced with a problem involving the modeling of a nonlinear system.

Describing Functions

We now turn to a discussion of a technique that has proved very useful in many applications involving nonlinear systems. This method, known as the *describing function method*, is important in many engineering systems where steady-state frequency response characteristics are of primary importance. For example, audio amplifiers comprised of active electronic devices are designed to exhibit linear characteristics over the range of audio frequencies. Yet the devices themselves are not linear over all operating conditions, and small nonlinear distortions are inevitably introduced by such systems. As another example, an automobile cruise control system must be designed to operate over a range of engine rpm's despite nonlinear dynamic characteristics of the engine. Finally, even in applications such as position control, where transient response characteristics are of primary importance, the describing function method can be fruitfully employed.

One powerful motivation for the technique is a wealth of accumulated engineering experience about the importance of understanding *amplitude dependence* of the response of a nonlinear system to inputs. Based on the use of frequency response characteristics for linear systems, it is quite natural to adopt the viewpoint that sinusoidal inputs comprise a class of inputs for which an appropriate notion of amplitude dependence should be formulated.

Fourier analysis, the principle of representing a signal as a superposition of sinusoidal components, provides one justification that these signals comprise a sufficiently rich class for the purpose at hand. That the response of a linear system to an input sinusoid can be described by its effects on amplitude and phase (and its utter lack of effects on input frequency) is a second. But from a nonlinear systems point of view, a third quite important and complementary justification arises. Instead of adopting the "linear" viewpoint that the signal $A \sin \omega t$ is characterized by its maximum amplitude, A, we instead think of the corresponding distribution of instantaneous amplitudes over time. We describe this distribution using a nonnegative function $p(a), -A \leq a \leq A$, called the *amplitude density function*, which is defined by

$$\int_{-A}^{a} p(\alpha) \, d\alpha = \text{the fraction of each period for which } A \sin \omega t < a \quad (4.120)$$

From the graph of the function $A \sin \omega t$ we have

$$\int_{-A}^{a} p(\alpha) \, d\alpha = \frac{1}{2} + \sin^{-1} \frac{(a/A)}{\pi} \quad (4.121)$$

from which $p(a)$ may be found by differentiation:

$$p(a) = \left[a\pi \sqrt{1 - \left(\frac{a}{A}\right)^2} \right]^{-1} \quad (4.122)$$

This shows that the sine function spends relatively large proportions of each period in the vicinity of $\pm A$ and a relatively small fraction of each period near 0 (Fig. 4.31). This agrees with intuition since the magnitude of the derivative of $A \sin \omega t$ is largest at values of t where the function itself is near zero; similarly, the magnitude of the derivative is small for the part of each period when the sine wave is at its peaks.

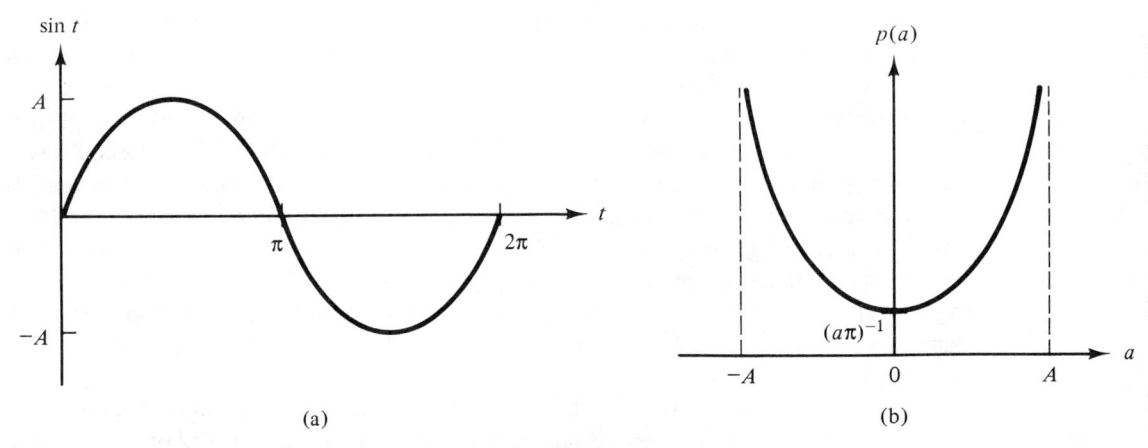

Figure 4.31 (a) One cycle of $A \sin t$; (b) Corresponding amplitude density function.

Sec. 4.5 Input-Output Analysis of Nonlinear Systems

The sine function's distribution of instantaneous amplitudes is typical of a wide class of smooth periodic waveforms. We thus think of the sine function as an idealized input in this sense; its particular characteristics permit analytical calculations that would otherwise be difficult or impossible. Thus we are led to use of a method for characterizing the important amplitude-dependent gain characteristics of nonlinear functions based on sine wave inputs, the describing function method.

The describing function method is derived from Fourier analysis. Recall that any periodic function can be expanded in a Fourier series consisting of sines and cosines at integer multiples of the fundamental frequency. If a nonlinear function N is applied to a sine function $A \sin \omega t$, then clearly $N(A \sin \omega t)$ is periodic with a fundamental frequency no lower than the input frequency ω. In its simplest form, the method of describing functions may be applied to systems whose nonlinearity may be viewed as arising from a single nonlinear function operating on a signal within an otherwise linear system. (Multiple nonlinear functions can be handled with suitable extensions of the approach.) Describing function analysis proceeds as follows: For modeling the characteristics of the system at a frequency ω, replace the nonlinear function with an amplitude-dependent linear gain, $D(A)$, which provides the same response to the input $A \sin \omega t$ at the fundamental frequency, ω. In effect, the method simply ignores the higher harmonic terms arising in the Fourier series representation of the output of the nonlinear function, using the amplitude-dependent linear gain to provide the correct response at the fundamental input frequency.

An assumption about the general form of the nonlinear function corresponding to many cases of practical significance and providing some of the justification of the method is that the nonlinear function N is *odd symmetric*, by which is meant $N(-x) = -N(x)$; this symmetry property is satisfied by linear functions and such practically important nonlinear functions as the following:

The *on/off function*:

$$N_{OO}(x) = \begin{cases} a; & x > 0 \\ -a; & x < 0 \end{cases} \tag{4.123}$$

The *dead zone function*:

$$N_{DZ}(x) = \begin{cases} 0; & |x| \leq a \\ k(x-a); & x > a \\ k(x+a); & x < -a \end{cases} \tag{4.124}$$

The *saturating amplifier*:

$$N_{SAT}(x) = \begin{cases} kx; & |x| \leq a \\ ka; & x > a \end{cases} \tag{4.125}$$

The *Coulomb friction-stiction function*:

$$N_{CS}(x) = \begin{cases} a_1 + a_2 e^{-kx}; & x > 0 \\ -a_1 - a_2 e^{kx}; & x < 0 \end{cases} \tag{4.126}$$

In these descriptions, the parameters a, a_1, a_2, and k are assumed to be positive (Fig. 4.32).

These nonlinearities are applied in a wide variety of models. For example, in considering the limitations of transistor audio amplifiers and of hydraulic amplifiers such as the ones employed in automobile power steering systems and aircraft flight control systems, saturation effects occur for large-amplitude input signals. The piecewise linear saturation function is a simple amplifier model with linear gain for small input signals and saturation effects. For systems involving an electromechanical relay, such as a thermostatically controlled resistive heating system, the need for a sufficiently large "triggering signal" to trip the relay is often modeled with the help of the piecewise linear dead-zone function, a common nonlinearity arising in mechanical systems involving friction. The discontinuity of this function at $x=0$ reflects the indeterminate force generated under conditions of zero relative velocity. Electric torque motors with belt drives (used in some robot position control systems) are modeled with stiction.

It is a straightforward exercise to compute the describing function approximations of the odd symmetric functions mentioned above. As examples, we find the following:

Linear function: $(N_L(x) = ax)$: $D_L(A) = a$.
On/off function: $D_{OO}(A) = 4a/(\pi A)$.
Dead zone function:

$$D_{DZ}(A) = \begin{cases} 0; & A < a \\ k - \dfrac{2kB}{\pi} - \dfrac{k}{\pi}\sin 2B; & A > a \end{cases} \quad (4.127)$$

where $B = \sin^{-1}(a/A)$.

Example

We will work out one example in detail to illustrate the method. For the on/off function, we write the Fourier series expansion as

$$N_{OO}(A \sin \omega t) = \sum_{i=1}^{\infty} b_i \sin i\omega t$$

(Note that $N_{OO}(A \sin \omega t)$ is a square wave of amplitude a and frequency ω.) The describing function is simply $D_{OO}(A) = b_1/A$. b_1 is computed using the integral formula for Fourier coefficients:

$$b_i = \frac{2}{T} \int_0^T N_{OO}(A \sin \omega t) \sin i\omega t \, dt$$

where $T = 2\pi/\omega$ is the fundamental period. Setting $i = 1$ to compute b_1, and using the form of N_{OO}, gives

$$b_1 = \frac{2}{T} \int_0^{T/2} a \sin \omega t \, dt - \frac{2}{T} \int_{T/2}^T a \sin \omega t \, dt = \frac{4a}{\pi}$$

Thus we find that $D_{OO}(A) = b_1/A = 4a/(\pi A)$.

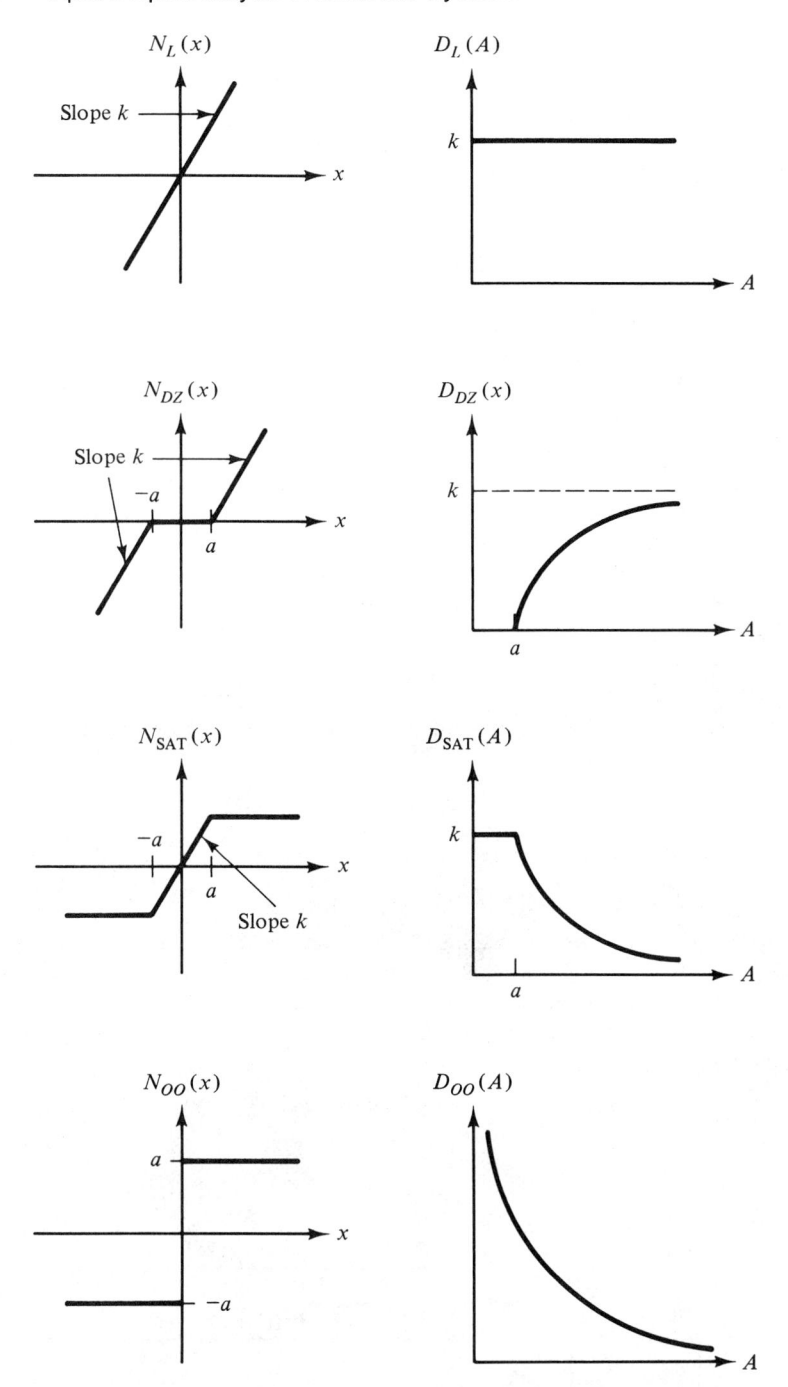

Figure 4.32 Nonlinearities and describing functions.

Notice that the two limiting values of $D_{00}(A)$, when $A \to 0$ and $A \to \infty$, can be determined from qualitative reasoning. For very, very small input amplitude A, the fixed output amplitude of a will require a very, very large gain, one tending to ∞ as A tends to 0. Similarly, for very, very large input amplitude A, the fixed output amplitude requires a very, very small gain, one that vanishes in the limit of infinite input amplitude.

The describing function method is justified by several considerations. First, in most practical cases, the largest frequency component of the output of the nonlinear function is at the fundamental frequency; that is, the Fourier coefficient of largest magnitude is typically the first one. Indeed, for odd symmetric N, there is no second harmonic term in the Fourier series expansion. Second, it may be argued that in most cases, the main qualitative features of the amplitude distribution function of the signal at the output of the nonlinearity are retained by the kind of approximation introduced in a describing function model. Finally, in applications of the method to the design of feedback control systems, it is often argued that in a well-designed system, higher-frequency terms in the output of the nonlinearity are subject to greater attenuation (damping) within the feedback loop and do not have as great an effect on the overall dynamics of the system. The lack of a second harmonic term in the Fourier series of the output of the nonlinear term also serves to improve the quality of the approximation from this standpoint.

Example

To illustrate a simple application of a describing function, consider a system described by the equation

$$y^{(3)}(t) + 6y^{(2)}(t) + 5y^{(1)}(t) = 10u(t)$$

together with the unity feedback control equation

$$u(t) = N(u_{ref}(t) - y(t))$$

(Fig. 4.33). The superscripts in the first equation denote the orders of differentiation, $u_{ref}(t)$ is a reference input, and N is a nonlinear function. For purposes of determining approximate conditions on nonlinear function N that ensure stability of this system, we use the describing function gain, $D(A)$. Thus we replace the last equation by the approximate one

$$u(t) = D(A)(u_{ref}(t) - y(t))$$

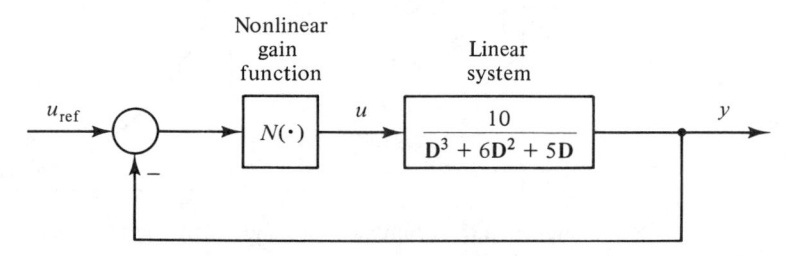

Figure 4.33 Feedback system with nonlinear gain.

Sec. 4.5 Input-Output Analysis of Nonlinear Systems

and this may be substituted in the differential equation for y to give

$$y^{(3)}(t) + 6y^{(2)}(t) + 5y^{(1)}(t) + 10D(A)y(t) = 10D(A)u_{ref}(t)$$

The conditions for stability of this linear system may be obtained by applying the Routh-Hurwitz test; the results are $D(A) > 0$ and $30 - 10D(A) > 0$. Thus we expect that the system will be unstable when $D(A) > 3$. Taking the dead-zone nonlinearity as an example, we expect that for $k < 3$ the system is always stable. For $k > 3$, the graph of $D(A)$ increases above the value 3 for amplitudes greater than a certain A_{crit}, and we expect the system to be unstable for sufficiently large reference input amplitudes. (The threshold amplitude, say A_θ, can be estimated as the one for which $u_{ref}(t) - y(t)$ has amplitude A_{crit}, using the last equation to relate $y(t)$ to $u_{ref} = A_\theta \sin \omega t$. Notice that the threshold amplitude will depend on ω since the amplitude of y is frequency dependent.)

Variations of the basic describing function technique have been developed to treat other situations. A phase shift term can be added to deal with more general nonlinearities such as ones that are not odd-symmetric and even the multivalued nonlinearities corresponding to hysteresis effects. Random input describing functions for stochastic system modeling have been studied. We will not take time to pursue such details. Rather, we want to emphasize the basic strategy of the describing function method by turning to problems that go beyond stability analysis for input-output models.

Quasilinearization

In dealing with problems such as periodic solutions and forced responses of nonlinear systems, the suggestive term *quasilinearization* is often used to describe the approach to be described next. The motivation stems from the fact that the idea of replacing a nonlinear function by a variable-gain linear function has limited applicability; it can really be applied only to system models consisting of linear subsystems and nonlinear interconnection effects. (The nonlinear feedback system of the previous example is such a system.) While this covers many systems of interest for applications, a broader view can be taken.

"Quasilinearization" is the term used in the following kind of context: A nonlinear system is to be modeled by an approximating system that displays one important property of a linear system, namely its response to a sinusoidal input is a sinusoid of the same frequency. However, the approximation need not display the most fundamental of all of the properties of linear systems, namely the superposition principle. Indeed, even homogeneity is not necessarily retained, since the output response to a sinusoid may have an amplitude that depends nonlinearly on the input amplitude.

Example

There are a number of applications of quasilinearization in the analysis of nonlinear signal processing systems. One commonly used filter is the *phase-locked loop*, or PLL for short, which is used for demodulation of FM signals and in other situations where the tracking of a time-varying phase is required. For a frequency-modulated (FM) input signal taking the form

$$u(t) = \sin(\omega_c t + \phi_u(t))$$

where the phase is related to a message waveform, $m(t)$, by

$$\frac{d}{dt}\phi_u(t) = m(t)$$

it is desired to obtain an output signal $y(t) \approx m(t)$.

The PLL consists of three parts: a loop filter, a voltage-controlled oscillator (or VCO), and a signal multiplier (Fig. 4.34). The loop filter is linear; its output is the estimated message, $y(t)$, and its input, $v(t)$, is the output of the signal multiplier. For a simple first-order filter, the input/output differential equation is

$$\dot{y}(t) = -ay(t) + bv(t)$$

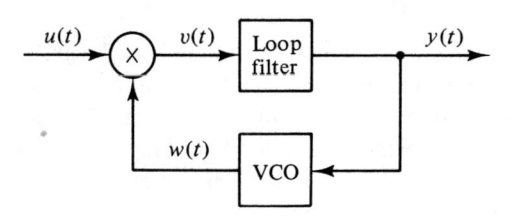

Figure 4.34 Block diagram of phase-locked loop.

The VCO generates a frequency-modulated signal, $w(t)$, with carrier frequency ω_c, using the loop filter output as its message signal:

$$w(t) = 2A\cos(\omega_c t + \phi_w(t))$$

where

$$\frac{d}{dt}\phi_w(t) = y(t)$$

Finally, the signal multiplier forms the product of $u(t)$ and $w(t)$, giving

$$v(t) = u(t)w(t)$$

Choosing $y(t)$, $\phi_w(t)$, and $\phi_u(t)$ as state variables leads to the following state equation description:

$$\dot{y}(t) = -a\,y(t) + bA\sin(\phi_u(t) - \phi_w(t)) + bA\sin(2\omega_c t + \phi_u(t) + \phi_w(t))$$

$$\dot{\phi}_w(t) = y(t)$$

$$\dot{\phi}_u(t) = m(t)$$

Notice that in the mathematical description of the system, the message signal $m(t)$ plays the role of the input.

In applications of the PLL, the carrier frequency ω_c is much greater than the highest frequency present in the message. This allows for some approximations that greatly simplify the analysis of this system. Consider the response of the system when the message is zero and the initial condition $\phi_u(0) = 0$ holds. In this case, $\phi_u(t) = 0$ for all t, and we have only the response of the remaining second-order system to consider:

$$\dot{y}(t) = -ay(t) - bA\sin\phi_w(t) + bA\sin(2\omega_c t + \phi_w(t))$$
$$= -ay(t) - bA\sin\phi_w(t)(1 - \cos 2\omega_c t) + bA\cos\phi_w(t)\sin 2\omega_c t$$

$$\dot{\phi}_w(t) = y(t)$$

Sec. 4.5 Input-Output Analysis of Nonlinear Systems

Notice that this is a periodic time-varying nonlinear system. It is natural to guess that there is a periodic solution, and this suggests trying a method known as *harmonic balancing*: determining coefficients in a truncated Fourier series expansion for the periodic solution by substituting into the differential equation and neglecting higher harmonic terms. For the problem at hand, we adopt the truncated Fourier series approximation

$$y(t) \approx A_0 + A_1 \cos 2\omega_c t + B_1 \sin 2\omega_c t$$

Integrating the differential equation for $\phi_w(t)$ using this approximation for $y(t)$ gives

$$\phi_w(t) \approx \phi_w(0) + A_0 t + \frac{A_1}{2\omega_c} \sin 2\omega_c t + \frac{B_1}{2\omega_c}(1 - \cos 2\omega_c t)$$

Thus $A_0 = 0$ is required to obtain a periodic solution for $\phi_w(t)$. The resulting expression for $\phi_w(t)$ may be substituted in the differential equation for $y(t)$ to obtain equations for the coefficients A_1 and B_1. (We will assume that using small-angle approximations for the $\sin \phi_w(t)$ and $\cos \phi_w(t)$ terms is justified; after all, if the PLL is to function correctly, $y(t) \approx 0$ and the solution obtained for $\phi_w(t)$, when $\phi_w(0) = 0$, is smaller still by a factor of $2\omega_c$.) Equating the coefficients of the $\sin 2\omega_c t$ terms, the following equation is found:

$$-2\omega_c A_1 = -aB_1 + bA - \frac{bAA_1}{2\omega_c}$$

and from the coefficients of the $\cos 2\omega_c t$ terms:

$$2\omega_c B_1 = -aA_1 - bA\phi_w(0)$$

The solution to these equations may be obtained explicitly, but the exact form is not important. It is enough to note that sizes of the expressions for A_1 and B_1 are dominated by at least one factor of ω_c in their denominators, assuming that the loop filter bandwidth, a, is much smaller than ω_c. Thus, at least for the fundamental frequency terms in their Fourier series, $y(t) \approx 0$ and $\phi_w(t) \approx 0$, as expected.

Notice that the analysis agrees with the behavior expected on intuitive grounds; the low-pass loop filter, whose bandwidth is chosen to be large enough to pass the message signal, essentially eliminates the high-frequency terms in the multiplier output. A more elaborate analysis is required to show that the high-frequency terms may also be neglected in the presence of a low bandwidth message signal such as $m(t) = \beta \cos \omega_m t$, but the qualitative result is the same.

By neglecting high-frequency terms, a simplified description of the PLL is obtained. Just two state variables are needed because only the phase difference term appears explicitly. Taking $\psi(t) = \phi_w(t) - \phi_u(t)$, the resulting state equations are

$$\dot{y}(t) = -ay(t) - bA \sin \psi(t)$$
$$\dot{\psi}(t) = y(t) - m(t)$$

This is equivalent to the second-order equation

$$\ddot{\psi}(t) + a\dot{\psi}(t) + bA \sin \psi(t) = -am(t) - \dot{m}(t)$$

which may be recognized as a damped pendulum equation with a forcing term.

One important application of quasilinearization involves predicting the existence of intrinsic periodic solutions (i.e., limit cycles) in nonlinear feedback systems.

Suppose that a unity feedback loop is closed around a nonlinear system, resulting in the model

$$\dot{\mathbf{x}}(t) = \mathbf{f}(\mathbf{x}(t), \mathbf{u}(t)) \tag{4.128}$$

$$\mathbf{y}(t) = \mathbf{h}(\mathbf{x}(t)) \tag{4.129}$$

$$\mathbf{u}(t) = -\mathbf{y}(t) \tag{4.130}$$

If a periodic solution exists and if the resulting input and output signals are well-approximated by sinusoidal functions so that quasilinearization may be applied, conditions are obtained that relate the system and the amplitude/frequency characteristics of the periodic solution. For this purpose, and for some others, a convenient way of displaying the characteristics of quasilinear systems is to use the parametrized family of frequency response functions, say $H(j\omega, A)$. Here

$$\mathbf{y}(t) \approx H(j\omega, A)\mathbf{u}(t) \tag{4.131}$$

when

$$\mathbf{u}(t) = Ae^{j\omega t} \tag{4.132}$$

This characterization can be determined analytically in simple cases, and it can also be obtained experimentally using a set of frequency response measurements.

With $H(j\omega, A)$ providing a quasilinear model for the nonlinear system in the forward branch of the feedback loop, when $\mathbf{u}(t) = A \sin \omega t$ we have the following equation arising from the interconnection constraint $\mathbf{u}(t) = -\mathbf{y}(t)$:

$$A \sin \omega t = -H(j\omega, A) A \sin \omega t \tag{4.133}$$

By considering the real and imaginary parts of this equation, the following two equations for the unknowns, ω and A, are obtained:

$$1 + \operatorname{Re} H(j\omega, A) = 0 \tag{4.134}$$

$$\operatorname{Im} H(j\omega, A) = 0 \tag{4.135}$$

The solutions to these equations provide estimates for the frequency and amplitude of periodic solutions to the equations for the feedback system.

Example

When the nonlinear system takes the form

$$\dot{\mathbf{x}}(t) = \mathbf{A}\mathbf{x}(t) + \mathbf{B} N(\mathbf{u}(t))$$

$$\mathbf{y}(t) = \mathbf{C}\mathbf{x}(t)$$

so that it is simply a linear system preceded by a memoryless nonlinear input transformation (assumed to have odd symmetry), the quasilinear system is characterized by the product of the linear system's frequency response function and the corresponding describing function,

$$H(j\omega, A) = \mathbf{C}(j\omega \mathbf{I} - \mathbf{A})^{-1} \mathbf{B} D(A)$$

Since $D(A)$ is real, this simplification means that the potential periodic solution frequencies may be obtained by solving

$$\operatorname{Im} \mathbf{C}(j\omega \mathbf{I} - \mathbf{A})^{-1} \mathbf{B} = 0$$

Nonlinear Phenomena in Forced Responses

Quasilinearization is also an important tool for the study of certain peculiarly nonlinear phenomena related to forced responses of nonlinear system. Rather than treat such problems in detail, which would require considerable sophistication of the user, we will simply present a rather qualitative overview here, using simple examples to convey the general ideas.

One well-studied nonlinear system capable of exhibiting an interesting variety of behavior is described by *Duffing's equation*,

$$\ddot{x}(t) + 2\zeta \dot{x}(t) + x(t) + k_1 x^3(t) = u(t) \tag{4.136}$$

The parameter 2ζ is a positive damping term (ζ is the damping ratio). If $k_1 = 0$, this is a damped linear oscillator (e.g., a spring-mass system with damping (or a RLC circuit), whose natural frequency is $\omega_n = 1$). For positive k_1, the spring has a nonlinear restoring force (growing much faster than proportional to displacement); the spring gets "stiffer" as k_1 gets larger.

We start with quasilinearization. If the input of the system is $A \cos \omega t$, we find that the first harmonic term in the solution of the equation is

$$x_q(t) = a \cos(\omega t - \phi) \tag{4.137}$$

where

$$\tan \phi = \frac{8\zeta}{3k_1 a^2 + 4(1 - \omega^2)} \tag{4.138}$$

$$a^2 \left(\frac{3k_1 a^2}{4} + (1 - \omega^2) \right)^2 + 4\zeta^2 \omega^2 a^2 = A^2 \tag{4.139}$$

This provides us with the necessary information to obtain the family of amplitude-parametrized frequency response functions, $H(j\omega, A)$; notice that both the magnitude and phase are amplitude dependent.

A rather unusual phenomenon is demonstrated by this example; for certain choices of the parameters ζ and k_1, and for fixed input amplitude, $|H(j\omega, A)|$ is multivalued over a range of input frequencies. This situation arises because the squared amplitude of the solution, a^2, is the solution of a cubic equation that has multiple roots for certain values of its coefficients. (To use some technical jargon, the cubic equation for a^2, where ω and k_1 are viewed as parameters, can be put in the form of the so-called *cusp catastrophe*; the surface defined by solutions of the cubic equation as a function of the two parameters has a "fold" in it.)

The physical manifestation of this multivalued frequency response characteristic is known as *jump resonance*; it may be described as follows. Take $\zeta = 0.5$. For the damped linear oscillator ($k_1 = 0$), a achieves its maximum value when the input is tuned to the natural frequency $\omega_n = 1$. However, for a sufficiently strong nonlinear spring (e.g., $k_1 = 5$), a will increase smoothly with ω beyond $\omega = \omega_n$ until a frequency ω_H (in the vicinity of 3) is reached where the amplitude drops dramatically and discontinuously. On the other hand, if a large value of ω is selected and then gradually decreased, the amplitude increases smoothly until a frequency ω_L, $\omega_L < \omega_H$ is reached where the amplitude discontinuously jumps up (Fig. 4.35). It is the interval between ω_L and ω_H where the squared amplitude function is multivalued.

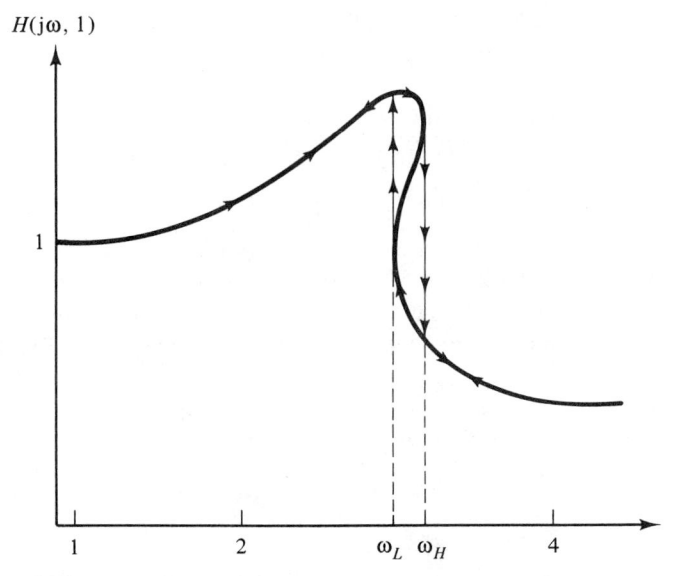

Figure 4.35 Amplitude of frequency response for Duffing's equation via quasilinearization; jump resonance effect indicated.

Another version of Duffing's equation serves as an example of a system exhibiting *subharmonic response*. This kind of behavior, which is never observed in linear systems, is simply the existence of periodic solutions of lower frequency than the frequency of system excitation. A modification of the quasilinearization function approach can be used to investigate the potential of a system for supporting subharmonic response. (Proving that subharmonic responses are actually manifested in the solution is more difficult and we shall not worry about it here.)

The particular example to be given involves an undamped system arising from a nonlinear feedback loop (Fig. 4.36). The equation takes the form

$$\ddot{x}(t) + x(t) + k_1 x^3(t) = k_1 A \sin \omega t \tag{4.140}$$

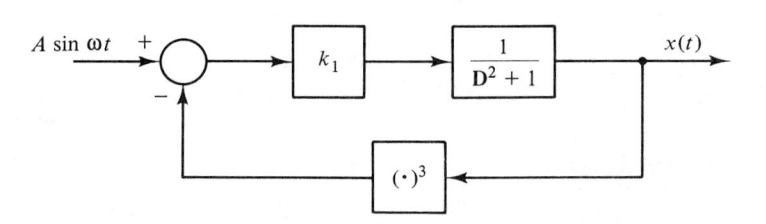

Figure 4.36 Feedback system: Duffing's equation with no damping.

Sec. 4.5 Input-Output Analysis of Nonlinear Systems

From a knowledge of trigonometric identities, we might be led to guess that $x(t)$ contains a subharmonic term at one-third frequency. As in our analysis of the PLL, we use harmonic balancing, in this case assuming that the solution can be approximated by a two-term expression

$$x(t) \approx a_0 \sin \frac{\omega t}{3} + a_1 \sin \omega t \tag{4.141}$$

(We have used some hindsight to avoid introducing phase terms, or equivalently, the cosine terms.) Substituting into the differential equation and ignoring terms involving other frequencies leads to two equations for the amplitudes in terms of the three variables A, ω, and k_1. From the coefficients of the subharmonic terms,

$$-\frac{a_0}{9}\omega^2 + a_0 + \frac{3k_1 a_0^3}{4} = 0 \tag{4.142}$$

and from the coefficients of the fundamental terms

$$-a_1 \omega^2 + a_1 + \frac{3k_1(a_0^3 + a_1^3)}{4} = k_1 A \tag{4.143}$$

For the first equation to have a solution, assuming that $k_1 > 0$, $\omega > 3$ is required. Then for the second equation to have a solution, A must not be too large. The details are left as an exercise.

In some nonlinear systems with internal oscillations arising from limit cycles, an appropriately chosen periodic input can suppress all traces of the limit cycle in the forced response. This is sometimes called "quenching" of the limit cycle. The effect can be demonstrated in the case of the van der Pol oscillator that was introduced earlier as an example in our discussion of limit cycles. Adding the input term appropriately, we obtain the equation

$$\ddot{z}(t) + \varepsilon(3z^2(t) - 1)\dot{z}(t) + z(t) = \varepsilon A \omega \cos \omega t \tag{4.144}$$

We assume that the driving frequency ω is unrelated to the limit cycle solution frequency ω_{lc}, and we follow the harmonic balancing approach used for subharmonic responses, assuming a two-sinusoid approximate solution and equating terms after substituting in the differential equation. The steps are left to be carried out as an exercise. (One hint: Use the approximate solution $a_0 \sin \omega_{lc} t + a_1 \sin(\omega t - \phi)$.) The resulting system of two equations can be analyzed to show that for sufficiently large amplitude inputs, there will be no output response term at the limit cycle frequency ω_{lc}.

Finally, we want to mention that periodic inputs to fairly simple nonlinear systems can sometimes produce very complex behavior. For example, in the harmonically forced nonlinear pendulum system

$$\ddot{\theta}(t) + d\,\dot{\theta}(t) + \sin \theta(t) = u_0 \cos t \tag{4.145}$$

there are choices of the damping coefficient d and input amplitude u_0 (e.g., $d = 0.22$ and $u_0 = 2.7$) leading to chaotic, not periodic, solutions (i.e., highly irregular and unpredictable behavior similar to the solutions of the Lorenz equations mentioned at the end of Section 4.3).

4.6 PIECEWISE LINEAR SYSTEMS

Piecewise linear functions are frequently used to approximate nonlinear functions of a single variable. They are most effectively used when a relatively few line segments provide a good global approximation to the function; the fitting of a piecewise linear approximation requires engineering judgment based on some criterion for measuring the goodness of fit. Because the endpoints of the approximating lines as well as the slopes and intercepts of the lines are subject to choice, there is considerable flexibility in choosing such an approximation. It is not to be expected that a good piecewise linear approximation corresponds to a set of local linearizations pieced together. In any event, we will not belabor the subjective issue of determining good approximations. Rather we will recall some common instances of piecewise linear models where piecewise linear functions provide simple models for certain qualitative properties.

We will restrict our attention to functions of a single variable. This will cover a broad class of applications because we retain the flexibility of choosing state variables appropriately when it comes to formulating a differential equation model for a system of interest. In addition, it turns out that quite general nonlinear functions of several variables can be expressed as sums of products of functions of one variable, so the approach may be used to obtain piecewise linear models (whose pieces are segments of hyperplanes) when functions of several variables are involved.

Simple electronic circuits involving semiconductor diodes comprise a class of very useful systems where piecewise linear models are often employed. A common diode model is represented by the current-voltage relationship $i_d = N_D(v_d)$, where v_d is the voltage across the diode terminals (using proper polarity) and i_d is the current through the diode. The function $N_D(v)$ is the piecewise linear function

$$N_D(v) = \begin{cases} 0; & v \leq v_0 \\ \dfrac{v - v_0}{R_d}; & v > v_0 \end{cases} \tag{4.146}$$

The diode cutoff voltage v_0 and the forward resistance R_d may be determined from experimental or theoretical considerations.

Several of the nonlinear functions introduced in our earlier discussion of describing functions are piecewise linear:

The on/off function:

$$N_{OO}(x) = \begin{cases} a; & x > 0 \\ -a; & x < 0 \end{cases} \tag{4.147}$$

The saturation function:

$$N_{SAT}(x) = \begin{cases} kx; & |x| \leq a \\ ka; & x > a \end{cases} \tag{4.148}$$

Sec. 4.6 Piecewise Linear Systems

The dead-zone function:

$$N_{DZ}(x) = \begin{cases} 0; & |x| \leq a \\ k(x-a); & x > a \\ k(x+a); & x < -a \end{cases} \quad (4.149)$$

A main advantage of piecewise linear systems from an analytical viewpoint is that the solution of piecewise linear differential equations can be obtained by joining together segments of solutions to linear differential equations. For time-invariant systems, this means that analytical solutions can be exploited, at least to some degree. As an example of how such an analysis proceeds, we investigate the (artificial) piecewise linear system described by the equations

$$\dot{x}_1(t) = x_2(t) \quad (4.150)$$

$$\dot{x}_2(t) = -g(x_2) - x_1(t) \quad (4.151)$$

where the function $g(x)$ is given by

$$g(x) = \begin{cases} x-2; & x > 1 \\ -x; & |x| \leq 1 \\ x+2; & x < -1 \end{cases} \quad (4.152)$$

This is a two-dimensional system, so a phase plane analysis is a convenient way to study its behavior. Since this is a linear oscillator with nonlinear damping, some intuitive observations can be made. The unstable equilibrium at the origin corresponds to the negative damping coefficient for small velocities. For large velocities, the damping is negative but there is a constant bias term acting as a system input. The effect is a stabilizing one, and thus we are led to expect the existence of a limit cycle. We will exploit the piecewise linear form of the system to construct the limit cycle (semi-explicitly).

An odd-symmetry condition makes the analysis a bit easier: Solutions of the system for $x_2 > 0$ are reflected through the origin of the phase plane to obtain solutions for $x_2 < 0$. Thus we need only find half a period of the presumed limit cycle. We will start at the point $(q, 1)$, where $q < 0$, and see if we can determine q so that the point lies on the limit cycle. A look at the equations tells us that we expect the solutions passing through $(q, 1)$ to move up and to the right in the phase plane (provided that we choose q sufficiently negative), to reach a "peak" where the slope of the graph of x_2 as a function of x_1 is zero, and then to move down and to the right until it again intersects the $x_2 = 1$ line. From the linear equation describing the system for $x_2 \geq 1$ (assuming that we start at $t = 0$) we obtain

$$x_1(t) = q e^{-t/2} \cos \frac{\sqrt{3}\, t}{2} + a(q) e^{-t/2} \sin \frac{\sqrt{3}\, t}{2} + 2t \quad (4.153)$$

from which $x_2(t)$ can be found by differentiation. The coefficient $a(q)$ is determined by the condition $x_2(0) = 1$; we have explicitly denoted the q dependence of this coefficient for emphasis.

Using the form of the solution, we determine the first positive time instant $t_1(q)$ when $x_2(t_1(q)) = 1$. (This must be done numerically for any trial choice of q, using

Newton's method, for example.) Then $x_1(t_1(q))$ can be found from the solution above. For the next segment of the solution, the differential equations for the region $|x_2| < 1$ must be used. The solution is again obtained analytically:

$$x_1(t) = b(q) e^{t/2} \cos \frac{\sqrt{3}\, t}{2} + c(q) e^{t/2} \sin \frac{\sqrt{3}\, t}{2} \qquad (4.154)$$

where the coefficients are determined by the boundary conditions on x_1 and x_2 at $t_1(q)$. Finally, the first time instant $t_2(q) > t_1(q)$, where $x_2(t_2(q)) = -1$, must be found, again using a method such as Newton's method. Then we must evaluate $x_1(t_2(q))$, and the condition for the constructed solution to be half of a limit cycle is

$$q = -x_1(t_2(q)) \qquad (4.155)$$

(Fig. 4.37). This nonlinear equation for q needs to be solved. Notice that the method of successive approximation is well suited to this task, even though the function $x_1(t_2(q))$ is not available in analytical form. We have just described how to evaluate it for any given q. This was made possible because of the piecewise linear form of the system.

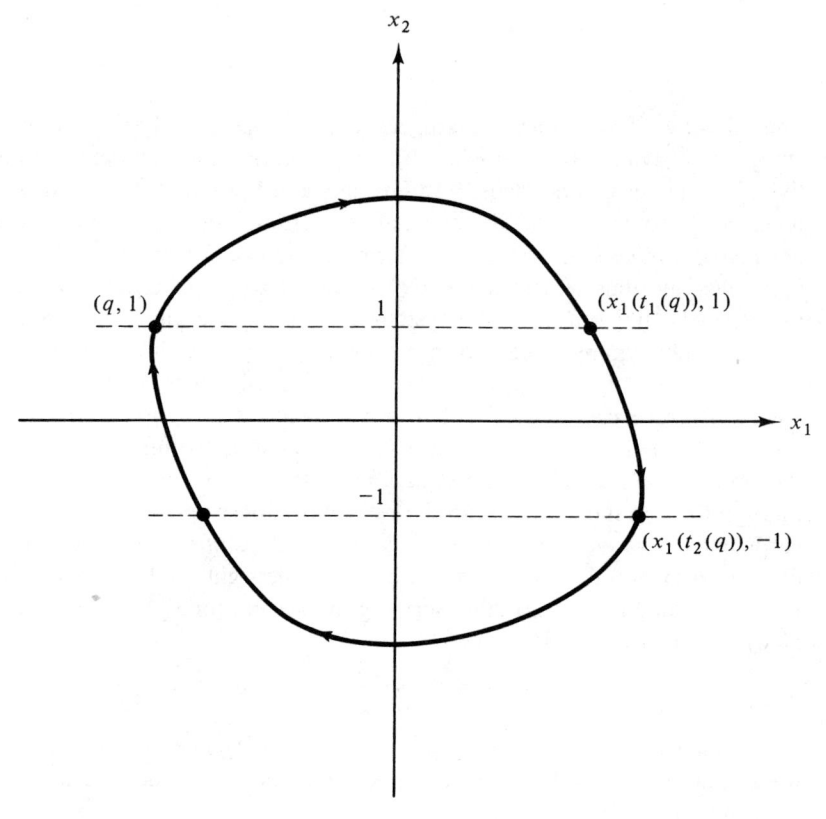

Figure 4.37 Phase plane plot of limit cycle for the piecewise linear system example.

The applications of piecewise linear systems often involve discontinuous models due to effects like stiction or because an on/off switching function is part of the system. Switched systems can provide stability without introducing damping (unwanted dissipation of energy as heat, for example) and are used in dc-ac conversion circuits for solar electric power applications. Switched systems also arise as solutions to optimal control problems with control magnitude bounds. An example is the minimum fuel orbit transfer used to place satellites in geosynchronous orbit from low Earth orbit.

More general kinds of piecewise linear models have found applications to analysis of hysteresis effects (another typical characteristic of systems involving relays). Discussions of this and related topics are left for the references.

4.7 NOTES AND REFERENCES

Nonlinear systems is a vast topic, and these few brief comments and short bibliography are by no means intended to be comprehensive. The books and articles listed are ones that were found to be particularly useful during the writing of this chapter. The book by Cook [4] was particularly helpful and contains more details about many of the topics covered in Sections 4.2, 4.3, 4.5, and 4.6. McClamroch's book [10] uses case studies to introduce many of the same topics; Beltrami's book [2] is aimed at a somewhat more mathematical reader. The basic mathematical material is found in the coverage of differential equations in Hirsch and Smale [6]. For a mathematical treatment with a particular emphasis on control, see Sontag's book [14].

Solution of nonlinear equations is of fundamental importance for steady-state nonlinear circuit analysis; some of the material from Section 4.1 appears in Chua and Lin [3] and in Mastascusa [9]. Use of Newton's method and successive approximation for dynamic circuit analysis is discussed in the book by White and Sangiovanni-Vincentelli [16].

The remaining references reflect some of the breadth of the nonlinear systems area. Hopfield's papers [7] [8] describe continuous-time and discrete-time neural network models and Hebbian learning; the paper by Personnaz, Guyon, and Dreyfus [11] describes the pseudo-inverse learning rule. Rugh's book [12] covers the many aspects of Volterra series models. A discussion of describing-function methods for control system design is given in the paper by Taylor [15]. The chaotic behavior of the forced, damped pendulum is discussed in the survey paper by Grebogi, Ott, and Yorke [5]. Connections with catastrophe theory are described by Saunders [13]. Control of the stick balancer using linearization approaches is the subject of the paper by Baumann and Rugh [1].

BIBLIOGRAPHY

[1] W.T. Baumann and W.J. Rugh, "Feedback Control of Nonlinear Systems by Extended Linearization," *IEEE Trans. Automat. Control*, **31** (1986), 40–46.

[2] E. Beltrami, *Mathematics for Dynamic Modeling*, Academic Press, Orlando, FL, 1987.

[3] L.O. Chua and P.-M. Lin, *Computer-Aided Analysis of Electronic Circuits: Algorithms and Computational Techniques*, Prentice-Hall, Englewood Cliffs, NJ, 1975.

[4] P.A. Cook, *Nonlinear Dynamical Systems*, Prentice-Hall International, London, 1986.

[5] C. Grebogi, E. Ott, and J.A. Yorke, "Chaos, Strange Attractors, and Fractal Basin Boundaries in Nonlinear Dynamics," *Science*, **238** (1987), 632–638.

[6] M.W. Hirsch and S. Smale, *Differential Equations, Dynamical Systems, and Linear Algebra*, Academic Press, New York, 1974.

[7] J.J. Hopfield, "Neural Networks and Physical Systems with Emergent Collective Computational Abilities," *Proc. Natl. Acad. Sci. U.S.A.*, **79** (1982), 2554–2558.

[8] J.J. Hopfield, "Neurons with Graded Response Have Collective Computational Properties Like Those of Two-State Neurons," *Proc. Natl. Acad. Sci. U.S.A.*, **81** (1984), 3088–3092.

[9] E.J. Mastascusa, *Computer-Assisted Network and System Analysis*, John Wiley & Sons, New York, 1988.

[10] N.H. McClamroch, *State Models of Dynamic Systems*, Springer-Verlag, New York, 1980.

[11] L. Personnaz, I. Guyon, and G. Dreyfus, "Collective Computational Properties of Neural Networks: New Learning Mechanism," *Phys. Rev. A*, **34** (1986), 4217–4228.

[12] W.J. Rugh, *Nonlinear System Theory*, The Johns Hopkins University Press, Baltimore, MD, 1981.

[13] P.T. Saunders, *An Introduction to Catastrophe Theory*, Cambridge University Press, New York, 1980.

[14] E.D. Sontag, *Mathematical Control Theory: Deterministic Finite Dimensional Systems*, Springer-Verlag, New York, 1990.

[15] J.H. Taylor, "A Systematic Nonlinear Controller Design Approach Based on Quasilinear System Models," *Proceedings of the 1983 American Control Conference*, San Francisco, pp. 141–145.

[16] J.K. White and A. Sangiovanni-Vincentelli, *Relaxation Techniques for the Simulation of VLSI Circuits*, Kluwer Academic Publishers, Boston, 1987.

PROBLEMS

1. Write down state equations for the stick balancer system described in the introductory section of this chapter. Use the state variables suggested: Take x_1 to be the horizontal displacement of the cart from some reference position, x_2 to be the cart velocity, $x_3 = \theta$, the angular displacement of the stick away from its desired vertical orientation, and $x_4 = \dot{\theta}$, the angular velocity of the stick.

2. Carry out the linearization of the stick balancer system around the equilibrium solution and around the "constant acceleration" solution described in the example in Section 4.1.

3. Linearize the nonlinear system

$$\dot{x}_1(t) = -x_1(t) + (x_1(t) + x_2(t) - 1)^2$$
$$\dot{x}_2(t) = x_1(t) - 2x_2(t) + 2 + x_1^3(t)$$

about the equilibrium point $x_1 = 0$, $x_2 = 1$. Is this an asymptotically stable equilibrium point? Why?

Problems

4. Suppose that you are told that a nonlinear system described by the equations

$$\dot{x}_1(t) = f_1(x_1(t), x_2(t))$$
$$\dot{x}_2(t) = f_2(x_1(t), x_2(t))$$

has a solution $x_1(t) = \phi_1(t)$, $x_2(t) = \phi_2(t)$, where $\phi_1(t)$ and $\phi_2(t)$ are specified time functions.

(a) Suppose that the functions $\phi_1(t)$ and $\phi_2(t)$ are periodic with common period. Describe how you would go about checking whether the given solution is a limit cycle for the system.

(b) Suppose instead that $\lim_{t \to \infty} \phi_i(t) = 1$ for $i = 1, 2$. Give conditions on the system equations for which this limit point is an asymptotically stable equilibrium solution of the system.

(c) Give an explicit example of a system (i.e., give nonlinear functions f_1 and f_2) that has an equilibrium solution $x_1(t) = 1$, $x_2(t) = 1$ which is a *saddle point*.

5. The time constant of a first-order linear system, say τ, can be computed by examining the system's unit step response

$$y(t) = c(1 - e^{-t/\tau}), \qquad t \geq 0$$

(a) Suppose that c is known. Show how Newton's method may be used to determine τ from a measurement of the time t at which $y(t)$ takes the value $c/2$.

(b) Suppose that c is unknown. Show how Newton's method may be used to determine τ and c from measurements of two times, say t_1 and t_2 for which $y(t_1) = y(t_2)/2$. You may assume that c is positive and that a nonzero lower bound for c is known.

6. Consider the nonlinear system

$$\dot{x}_1(t) = -x_1(t) + (x_1(t) + x_2(t))^2$$
$$\dot{x}_2(t) = x_1(t) - 2x_2(t) + (x_1(t) + x_2(t))^2$$

Construct an approximate backward difference discretization method and apply it to obtain a phase plane portrait for the system in the vicinity of the equilibrium point $x_1 = 0$, $x_2 = 0$.

7. Compare the performance of Newton's method and the successive approximation method for finding the point of intersection of the graphs of the two equations $y = x - 1$ and $y = \tanh x$. (A programmable calculator will suffice to carry out the computations.)

8. Use the method of successive approximation to obtain a sequence of approximate solutions of the differential equation

$$\dot{\mathbf{x}}(t) = \mathbf{A}\mathbf{x}(t), \qquad \mathbf{x}(0) = \mathbf{x}_0$$

on the interval $0 \leq t \leq 1$. Use $\mathbf{x}^{[1]}(t) = \mathbf{x}_0 + \mathbf{A}^2 \mathbf{x}_0 t^2 / 2$ as the first approximate solution and determine the second and third approximate solutions generated by the method. Repeat for $\mathbf{x}^{[1]}(t) = \mathbf{x}_0 e^{-t}$.

9. For the system

$$\dot{x}_1(t) = x_2^2(t)$$
$$\dot{x}_2(t) = \frac{1}{3}$$

find an analytical expression for the phase plane trajectories and sketch the phase plane portrait.

10. Show that the nonlinear system

$$\dot{x}_1(t) = x_2(t) + x_1(t)x_2^2(t)$$
$$\dot{x}_2(t) = -x_1(t) + x_1^2(t)x_2(t)$$

has no limit cycles.

11. Complete the analysis, with the help of parameter values chosen on the basis of numerical calculation and experiments if necessary, for the predator-prey population models described in Section 4.3. Show that the self-limiting prey model leads to an asymptotically stable equilibrium for suitable parameter choices. Show that a limit cycle can arise when delay in the predator equation is modeled as described.

12. For the linear oscillator, $\ddot{x}(t) + \omega^2 x(t) = 0$, solutions evolve on an ellipse in the phase plane. When $\omega = 1$ the ellipse is a circle. Three-dimensional systems which evolve on a sphere in 3-space (or some other ellipsoid) are important in applications involving purely angular motion, such as the orientation of a satellite in orbit. Each point on the sphere in 3-space can be described by two angular coordinates (usually called θ and ϕ in discussions of spherical coordinates). Find the conditions on the 3×3 matrix $\mathbf{A}$ which guarantee that the trajectories of the system

$$\dot{\mathbf{x}}(t) = \mathbf{A}\mathbf{x}(t)$$

satisfy $\|\mathbf{x}(t)\|^2 = C$, where C is a constant. Show that when the trapezoidal method of discretization is applied to such a system, the discretized system still evolves (in discrete steps) on the same sphere.

13. For the Lorenz model, equations (4.74–4.76), use simulation to show that in the case when $\beta = 10$, the basin of attraction of the equilibrium has shrunk so much that initial condition vectors whose components agree with the equilibrium values to two decimal places still tend to the chaotic attractor.

14. In the Lorenz model, modifying x_2 to δx_2 and then choosing $\delta = 0$ gives a system having a limit cycle instead of a chaotic attractor. Investigate the system for various values of δ to explore the transition between periodic solutions and chaotic ones.

15. The Rössler system has been used as a model for a variety of systems which seem to demonstrate evolution from well-behaved limit cycle solutions to chaotic solutions as a parameter value is increased (or decreased). Examples are: a dripping water faucet as the flow is increased; onset of cardiac fibrillation as levels of chemical neuromuscular transmitters vary; onset of rhythmic EEG as the level of cocaine intoxication increases. The system is described by the equations

$$\dot{x}_1 = x_1 - x_1 x_2 - x_3$$
$$\dot{x}_2 = x_1^2 - a x_2$$
$$\dot{x}_3 = b x_1 - c x_3$$

Use linearization to investigate stability of the equilibria as a function of the parameters a, b, and c. Guided by the results, investigate possible chaotic solutions experimentally.

16. For the nonlinear function

$$N(x) = \begin{cases} x + 1 & \text{for} \quad x > 0 \\ x - 1 & \text{for} \quad x < 0 \end{cases}$$

find the describing function gain for sinusoidal input $A \sin \omega t$.

17. Let $N_1(x)$ and $N_2(x)$ be two nonlinear functions having odd symmetry (i.e., $N_1(-x) = N_1(x)$ and $N_2(-x) = -N_2(x)$). Let $D_1(A)$ and $D_2(A)$ be the corresponding describing function gains for a sinusoidal input $A \sin \omega t$. Determine the describing function gain for the nonlinear function

$$N(x) = 5x + 2N_1(x) - 3N_2(x)$$

18. Let $N(x)$ be a nonlinear function having odd symmetry (i.e., $N(-x) = -N(x)$). Let $D_N(A)$ be the corresponding describing function gain for a sinusoidal input $A \sin \omega t$. Show that the describing function gain for the nonlinearity $2N(x)$ is equal to $2D_N(A)$.

19. Give specific examples of odd-symmetric nonlinear functions whose describing function gains have the following limiting behavior:

(a) $\lim_{A \to \infty} D_N(A) = 1$

(b) $\lim_{A \to \infty} D_N(A) = \infty$

(c) $\lim_{A \to \infty} D_N(A) = 0$ and $\lim_{A \to 0} D_N(A) = 1$

20. Use SPICE, or design a simulation using other available software, to investigate the AM demodulator circuit described in the examples at the beginning of this chapter. (Use realistic circuit element values corresponding to a carrier frequency of 1.21 MHz and bandwidth of 10 kHz.) Determine the mean-squared error from ideal demodulation when the message is a sum of two given sine waves.

21. Experimentally verify the jump resonance phenomenon for the Duffing equation model, equation (4.136) in Section 4.5, by carrying out simulation experiments.

22. Carry out the analysis and give an experimental verification of limit cycle quenching in the van der Pol oscillator; see equation (4.144) in Section 4.5 for the model. Repeat for the piecewise linear "approximate van der Pol" system analyzed in the Section 4.6, equations (4.150–4.152).

23. Use SPICE, or design a simulation using other available software, to investigate the two-transistor clock waveform generator described in the examples at the beginning of this chapter. (Use realistic circuit element values corresponding to a frequency of about 2 MHz.) Analyze the system and show (or give supporting evidence) that the system has a limit cycle.

CHAPTER 5

Optimization

The process of engineering design is (implicitly or explicitly) based on optimization. System performance, whether described in terms of some kind of "figure of merit" such as "rapid step response with no overshoot" or in more a more mathematical form such as "minimize the time required to carry out a particular task," is a criterion to be optimized by choice of certain parameters or controls that may be chosen by the system designer. It is generally an art, rather than a science, for a designer to translate all of the desired system performance specifications and constraints into a mathematically tractable form to which optimization techniques may be applied. Thus we will have little to say about this important issue. However, it should be clear that computer-aided design methods greatly extend the capability of a designer to incorporate considerable trial-and-error experimentation in all stages of the design process, beginning with problem formulation.

The kinds of optimization methods that are available to the designer include the following general categories: unconstrained parameter optimization, constrained parameter optimization, and path optimization (optimal control). The first two methods share the distinction of applying to problems where there are a finite number of parameters (degrees of freedom) that must be chosen to minimize a "cost function" or to maximize a "performance measure." Since a maximization problem can be turned into an equivalent minimization problem by changing the sign of the quantity to be maximized, we will always assume that our optimization problems are posed in the form of minimization problems.

5.1 PARAMETER OPTIMIZATION

Unconstrained Parameter Optimization

The most basic type of parameter optimization problem involves minimizing a real-valued "cost function" by the choice of some free variables. The mathematical tools that are used to solve unconstrained parameter optimization problems come directly from multivariable calculus. To minimize a cost function $C(\mathbf{u})$, where $\mathbf{u} = [u_1, \ldots, u_m]^T$, some familiar necessary conditions are obtained by setting the derivative of C with respect to $\mathbf{u}$ equal to $\mathbf{0}$. This derivative takes the form of a *row* vector of partial derivatives, familiar as the gradient of the scalar function C with respect to its arguments; necessary conditions for the value $\mathbf{u}_0$ to be a local minimum of C are

$$C'(\mathbf{u}_0) = \nabla C(\mathbf{u}_0) = \left[\frac{\partial C}{\partial u_1} \quad \frac{\partial C}{\partial u_2} \quad \cdots \quad \frac{\partial C}{\partial u_m} \right] = \mathbf{0} \tag{5.1}$$

where all of the partial derivatives with respect to the components of $\mathbf{u}$ are evaluated at $\mathbf{u}_0$. This gives m equations for the m unknown components of $\mathbf{u}_0$. Generally, the equations are nonlinear, and neither existence nor uniqueness of solutions to these equations holds in general. For most cases encountered in applications, at least existence can be argued from the form of the cost function C. When the derivative equations are nonlinear, a numerical technique such as Newton's method is usually required to find a solution.

Of course, these necessary conditions are far from sufficient. Indeed, these are the same conditions satisfied by all *critical points*: local minima, local maxima, and saddle points. To gain some insight into these conditions, we will generalize the discussion given earlier about the connections between derivatives and linear approximation of functions to consider quadratic approximation. The intuition behind this analysis is that in the vicinity of a local minimum of a "sufficiently smooth" function, the "bowl-like" character of the function is well approximated by a quadratic function.

We will adopt the following convenient notational convention for partial derivatives of a real-valued function: when a variable is used as a subscript, it denotes the corresponding partial derivative. Thus

$$C_{u_i} = \frac{\partial C}{\partial u_i} \quad \text{and} \quad C_{u_i u_j} = \frac{\partial^2 C}{\partial u_i \, \partial u_j} \tag{5.2}$$

An argument is included when it is desired to indicate the point at which the partial derivatives are evaluated (e.g., $C_{u_i}(\mathbf{u}_0)$). The convention is naturally extended to provide a compact notation for the derivative or gradient by using an m-vector-valued variable as a subscript:

$$\nabla C(\mathbf{u}_0) = C_\mathbf{u}(\mathbf{u}_0) = [C_{u_1}(\mathbf{u}_0) \; C_{u_2}(\mathbf{u}_0) \; \cdots \; C_{u_m}(\mathbf{u}_0)] \tag{5.3}$$

A similar notation is used to denote the $m \times m$ matrix of second partial derivatives:

$$C_{\mathbf{uu}}(\mathbf{u}_0) = \begin{bmatrix} C_{u_1 u_1} & \cdots & C_{u_1 u_m} \\ \vdots & & \vdots \\ C_{u_m u_1} & \cdots & C_{u_m u_m} \end{bmatrix} \qquad (5.4)$$

The matrix of second partial derivatives, $C_{\mathbf{uu}}(\mathbf{u}_0)$ is sometimes called the *Hessian matrix* of the function $C(\mathbf{u})$ evaluated at $\mathbf{u}_0$.

The mathematical formulation that will be used to analyze quadratic approximation is the *second-order Taylor series expansion (with remainder)*. For a real-valued function $C(\mathbf{u})$, this is the expression

$$C(\mathbf{u}) = C(\mathbf{u}_0) + C_{\mathbf{u}}(\mathbf{u}_0)(\mathbf{u} - \mathbf{u}_0) + \tfrac{1}{2}(\mathbf{u} - \mathbf{u}_0)^T C_{\mathbf{uu}}(\mathbf{u}_0)(\mathbf{u} - \mathbf{u}_0) + E(\mathbf{u} - \mathbf{u}_0) \qquad (5.5)$$

The remainder term in the second-order Taylor series, E, is a quantity that goes to zero "faster than quadratically," assuming that C is twice continuously differentiable:

$$\lim_{\|\mathbf{u} - \mathbf{u}_0\| \to 0} \frac{E(\mathbf{u} - \mathbf{u}_0)}{\|\mathbf{u} - \mathbf{u}_0\|^2} = 0 \qquad (5.6)$$

We will use the second-order Taylor series expansion to arrive at a complete characterization of the local minima of the function C.

First we show that at any point $\mathbf{u}_0$, the derivative, or gradient, $C_{\mathbf{u}}(\mathbf{u}_0)$, is a row vector with an important geometric significance: its transpose, the column vector $(C_{\mathbf{u}}(\mathbf{u}_0))^T$, points in the direction of local maximum increase of the function C and the direction of local maximum decrease is specified by $-(C_{\mathbf{u}}(\mathbf{u}_0))^T$ (Fig. 5.1). To verify these facts, we start with the defining property of the derivative:

$$\lim_{\|\mathbf{u} - \mathbf{u}_0\| \to 0} \frac{\|C(\mathbf{u}) - C(\mathbf{u}_0) - C_{\mathbf{u}}(\mathbf{u}_0)(\mathbf{u} - \mathbf{u}_0)\|}{\|\mathbf{u} - \mathbf{u}_0\|} = 0 \qquad (5.7)$$

We will examine this limit for the particular case when the the direction of $(\mathbf{u} - \mathbf{u}_0)$ remains fixed while its norm tends to zero. Take any point $\mathbf{u} \neq \mathbf{u}_0$ and define r by $\|\mathbf{u} - \mathbf{u}_0\| = r$. Let $\|C_{\mathbf{u}}(\mathbf{u}_0)\| = R$. Then the term $C_{\mathbf{u}}(\mathbf{u}_0)(\mathbf{u} - \mathbf{u}_0)$ is simply the inner product of the two vectors $(C_{\mathbf{u}}(\mathbf{u}_0))^T$ and $(\mathbf{u} - \mathbf{u}_0)$, so we have

$$C_{\mathbf{u}}(\mathbf{u}_0)(\mathbf{u} - \mathbf{u}_0) = R\, r \cos\theta \qquad (5.8)$$

where θ is the angle between the two vectors. Letting $\mathbf{u}$ tend toward $\mathbf{u}_0$ while the direction of $(\mathbf{u} - \mathbf{u}_0)$ remains fixed, we obtain the following simplification:

$$\lim_{r \to 0} \left| \frac{C(\mathbf{u}) - C(\mathbf{u}_0)}{r} - R \cos\theta \right| = 0 \qquad (5.9)$$

In the limit, the difference quotient $(C(\mathbf{u}) - C(\mathbf{u}_0))/r$, which is the rate of increase of C at $\mathbf{u}_0$ in the direction $(\mathbf{u} - \mathbf{u}_0)$, is thus maximized when $\theta = 0$ and minimized when $\theta = \pi$. In other words, the gradient direction at $\mathbf{u}_0$ points in the direction of greatest local increase and opposite to the direction of greatest local decrease.

The familiar necessary conditions for C to have a local minimum at $\mathbf{u}_0$ follow directly from this analysis. Unless the derivative is the $\mathbf{0}$ vector, the value of $C(\mathbf{u})$ will

Sec. 5.1 Parameter Optimization

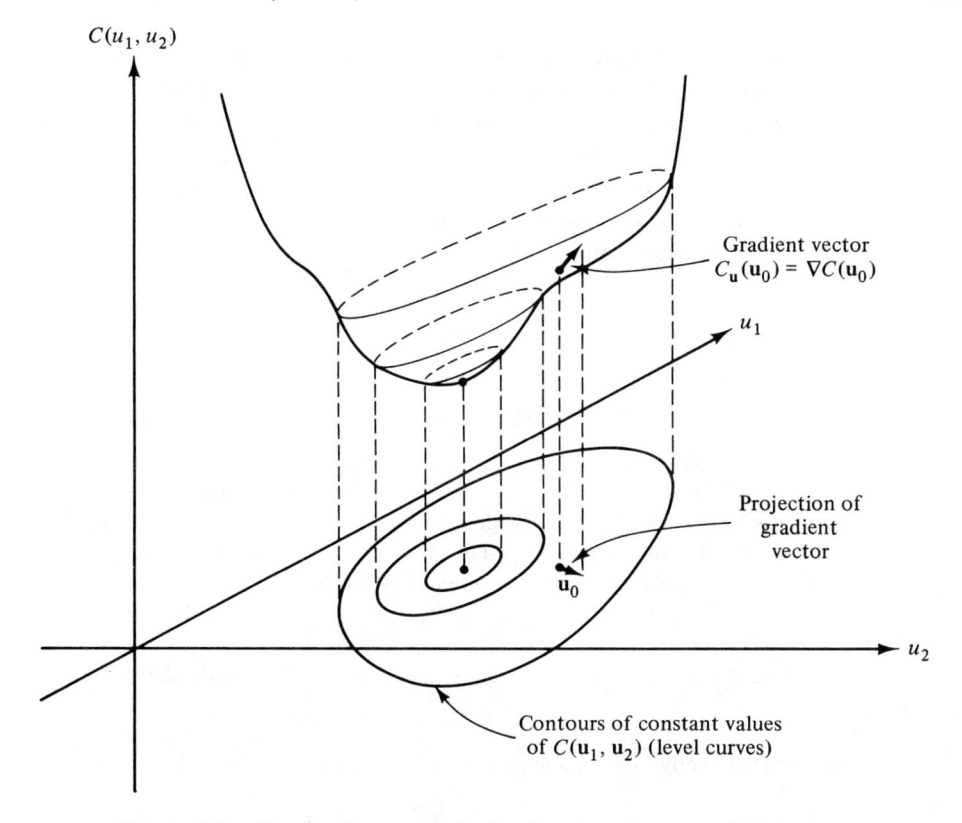

Figure 5.1 The gradient points in the direction of most rapid increase.

be less than the value of $C(\mathbf{u}_0)$ for points sufficiently close to $\mathbf{u}_0$ in the negative gradient direction. The second-order Taylor series expansion can also be used to draw this conclusion in a slightly different way. For small enough $\|\mathbf{u} - \mathbf{u}_0\|$, the linear term will be much larger than the quadratic and higher-order terms. If $C_\mathbf{u}(\mathbf{u}_0)$ were to have some nonzero component, varying only the corresponding component of $\mathbf{u}_0$ will enable both increases and decreases in the value of C to be obtained (depending on the sign of the variation), so $\mathbf{u}_0$ could not be a local minimum.

Once it is realized that it is necessary for the derivative of C to vanish at local extrema, analysis of the second-order terms in the Taylor series approximation leads to higher-order necessary conditions. In particular, it is clearly necessary that the second-order term be nonnegative. This is the condition that $C_{\mathbf{uu}}(\mathbf{u}_0)$ is a *nonnegative definite* matrix. (Notice that the matrix is symmetric, so it has real eigenvalues. It will be nonnegative definite if and only if all of its eigenvalues are nonnegative.)

Sufficient conditions for $\mathbf{u}_0$ to be a local minimum may also be obtained from a consideration of the Taylor series. If the second-order term is always positive for a point $\mathbf{u}_0$ where $C_\mathbf{u}(\mathbf{u}_0)$ is zero, then that point is a local minimum. This positivity condition is simply the condition that $C_{\mathbf{uu}}(\mathbf{u}_0)$ is a *positive definite* matrix. This is equivalent to the condition that all eigenvalues of $C_{\mathbf{uu}}(\mathbf{u}_0)$ are positive.

In summary, the unconstrained minimum of a function is found by setting its partial derivatives (with respect to the parameters that may be varied) equal to zero and solving for the parameter values. Among the sets of parameter values obtained, those at which the matrix of second partial derivatives of the cost function is positive definite are local minima. If there is a single local minimum, it is also the global minimum; otherwise, the cost function must be evaluated at each of the local minima to determine which one is the global minimum.

Examples

Quadratic functions can be thoroughly analyzed as described above. Starting with a very simple case, notice that

$$Q(u_1, u_2) = \tfrac{1}{2} q_1 u_1^2 + \tfrac{1}{2} q_2 u_2^2 + m_1 u_1 + m_2 u_2 + q_0$$

is already expressed as a second-order Taylor series. The derivative vanishes at the point $u_1 = -m_1/q_1$, $u_2 = -m_2/q_2$. Since the matrix of second partial derivatives is diagonal, it is easily checked for positive definiteness. This holds if and only if $q_1 > 0$ and $q_2 > 0$. These conditions are necessary and sufficient for a global minimum.

For a general quadratic function of n variables,

$$Q(\mathbf{u}) = \tfrac{1}{2} \mathbf{u}^T \mathbf{Q} \mathbf{u} + \mathbf{m}^T \mathbf{u} + q_0$$

where $\mathbf{Q}$ is a symmetric, $n \times n$ matrix and $\mathbf{m}$ is an n-vector, the derivative is

$$Q_\mathbf{u}(\mathbf{u}_0) = \mathbf{u}_0^T \mathbf{Q} + \mathbf{m}^T$$

so the derivative vanishes at points $\mathbf{u}_0$ that are solutions to the linear equations

$$\mathbf{Q} \mathbf{u}_0 = -\mathbf{m}$$

The matrix of second partial derivatives is $Q_{\mathbf{uu}} = \mathbf{Q}$. Thus, the necessary and sufficient condition for there to be a unique minimizing $\mathbf{u}_0$ is that $\mathbf{Q}$ be a positive definite matrix.

The simplicity of minimization problems for quadratic functions is not merely an academic nicety. A wide range of important applications lead to a problem of this form; for example, linear least squares approximation problems are included in this class. For special classes of quadratic minimization problems, where the matrix $\mathbf{Q}$ is sparse or has a structure that may be exploited in solving the linear equations that determine the minimizing $\mathbf{u}_0$, very large numbers of parameters may be handled efficiently.

Example

Linear prediction filters are applied to a large number of signal processing problems, and the linear prediction filter design problem is usually posed as a quadratic minimization problem. We will describe a simple case. Suppose $\{s_k : 0 \leq k \leq N\}$ is a set of consecutive samples of a discrete signal (e.g., a sampled speech waveform). A linear prediction filter takes a weighted sum of p consecutive samples to produce an output that is intended to approximate the next sample; denoting the filter output by y_k, the filter equation is

$$y_k = u_1 s_{k-1} + u_2 s_{k-2} + \cdots + u_p s_{k-p}$$

where the coefficients, $u_1, \ldots, u_p$, are to be determined. For speech processing applications, N may be 250 or more while p might be 10 or 20. Ideally, $y_k = s_k$, but if equality is

Sec. 5.1   Parameter Optimization

to hold for all the available samples (i.e., for $p \leq k \leq N$ and $N - p > p$), the equations will typically be inconsistent.

A least-squares approximate solution to the equations may be used to determine the filter coefficients. The approximate solution is the one that minimizes the sum of squared errors,

$$Q(\mathbf{u}) = \sum_{k=p}^{N} (s_k - y_k)^2$$

$$= \sum_{k=p}^{N} (s_k - \sum_{l=1}^{p} u_l s_{k-l})^2$$

This is a quadratic expression for $\mathbf{u}$ and setting its gradient to zero leads to a set of linear equations to be solved as the design procedure for the linear prediction filter. The components of the gradient are the partial derivatives

$$Q_{u_i}(\mathbf{u}) = \frac{\partial Q(\mathbf{u})}{\partial u_i} = 2 \sum_{k=p}^{N} (s_k - \sum_{l=1}^{p} u_l s_{k-l}) s_{k-i}, \quad 1 \leq i \leq p$$

and setting these expressions to zero gives, after some rearrangement of the sums,

$$\sum_{l=1}^{p} \left(\sum_{k=p}^{N} s_{k-i} s_{k-l} \right) u_l = \sum_{k=p}^{N} s_{k-i} s_k, \quad 1 \leq i \leq p$$

These are p linear equations for the p components of the parameter vector $\mathbf{u}$; the matrix $\mathbf{Q}$ for this quadratic minimization problem is given elementwise as

$$(\mathbf{Q})_{il} = \sum_{k=p}^{N} s_{k-i} s_{k-l}$$

and $\mathbf{Q}$ may be verified to be the Gram matrix of the set of p vectors

$$\begin{bmatrix} s_{p-1} \\ s_p \\ \cdot \\ \cdot \\ \cdot \\ s_{N-1} \end{bmatrix} \begin{bmatrix} s_{p-2} \\ s_{p-1} \\ \cdot \\ \cdot \\ \cdot \\ s_{N-2} \end{bmatrix} \cdots \begin{bmatrix} s_0 \\ s_1 \\ \cdot \\ \cdot \\ \cdot \\ s_{N-p} \end{bmatrix}$$

Thus, linear independence of these vectors is the necessary and sufficient condition for existence of a unique minimum; this is equivalent to the condition $\det \mathbf{Q} \neq 0$.

The calculations required in these quadratic examples are hardly typical of more general cases; in many cases, setting the derivative of the cost function to zero gives coupled nonlinear equations that can be solved only with a numerical method such as Newton's method.

Example

Various techniques for digital filter design based on desired frequency response characteristics have been formulated as optimization problems. We will describe one approach employing a series connection of second-order filters. When compared to a realization employing the numerator and denominator polynomial coefficients directly, this filter structure turns out to reduce significantly, with respect to variations in the design parameters, the sensitivity of the frequency response achieved. This property is particularly advantageous when the designed filter is to be implemented using custom digital hardware because coefficient accuracy requirements are relaxed.

We suppose that a discrete-time transfer function $H(z)$ is to be designed to provide a good fit to a nominal frequency response function $H_0(e^{jv})$. For this example we will suppose that it is the magnitude of H_0 that is to be matched, although the same kind of approach can handle general magnitude and phase design problems. To reduce the design problem to a parameter optimization problem, we assume that the designed filter takes the form

$$H(z) = b_0 \prod_{k=1}^{K} \frac{z^2 + b_{1,k} z + b_{2,k}}{z^2 + a_{1,k} z + a_{2,k}}$$

where the filter order, $2K$, is selected ahead of time. This expression corresponds to the series connection of an amplitude scaling factor b_0 and K second-order filters. We let

$$H_k(z) = \frac{z^2 + b_{1,k} z + b_{2,k}}{z^2 + a_{1,k} z + a_{2,k}}$$

denote the transfer function of the kth term. The squared magnitude of the frequency response of $H(z)$ is then given by

$$|H(e^{jv})|^2 = |b_0|^2 \prod_{k=1}^{K} |H_k(e^{jv})|^2$$

and from the form of $H_k(z)$ we obtain

$$|H_k(e^{jv})|^2 = \frac{1 + b_{1,k}^2 + b_{2,k}^2 + 2b_{1,k}(1+b_{2,k})\cos v + 2b_{2,k}\cos 2v}{1 + a_{1,k}^2 + a_{2,k}^2 + 2a_{1,k}(1+a_{2,k})\cos v + 2a_{2,k}\cos 2v}$$

The filter is designed by requiring the coefficients to provide a good fit to the nominal frequency response at a prespecified set of frequencies in the range from 0 to π radians per sample, say v_n, $1 \leq n \leq N$; the nominal data given are denoted by

$$M_n = |H_0(e^{jv_n})|$$

Finally, the measure of goodness of fit is the so-called *p-norm* of the difference:

$$C_p = \sum_{n=1}^{N} \big| |H(e^{jv_n})| - M_n \big|^p$$

The parameter p is a fixed positive integer, moderately large in most cases. (Designs are sometimes carried out for an increasing sequence of p values, motivated by the fact that as $p \to \infty$, the limit of cost function $C_p^{1/p}$ tends to the maximum value of the difference between the samples of the designed and nominal magnitudes.) To perform the minimization of C with respect to the $4K+1$ design parameters, the gain b_0, and the $\{a_{1,k}, a_{2,k}, b_{1,k}, b_{1,k}\}$ coefficients, numerical methods must be employed because of the highly nonlinear form of the parametric dependence of C and its gradient.

The preceding example shows that when a design problem is posed as an optimization problem, it may lead to a set of nonlinear equations to be solved for the design parameters. A second example will be given to illustrate a case where nonlinear equations arise from the choice of a cost function that imposes desirable properties on the corresponding optimized solution.

Sec. 5.1 Parameter Optimization

Example

The use of squared error as a cost associated with approximation errors has certain undesirable consequences. An instance of a "classic" experimental problem is to determine the resistance value that is characteristic of a batch of resistors, based on a set of N measurements of individual resistors. This problem can be formulated as seeking the minimum of the quadratic function

$$Q(R) = \sum_{i=1}^{N} (R_i - R)^2$$

where the R_i are the measured resistances. The minimizing value of R is the average of the R_i, but this gives a nonsensical answer if one of the samples is defective and behaves as an open circuit.

Clearly, the problem arises from the quadratic form of the terms in the cost function $Q(R)$. One way of providing an answer that could be called *resistant to outliers in the measurements* would be to use a modified cost function such as

$$C(R) = \sum_{i=1}^{N} \phi(R_i - R)$$

where $\phi(r)$ is a function that is nearly quadratic for small r but that tends to 0 for large r, such as

$$\phi(r) = \frac{r^2}{1 + (r/R_{max})^{10}}$$

where R_{max} is an upper bound on the largest resistance value for nondefective resistors (Fig. 5.2). In adopting this cost function, computational simplicity is sacrificed. Newton's method or some other numerical technique is required to determine the value R for which $C_R(R) = 0$.

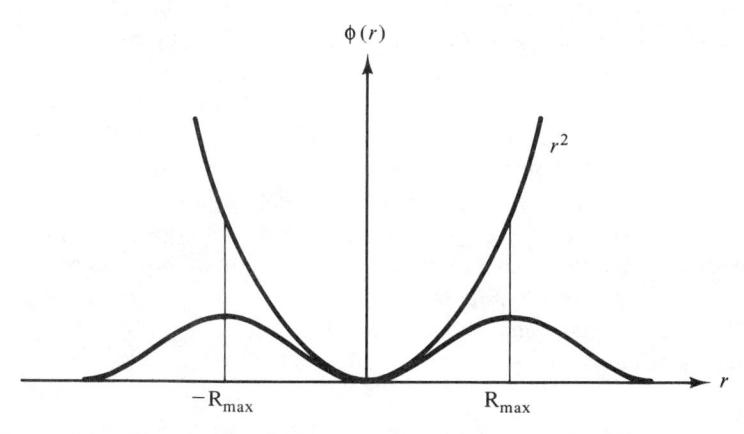

Figure 5.2 Function used in cost to provide estimate that is resistant to outliers.

In nonlinear optimization problems, there may be more than one value of $\mathbf{u}_0$ at which the gradient $C_\mathbf{u}$ is zero, thus complicating the determination of local and global minima. For a modest number of parameters, the complexity of evaluating the second derivative matrix as a function of **u** can even be daunting, and this matrix must be evaluated (and essentially inverted) at each step of Newton's method for finding zeros of the derivative. It is not surprising, therefore, that alternative numerical methods for performing function minimization are usually employed in applications. Such methods are discussed in the next section of this chapter.

Constrained Parameter Optimization: Equality Constraints

We now turn to minimization problems involving equality constraints. This type of problem arises when there are functional dependencies among the parameters to be chosen. A variety of geometrical problems fall in this class; for example, determine the shortest distance between the sphere of unit radius centered at the origin and the plane passing through the point (1,2,3) and normal to some given vector **n**. Another example: What triangle with unit perimeter has maximum area?

Such simple problems may be solved with an approach that has considerable generality and that provides important insights into the properties of the solutions. Furthermore, extensions to more general kinds of constrained optimization problems follow along similar lines. A brief summary of the method is the following. *Lagrange multipliers* are introduced to provide an augmented cost function; then setting partial derivatives to zero again leads to equations for the parameters and the multipliers. Also, a suitable matrix of second partial derivatives provides a sufficiency test for local minima. The Lagrange multipliers themselves have the interesting interpretation as *sensitivity functions*, giving the rates of change of the optimized cost with respect to changes in the constraint equations.

First we start with some notation. Let **x** be a column vector of dimension n and let **u** be a column vector of dimension m; let **f(x,u)** be an n-component vector-valued function. The components of **u** are variables to be chosen to make the function $L(\mathbf{x},\mathbf{u})$ as small as possible, subject to the n constraints **f(x,u)=0**. We will find it helpful to think of **u** as "control variables" that produce "state variables" **x** via the n "state equations" **f(x,u)=0**. When formulating problems arising in applications, the choice of which quantities are selected as the control variables is limited by a technical condition assuring that the minimizing choice of control variables determines a corresponding set of state variables through a linearized form of the constraint function **f**, although it is usually the case that the constraints take the form where **x** is an implicit function of **u**. Thus it is often not possible to use the constraints to eliminate the explicit dependence of L on **x** to obtain an unconstrained minimization problem. The following example will help to clarify the problem formulation.

Example

Consider the minimum sphere-to-plane distance problem described above (Fig. 5.3). We introduce six variables in order to specify the coordinates of a pair of points, one on the sphere and one on the plane. We write the equation of the unit sphere as

$$u_1^2 + u_2^2 + u_3^2 - 1 = 0$$

and the equation of the plane as

Sec. 5.1 Parameter Optimization 243

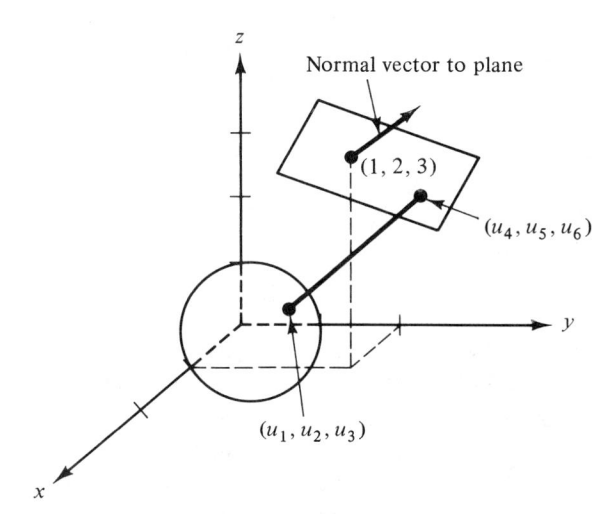

Figure 5.3 Unit sphere and plane passing through the point (1, 2, 3).

$$n_1(u_4 - 1) + n_2(u_5 - 2) + n_3(u_6 - 3) = 0$$

For the first constraint equation, each variable is an implicit function of the other two; we will arbitrarily let $x_1 = u_3$ so that the equation for the sphere becomes

$$u_1^2 + u_2^2 + x_1^2 - 1 = 0$$

For the second constraint equation, each variable can be expressed explicitly as a function of the other two. This is immaterial to the formulation, and we proceed by making another arbitrary choice, say $x_2 = u_5$. The equation of the plane is then

$$n_1(u_4 - 1) + n_2(x_2 - 2) + n_3(u_6 - 3) = 0$$

The latter two equations specify $\mathbf{f}(\mathbf{x},\mathbf{u})$, a two-component vector-valued function of $\mathbf{x} = (x_1, x_2)^T$ and $\mathbf{u} = (u_1, u_2, u_4, u_6)^T$. The function to be minimized is the squared distance, given by

$$L(\mathbf{x},\mathbf{u}) = (u_1 - u_4)^2 + (u_2 - x_2)^2 + (x_1 - u_6)^2$$

For this problem, it is intuitively clear that x_1 could just as well have been chosen as u_1 or u_2 and that x_2 could just as well have been chosen as u_4 or u_6. The validity of other choices, such as $x_1 = u_1$ and $x_2 = u_2$, may not be intuitively clear. We will return to this point shortly.

As motivation for the Lagrange multiplier approach, we look at the behavior of L in the vicinity of a constrained local minimum; we also take into consideration how the constraint vector $\mathbf{f}$ varies. Thus, let $\mathbf{u}_0$ and $\mathbf{x}_0$ be values at which L achieves a local minimum subject to the constraints $\mathbf{f}(\mathbf{x}_0, \mathbf{u}_0) = \mathbf{0}$. Using derivatives to provide linear approximations for L and $\mathbf{f}$ (i.e., writing the first order Taylor series expansions), we obtain

$$L(\mathbf{x},\mathbf{u}) \approx L(\mathbf{x}_0, \mathbf{u}_0) + L_\mathbf{x}(\mathbf{x}_0, \mathbf{u}_0)(\mathbf{x} - \mathbf{x}_0) + L_\mathbf{u}(\mathbf{x}_0, \mathbf{u}_0)(\mathbf{u} - \mathbf{u}_0) \qquad (5.10)$$

$$\mathbf{f}(\mathbf{x},\mathbf{u}) \approx \mathbf{f}(\mathbf{x}_0, \mathbf{u}_0) + \mathbf{f}_\mathbf{x}(\mathbf{x}_0, \mathbf{u}_0)(\mathbf{x} - \mathbf{x}_0) + \mathbf{f}_\mathbf{u}(\mathbf{x}_0, \mathbf{u}_0)(\mathbf{u} - \mathbf{u}_0) \qquad (5.11)$$

In these expressions, the **x** and **u** subscripts again denote partial derivatives; in the case of the vector-valued function **f(x,u)** the partial differentiation is done componentwise, so that

$$\mathbf{f_x} = [\ \mathbf{f}_{x_1}\ \cdots\ \mathbf{f}_{x_n}\ ] \tag{5.12}$$

and

$$\mathbf{f_u} = [\ \mathbf{f}_{u_1}\ \cdots\ \mathbf{f}_{u_m}\ ] \tag{5.13}$$

$\mathbf{f_x}$ is an $n \times n$ matrix and $\mathbf{f_u}$ is an $n \times m$ matrix. The approximations are valid to first-order in $\|\mathbf{x}-\mathbf{x}_0\| + \|\mathbf{u}-\mathbf{u}_0\|$. (That is, in either case the norm of the difference between the right- and left-hand sides of the expression tends to zero "faster than linearly.")

Assuming that $\mathbf{u}_0$ and $\mathbf{x}_0$ are values at which L achieves a constrained local minimum, we combine the two expressions to obtain some first-order necessary conditions. If **x** and **u** are also chosen to satisfy the constraints, so that both $\mathbf{f(x,u)} = \mathbf{0}$ and $\mathbf{f(x_0,u_0)} = \mathbf{0}$, then (to first order) we solve equation (5.11) to obtain

$$(\mathbf{x} - \mathbf{x}_0) = -\mathbf{f}_\mathbf{x}^{-1}\, \mathbf{f}_\mathbf{u}\, (\mathbf{u} - \mathbf{u}_0) \tag{5.14}$$

Here we have dropped the arguments of the partial derivatives of **f** for simplicity, and we have made the assumption that $\mathbf{f_x}(\mathbf{x}_0,\mathbf{u}_0)$ is nonsingular. This is the precise form of the assumption mentioned earlier about **u** determining **x** for a linearized version of the constraints. In fact, this assumption, through the *Inverse Function Theorem* of multivariable calculus, assures a well-behaved (i.e., differentiable) functional relationship between **u** and **x** in a neighborhood of the constrained local minimum point.

Having used the constraint equations to determine how changes in $\mathbf{u}_0$ and $\mathbf{x}_0$ must be related in order that the constraints continue to hold, we may substitute this result into the first-order approximation for the resulting change in L. This gives

$$L(\mathbf{x},\mathbf{u}) \approx L(\mathbf{x}_0,\mathbf{u}_0) + (L_\mathbf{u} - L_\mathbf{x}\, \mathbf{f}_\mathbf{x}^{-1}\, \mathbf{f}_\mathbf{u})(\mathbf{u}-\mathbf{u}_0) \tag{5.15}$$

Since this expression looks much like a first-order Taylor series, we introduce the following suggestive notation for the second term on the right side of this equation:

$$L_\mathbf{u} - L_\mathbf{x}\, \mathbf{f}_\mathbf{x}^{-1}\, \mathbf{f}_\mathbf{u} = L_{\mathbf{u}|\mathbf{f}=\mathbf{0}} \tag{5.16}$$

which we read as "the derivative of L with respect to **u** *along the constraints* $\mathbf{f} = \mathbf{0}$." In order for $L(\mathbf{x}_0,\mathbf{u}_0)$ to be locally minimized, it is therefore necessary that the coefficient of $(\mathbf{u}-\mathbf{u}_0)$ vanish; in other words, the derivative must vanish:

$$L_{\mathbf{u}|\mathbf{f}=\mathbf{0}} = L_\mathbf{u} - L_\mathbf{x}\, \mathbf{f}_\mathbf{x}^{-1}\, \mathbf{f}_\mathbf{u} = 0 \tag{5.17}$$

The m equations obtained by setting $L_{\mathbf{u}|\mathbf{f}=\mathbf{0}}$ to zero, together with the n constraint equations, must be solved to obtain the $n+m$ quantities $\mathbf{u}_0$ and $\mathbf{x}_0$. Being optimistic, since there are as many equations as unknowns we expect that we will usually be able to find a solution. Generally, the equations are nonlinear so that neither existence and uniqueness of solutions can be assured. Also, a numerical technique such as Newton's method will usually be required to determine a solution. Most important, these equations are only *necessary conditions*, and any solution that satisfies the invertibility assumption on

Sec. 5.1   Parameter Optimization   **245**

$\mathbf{f_x}$ is only guaranteed to be a *constrained critical point* (i.e., a constrained local minimum, local maximum, or saddle point). Finally, the approach provides no help in determining any constrained minima at points where $\mathbf{f_x}$ is not invertible.

Example

For the minimum sphere-to-plane distance problem formulated above, the following quantities are easily computed:

$$L_\mathbf{u} = 2(u_1 - u_4, u_2 - x_2, u_4 - u_1, u_6 - x_1)$$

$$L_\mathbf{x} = 2(x_1 - u_6, x_2 - u_2)$$

$$\mathbf{f_x} = \begin{bmatrix} 2x_1 & 0 \\ 0 & n_2 \end{bmatrix}$$

$$\mathbf{f_u} = \begin{bmatrix} 2u_1 & 2u_2 & 0 & 0 \\ 0 & 0 & n_1 & n_3 \end{bmatrix}$$

Clearly, $\mathbf{f_x}$ is invertible except when $x_1 = 0$ or $n_2 = 0$. It is easily shown that these conditions are not satisfied for any of the three of the "easy cases," namely when the planes are $u_4 = 1$, $x_2 = 2$, and $u_6 = 3$. This kind of drawback must always be kept in mind when working with *necessary*, but not *sufficient*, conditions of the type developed above.

The question raised at the end of the preceding example, whether other choices of $\mathbf{x}$ could be used in the formulation of the problem, can now be answered in the negative by noting that selecting x_1 and x_2 to be variables in the same constraint equation will automatically produce a zero row in the corresponding $\mathbf{f_x}$, and hence the formulation will never lead to a solution. At least in the original formulation the cases where the formulation fails are essentially "isolated cases" involving very special choices for the plane.

Lagrange Multipliers

There is a much easier-to-remember formulation of the solution to constrained minimization problems that also provides some additional insights of its own. The Lagrange multiplier approach to constrained minimization problems amounts to the following. Introduce an *n*-vector λ of undetermined quantities, and form an augmented cost function

$$H(\mathbf{x},\mathbf{u},\lambda) = \lambda^T \mathbf{f}(\mathbf{x},\mathbf{u}) + L(\mathbf{x},\mathbf{u}) \tag{5.18}$$

A straightforward calculation shows that the constrained local minima of L occur at (unconstrained!) critical points of H, provided that the constrained local minima occur at *regular points* of the constraint functions (i.e., at points where $\mathbf{f_x}$ is invertible). The resulting necessary conditions for constrained local minima of L are the following:

$$H_\mathbf{x} = \lambda^T \mathbf{f_x} + L_\mathbf{x} = \mathbf{0} \tag{5.19}$$

$$H_\mathbf{u} = \lambda^T \mathbf{f_u} + L_\mathbf{u} = \mathbf{0} \tag{5.20}$$

$$H_\lambda = \mathbf{f}^T = \mathbf{0} \tag{5.21}$$

This gives $m + 2n$ equations for the same number of unknowns, so we expect to be able to find solutions. Notice that the second of these sets of equations may be solved for λ^T, and substitution of the result into the first set of equations gives exactly the necessary conditions derived earlier.

For unconstrained problems, $\mathbf{f} = \mathbf{0}$ and L does not depend on $\mathbf{x}$, so only the first set of equations is nontrivial; this is exactly the familiar procedure of setting the derivative to zero to find possible local minima. In the remainder of the chapter, we will use L to denote the cost function even for unconstrained problems where no $\mathbf{x}$ variables appear.

The geometrical interpretation of the first two equations provides a valuable insight about constrained local minima. These equations may be combined into a single one,

$$[H_\mathbf{x}, H_\mathbf{u}] = \lambda^T [\mathbf{f_x}, \mathbf{f_u}] = [L_\mathbf{x}, L_\mathbf{u}] \tag{5.22}$$

Using the gradient symbol ∇ to denote the derivative with respect to both the $\mathbf{x}$ and $\mathbf{u}$ variables, for example

$$\nabla L = [L_{x_1} \cdots L_{x_n} L_{u_1} \cdots L_{u_m}] \tag{5.23}$$

equation (5.22) may be expressed in terms of the components of the constraint function as

$$[H_\mathbf{x}, H_\mathbf{u}] = \lambda^T \begin{bmatrix} \nabla \mathbf{f}_1 \\ \nabla \mathbf{f}_2 \\ \vdots \\ \nabla \mathbf{f}_n \end{bmatrix} + \nabla L = \mathbf{0} \tag{5.24}$$

In words, equation (5.24) says that at points where the necessary conditions are satisfied, the gradient vector of L must lie in the space spanned by the gradient vectors of the components of the constraint function. (In the case of a single constraint, the gradients of L and $\mathbf{f}$ must be scalar multiples of each other at a constrained local minimum; see Fig. 5.4.) This is intuitively reasonable because it means that the direction in which L is most rapidly decreasing is also a direction in which the values of one or more of the constraints are changing; thus, any small change along the direction of greatest decrease in L would produce a violation of at least one of the constraints.

We have made a point of distinguishing $\mathbf{u}$ and $\mathbf{x}$, the "control variables" and the "state variables," but equation (5.22) makes clear that they are treated the same in the necessary conditions. We will continue to make the distinction in the remainder of this chapter, but it is worth pointing out that many books formulate constrained optimization problems as "minimize $L(\mathbf{w})$ subject to $\mathbf{f}(\mathbf{w}) = \mathbf{0}$." From (5.24), it may be seen that the underlying regularity assumption for this formulation is that the matrix $\mathbf{f_w}$ must have linearly independent rows.

Example

When minimizing a quadratic function subject to linear constraints the Lagrange multiplier method provides a "closed-form" solution. Consider the problem of minimizing

$$L(\mathbf{x}, \mathbf{u}) = \tfrac{1}{2} \mathbf{x}^T \mathbf{Q} \mathbf{x} + \tfrac{1}{2} \mathbf{u}^T \mathbf{R} \mathbf{u}$$

where $\mathbf{Q}$ and $\mathbf{R}$ are symmetric matrices, subject to the constraints

Sec. 5.1 Parameter Optimization

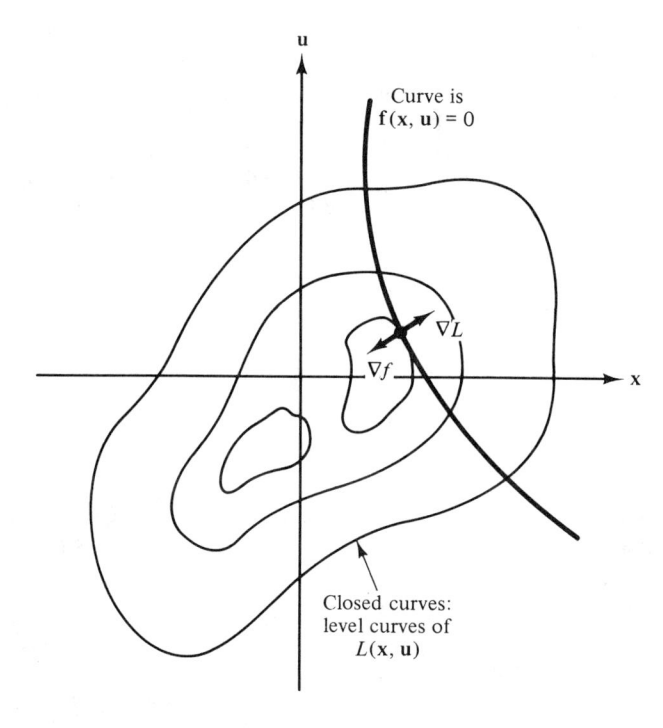

Figure 5.4 For a single constraint, the gradient vectors of $L(\mathbf{x}, \mathbf{u})$ and $\mathbf{f}(\mathbf{x}, \mathbf{u})$ are collinear at point of constrained local minimum. The case shown is for $\nabla L / \|\nabla L\| = -\nabla \mathbf{f}/\|\nabla \mathbf{f}\|$.

$$\mathbf{f}(\mathbf{x},\mathbf{u}) = \mathbf{A}\mathbf{x} + \mathbf{B}\mathbf{u} + \mathbf{y} = \mathbf{0}$$

Thus

$$H(\mathbf{x},\mathbf{u},\lambda) = \lambda^T \mathbf{A}\mathbf{x} + \lambda^T \mathbf{B}\mathbf{u} + \lambda^T \mathbf{y} + \tfrac{1}{2}\mathbf{x}^T \mathbf{Q}\mathbf{x} + \tfrac{1}{2}\mathbf{u}^T \mathbf{R}\mathbf{u}$$

from which we obtain

$$H_\mathbf{x} = \mathbf{0} = \lambda^T \mathbf{A} + \mathbf{x}^T \mathbf{Q}$$
$$H_\mathbf{u} = \mathbf{0} = \lambda^T \mathbf{B} + \mathbf{u}^T \mathbf{R}$$
$$H_\lambda = \mathbf{0} = \mathbf{x}^T \mathbf{A}^T + \mathbf{u}^T \mathbf{B}^T + \mathbf{y}^T$$

Since $\mathbf{f}_\mathbf{x} = \mathbf{A}$, the regularity condition for applying the Lagrange multipliers is that $\mathbf{A}$ is invertible. (Also, recall from the properties of matrix inversion and transposition that $(\mathbf{A}^T)^{-1} = (\mathbf{A}^{-1})^T$.) Hence the first and third equations may be solved to give

$$\lambda^T = -\mathbf{x}^T \mathbf{Q}\mathbf{A}^{-1} \quad \text{or} \quad \lambda = -(\mathbf{A}^{-1})^T \mathbf{Q}\mathbf{x}$$

and

$$\mathbf{x}^T = -(\mathbf{u}^T \mathbf{B}^T + \mathbf{y}^T)(\mathbf{A}^T)^{-1} \quad \text{or} \quad \mathbf{x} = -\mathbf{A}^{-1}(\mathbf{B}\mathbf{u} + \mathbf{y})$$

and these results may be substituted in the $H_\mathbf{u} = \mathbf{0}$ equation to give

$$\mathbf{0} = (\mathbf{u}^T \mathbf{B}^T + \mathbf{y}^T)(\mathbf{A}^{-1})^T \mathbf{Q}\mathbf{A}^{-1}\mathbf{B} + \mathbf{u}^T \mathbf{R}$$

After transposing and rearranging, this equation takes the form

$$(\mathbf{B}^T(\mathbf{A}^{-1})^T \mathbf{Q}\mathbf{A}^{-1}\mathbf{B} + \mathbf{R})\mathbf{u} = -\mathbf{B}^T(\mathbf{A}^{-1})^T \mathbf{Q}\mathbf{A}^{-1}\mathbf{y}$$

These linear equations will always have a solution when the coefficient matrix $(\mathbf{B}^T(\mathbf{A}^{-1})^T\mathbf{Q}\mathbf{A}^{-1}\mathbf{B} + \mathbf{R})$ is invertible; this solution determines $\mathbf{x}$ through the constraint equation, and λ is then determined from $\mathbf{x}$. Thus a constrained critical point can be found by applying the Lagrange multiplier method; whether or not the solution is a constrained local minimum can only be determined after some further analysis, to be described shortly.

This problem can be reduced to an unconstrained problem by using the constraint equation to write $\mathbf{x}$ in terms of $\mathbf{u}$ and substituting the result into $L(\mathbf{x},\mathbf{u})$ to obtain a function of $\mathbf{u}$ alone. We find that

$$\mathbf{x} = -\mathbf{A}^{-1}(\mathbf{B}\mathbf{u} + \mathbf{y})$$

so

$$\begin{aligned}L(\mathbf{x},\mathbf{u}) &= \tfrac{1}{2}(\mathbf{A}^{-1}(\mathbf{B}\mathbf{u}+\mathbf{y}))^T \mathbf{Q}(\mathbf{A}^{-1}(\mathbf{B}\mathbf{u}+\mathbf{y})) + \tfrac{1}{2}\mathbf{u}^T\mathbf{R}\mathbf{u} \\ &= \tfrac{1}{2}\mathbf{u}^T(\mathbf{B}^T(\mathbf{A}^T)^{-1}\mathbf{Q}\mathbf{A}^{-1}\mathbf{B}+\mathbf{R})\mathbf{u} + \mathbf{y}^T(\mathbf{A}^{-1})^T\mathbf{Q}\mathbf{A}^{-1}\mathbf{B}\mathbf{u} \\ &\quad + \tfrac{1}{2}\mathbf{y}^T(\mathbf{A}^{-1})^T\mathbf{Q}\mathbf{A}^{-1}\mathbf{y}\end{aligned}$$

Our previous analysis of unconstrained quadratic minimization problems leads to the same set of linear equations for $\mathbf{u}$ obtained above together with the condition that the matrix $(\mathbf{B}^T(\mathbf{A}^{-1})^T\mathbf{Q}\mathbf{A}^{-1}\mathbf{B} + \mathbf{R})$ be positive definite, not simply invertible, so that the critical point be a local minimum.

Sufficient Conditions for a Constrained Local Minimum

We now want to determine some sufficient conditions for a constrained critical point to be a constrained local minimum. Our route to these conditions proceeds via an analysis of *neighboring minima*. The idea behind the development is to determine how a small change in the constraint equation produces small changes in the variables $\mathbf{x}$, $\mathbf{u}$, and λ. Thus, suppose that the constraint equation is "perturbed" and the "neighboring" optimization problem to be considered is to minimize L subject to the constraints $\mathbf{f}(\mathbf{x},\mathbf{u})=\Delta_\mathbf{f}$. Here we use $\Delta_\mathbf{f}$ to indicate a small change in $\mathbf{f}$; Δ with other subscripts will indicate other small changes (Fig. 5.5).

The first equation we have that relates the small change in the constraint equation with changes in the problems variables is simply the first-order Taylor series expression for $\mathbf{f}(\mathbf{x},\mathbf{u})$. This was encountered earlier in connection with the necessary conditions for a local minimum. Written in terms of changes (Δ's) we have

$$\Delta_\mathbf{f} = \mathbf{f}(\mathbf{x}+\Delta_\mathbf{x},\mathbf{u}+\Delta_\mathbf{u}) - \mathbf{f}(\mathbf{x},\mathbf{u}) \approx \mathbf{f}_\mathbf{x}(\mathbf{x},\mathbf{u})\Delta_\mathbf{x} + \mathbf{f}_\mathbf{u}(\mathbf{x},\mathbf{u})\Delta_\mathbf{u} \qquad (5.25)$$

which we may solve (to first order) under the assumption that the changes produce another local minimum, giving

$$\Delta_\mathbf{x} = \mathbf{f}_\mathbf{x}^{-1}(\Delta_\mathbf{f} - \mathbf{f}_\mathbf{u}\Delta_\mathbf{u}) \qquad (5.26)$$

There are also two equations to be had from the assumption that a neighboring local minimum is achieved. Looking at the first-order Taylor series expansions for $H_\mathbf{u}$ and $H_\mathbf{x}$ leads to two equations by exploiting the following: These quantities must vanish (to first-order) at both the original and the neighboring minima. The following equations are obtained after careful attention to notational conventions (basically, to expand the

Sec. 5.1   Parameter Optimization   249

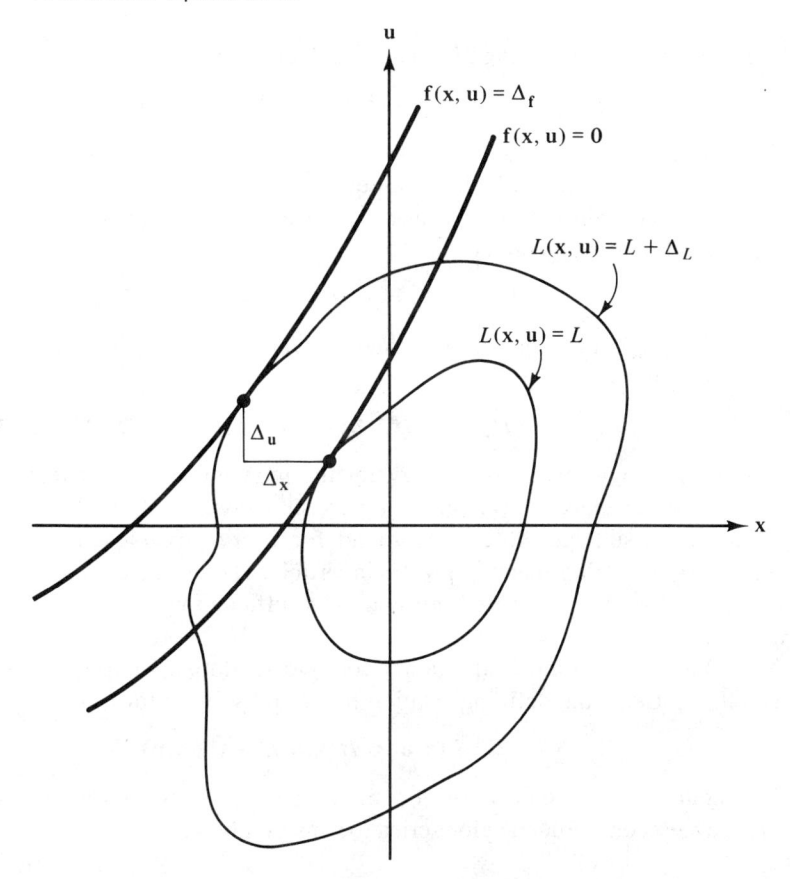

Figure 5.5 Neighboring constrained minima.

derivatives H_u and H_x in Taylor series, we must first transpose the quantities to turn them into column vectors).

$$0 = H_x^T(x+\Delta_x, u+\Delta_u, \lambda+\Delta_\lambda) - H_x^T(x, u, \lambda) = H_{xx}\Delta_x + H_{xu}\Delta_u + f_x^T\Delta_\lambda \quad (5.27)$$

$$0 = H_u^T(x+\Delta_x, u+\Delta_u, \lambda+\Delta_\lambda) - H_u^T(x, u, \lambda) = H_{ux}\Delta_x + H_{uu}\Delta_u + f_u^T\Delta_\lambda \quad (5.28)$$

In these equations, the doubly subscripted quantities are matrices of second partial derivatives; for example,

$$H_{ux} = (H_u^T)_x = \begin{bmatrix} H_{u_1 x_1} & \cdots & H_{u_1 x_n} \\ \vdots & & \vdots \\ H_{u_m x_1} & \cdots & H_{u_m x_n} \end{bmatrix} \quad (5.29)$$

The $2n+m$ linear equations in Δ_x, Δ_u, and Δ_λ obtained by this analysis of neighboring minima can be simplified by first solving for Δ_x in terms of Δ_f and Δ_u:

$$\Delta_x = f_x^{-1}(\Delta_f - f_u\Delta_u) \quad (5.30)$$

Then the equation involving H_x^T may be solved for Δ_λ:

$$\Delta_\lambda = -(\mathbf{f}_x^T)^{-1}(H_{xx}\Delta_x + H_{xu}\Delta_u) \tag{5.31}$$

Combining these two equations allows us to solve for Δ_λ in terms of Δ_f and Δ_u. (We omit the explicit form.) Finally, the equation involving H_u^T may be used to produce an explicit relationship between Δ_f and Δ_u, since both Δ_λ and Δ_x may be eliminated. The result of all of this algebra is

$$L_{uu|f=0}\Delta_u = -(H_{ux} - \mathbf{f}_u^T(\mathbf{f}_x^T)^{-1}H_{xx})\mathbf{f}_x^{-1}\Delta_f \tag{5.32}$$

where $L_{uu|f=0}$ is the "matrix of second derivatives of L with respect to the **u** variables, *along the constraints* $\mathbf{f}=\mathbf{0}$"

$$L_{uu|f=0} = H_{uu} - H_{ux}\mathbf{f}_x^{-1}\mathbf{f}_u - \mathbf{f}_u^T(\mathbf{f}_x^T)^{-1}H_{xu} + \mathbf{f}_u^T(\mathbf{f}_x^T)^{-1}H_{xx}\mathbf{f}_x^{-1}\mathbf{f}_u \tag{5.33}$$

Ignoring the suggestive terminology for the moment, it is clear that a sufficient condition for the existence of a neighboring constrained critical point is provided by guaranteeing that a suitable Δ_u can be found for every possible choice of Δ_f, and this is attained by requiring that $L_{uu|f=0}$ be invertible. Having determined this, we now return to the problem that motivated our analysis: sufficient conditions for a constrained local minimum.

Our interpretation of the rather complicated-looking expression $L_{uu|f=0}$ is verified as follows. Using the defining relation for H and solving for L gives

$$L(\mathbf{x},\mathbf{u}) = H(\mathbf{x},\mathbf{u},\lambda) - \lambda^T \mathbf{f}(\mathbf{x},\mathbf{u}) \tag{5.34}$$

This holds for any choice of the multipliers λ. Now, letting $\Delta_L = L(\mathbf{x}+\Delta_x, \mathbf{u}+\Delta_u) - L(\mathbf{x},\mathbf{u})$ a second-order Taylor series expansion gives

$$\Delta_L \approx H_x\Delta_x + H_u\Delta_u - \lambda^T\mathbf{f}_x\Delta_x - \lambda^T\mathbf{f}_u\Delta_u + \tfrac{1}{2}(\Delta_x^T \ \Delta_u^T)\begin{bmatrix} H_{xx} & H_{xu} \\ H_{ux} & H_{uu} \end{bmatrix}\begin{pmatrix} \Delta_x \\ \Delta_u \end{pmatrix} \tag{5.35}$$

Using the necessary conditions for a constrained minimum, and using the equation relating Δ_x, Δ_u, and Δ_f gives

$$\Delta_L \approx -\lambda^T\Delta_f + \tfrac{1}{2}(\Delta_u)^T L_{uu|f=0}\Delta_u \tag{5.36}$$

Setting $\Delta_f = \mathbf{0}$ in this expression, it follows that the sufficient condition for a constrained critical point to be a constrained local minimum is that the matrix $L_{uu|f=0}$ be positive definite.

Example

Continuing the preceding example where

$$H(\mathbf{x},\mathbf{u},\lambda) = \lambda^T\mathbf{A}\mathbf{x} + \lambda^T\mathbf{B}\mathbf{u} + \lambda^T\mathbf{y} + \tfrac{1}{2}\mathbf{x}^T\mathbf{Q}\mathbf{x} + \tfrac{1}{2}\mathbf{u}^T\mathbf{R}\mathbf{u}$$

we calculate the following quantities:

$$H_{ux} = \mathbf{0} = H_{xu}^T$$

$$H_{uu} = \mathbf{R}$$

$$H_{xx} = \mathbf{Q}$$

Sec. 5.1   Parameter Optimization

We have previously determined that $\mathbf{f_u} = \mathbf{B}$ and $\mathbf{f_x} = \mathbf{A}$. Hence

$$L_{\mathbf{uu}|\mathbf{f}=0} = \mathbf{R} + \mathbf{B}^\mathrm{T}(\mathbf{A}^{-1})^\mathrm{T}\mathbf{Q}\mathbf{A}^{-1}\mathbf{B}$$

and the positive definite condition on this matrix is precisely the sufficient condition for a local minimum determined previously by using the constraints to eliminate $\mathbf{x}$ and turn the constrained problem into an unconstrained one.

This sufficient condition means that $L_{\mathbf{uu}|\mathbf{f}=0}$ is invertible, so we are assured that neighboring constrained local minima exist, too. It is also of interest to determine how changes in the constraints affect the corresponding minimized value of L, and this is also revealed by the equation for Δ_L. The vector of Lagrange multipliers chosen to satisfy the necessary conditions has the following important interpretation: It gives the derivative (or sensitivity) of the minimum cost to changes in the constraint level; that is, if the constraints are given by the equations $\mathbf{f}=\mathbf{c}$ (for some constant vector $\mathbf{c}$), then, using obvious notation for the minimizing choice of λ, we have

$$\frac{\partial L_{\min}}{\partial \mathbf{c}} = -\lambda^{*\mathrm{T}} \tag{5.37}$$

This result is particularly important for applications involving *parametric modeling uncertainty*. Suppose that the constraint equation depends on a parameter vector $\mathbf{p}$ and takes the form $\mathbf{f}(\mathbf{x},\mathbf{u},\mathbf{p})=\mathbf{0}$. If $\mathbf{p}$ represents "nominal" parameter values (e.g., circuit component values) rather than "true" ones, it is important to evaluate how variations from the nominal values affect the minimized cost. Using the partial derivative of $\mathbf{f}$ with respect to $\mathbf{p}$ to obtain an estimate of $\Delta_\mathbf{f}$ gives

$$\Delta_\mathbf{f} \approx \mathbf{f_p}\Delta_\mathbf{p} \tag{5.38}$$

The corresponding change in the minimized cost is then obtained by premultiplying by $-\lambda^{*\mathrm{T}}$. We may also write this result as

$$\frac{\partial L_{\min}}{\partial \mathbf{p}} = -\lambda^{*\mathrm{T}}\mathbf{f_p} \tag{5.39}$$

Since it is not always easy to solve the parametric form of the constrained optimization problem, from which the derivative of the minimized cost with respect to parameter changes could be obtained, this expression involving the Lagrange multipliers is important for determining how modeling uncertainties are manifested.

Example

Consider the problem of finding the rectangle of maximum perimeter that can be inscribed in a circle of unit radius. We may turn this into a constrained minimization problem by minimizing the negative of the perimeter. By selecting the circle to be centered at the origin of the Cartesian plane and using x and u as coordinates (Fig. 5.6), the following mathematical formulation results:

$$\min -4(x+u)$$
$$\text{subject to } x^2 + u^2 - 1 = 0$$

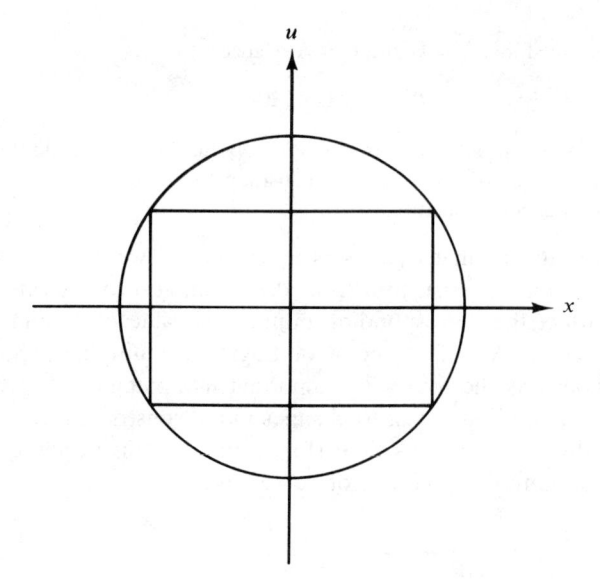

Figure 5.6 Unit circle $x^2 + u^2 = 1$ with inscribed rectangle.

Thus we have the identifications $L(x,u) = -4(x+u)$ and $f(x,u) = x^2 + u^2 - 1$; as usual, $H = \lambda f + L$. The necessary conditions are

$$H_u = 0 = 2\lambda u - 4$$
$$H_x = 0 = 2\lambda x - 4$$
$$H_\lambda = 0 = x^2 + u^2 - 1$$

The first two equations may be solved to obtain

$$u^* = x^* = \frac{2}{\lambda^*}$$

and since the point (x^*, u^*) must satisfy the third equation (to lie on the unit circle), we find that the inscribed rectangle is a square, with

$$u^* = x^* = \frac{1}{\sqrt{2}}$$

and the resulting perimeter is $4\sqrt{2}$. We also have

$$\lambda^* = 2\sqrt{2}$$

The sufficient conditions are easily checked, since $H_{uu} = H_{xx} = 2\lambda$ and $H_{ux} = H_{xu} = 0$. Thus

$$L_{uu|f=0} = 4\lambda^* = 8\sqrt{2}$$

which is positive, and the square is indeed the inscribed rectangle with maximum perimeter.

Notice that if the constraint were that the rectangle be inscribed in a circle of radius r, the constraint equation would be $f(x,u,r) = x^2 + u^2 - r^2$. If $r = 1$ is viewed as the nominal value of the radius, then from our discussion above about parametric modeling uncertainties, the derivative of the maximized perimeter with respect to r is equal to the product of $-\lambda^*$ times the derivative of f with respect to r. In this case, the derivative of f takes the value -2 at $r=1$. Since $\lambda^* = 2\sqrt{2}$, the result is $4\sqrt{2}$. This checks with a direct calculation that is easily done once the maximized perimeter is found to be $4r\sqrt{2}$.

For this problem, it is intuitively clear that the values of x^*, u^* are proportional to r, and the resulting maximized perimeter must be also. Another variation on the problem is to allow the circumscribing circle to be replaced with an ellipse by taking $f(x,u,a,b) = (x/a)^2 + (u/b)^2 - 1$. Now $a = b = 1$ are the nominal parameter values. The derivative of the maximized perimeter with respect to the parameter vector $\mathbf{p} = [\,a\ b\,]^T$, given by $-\lambda^* \mathbf{f_p}$, is $[\,2\sqrt{2}\ \ 2\sqrt{2}\,]$. This agrees, as it should, with the result obtained by solving the parametric version of the problem to find the maximum perimeter, which is $4\sqrt{a^2 + b^2}$, and calculating the derivative directly.

Constrained Parameter Optimization: Inequality Constraints

Many optimization problems arising in design applications involve one or more *inequality constraints* instead of, or in addition to, equality constraints of the type considered up to now. One common inequality constraint to be imposed on elements of $\mathbf{u}$ is nonnegativity; when optimization is used to determine physical quantities such as the mass of iron to be used in a transformer core, nonnegativity constraints arise. *Box constraints*, involving upper and lower bounds on variables, are perhaps even more common. Physical limitations on the values of resistors and capacitors impose box constraints in circuit design problems; optimum resource allocation problems (e.g., scheduling jobs on machines in a flexible manufacturing facility or controlling electric power production in a network of generators and loads) typically involve ranges of allowed parameter values. Physically important combinations of the $\mathbf{x}$ and $\mathbf{u}$ variables may also be subject to inequality constraints; in amplifier design problems, such quantities might include power dissipation and harmonic distortion.

A quite general form for describing constraints is the equation $\mathbf{f}(\mathbf{x},\mathbf{u}) \geq \mathbf{0}$. This is to interpreted as a set of inequalities: Each component of the vector $\mathbf{f}$ is nonnegative. A set of box constraints such as $\mathbf{g}(\mathbf{x},\mathbf{u}) \geq \mathbf{g}_1$ and $\mathbf{g}(\mathbf{x},\mathbf{u}) \leq \mathbf{g}_2$ can be incorporated in the general form as

$$\begin{bmatrix} \mathbf{g}(\mathbf{x},\mathbf{u}) - \mathbf{g}_1 \\ \mathbf{g}_2 - \mathbf{g}(\mathbf{x},\mathbf{u}) \end{bmatrix} \geq \mathbf{0} \tag{5.40}$$

An equality constraint can be written as a degenerate box constraint: $\mathbf{g}_1 = \mathbf{g}_2$ ensures that $\mathbf{g}(\mathbf{x},\mathbf{u}) - \mathbf{g}_1 = \mathbf{0}$.

In carrying out an analysis leading to first-order necessary conditions for constrained minimization, it is important to distinguish, at all potential solution points $(\mathbf{x}_0, \mathbf{u}_0)$, between the constraints that hold with equality, the so-called *active* or *binding* constraints, and the remaining constraints, where the required inequalities are strict. Intuitively, the strict inequalities do not impose conditions limiting the range of small changes in $\mathbf{u}$ and $\mathbf{x}$ for which a zero-derivative kind of condition must hold; the active constraints force the small changes in $\mathbf{u}$ and $\mathbf{x}$ to be related through certain linearized inequalities. Rather than carry out all of the details, we will simply summarize the end result, which takes the form of the *Kuhn-Tucker* necessary conditions:

$$H_\mathbf{u} = \lambda^T \mathbf{f_u} + L_\mathbf{u} = \mathbf{0} \tag{5.41}$$

$$H_\mathbf{x} = \lambda^T \mathbf{f_x} + L_\mathbf{x} = \mathbf{0} \tag{5.42}$$

$$H_\lambda = \mathbf{f}^T \geq \mathbf{0} \tag{5.43}$$

$$\lambda_i \mathbf{f}_i = 0 \tag{5.44}$$

$$\lambda \leq \mathbf{0} \tag{5.45}$$

Conditions (5.43) to (5.45) express mathematically two easily described conditions on the Lagrange multipliers. When the ith constraint is inactive, so that $\mathbf{f}_i(\mathbf{x},\mathbf{u}) > 0$, the corresponding multiplier is zero (i.e., $\lambda_i = 0$). On the other hand, when the ith constraint is active, all that can be said about the ith multiplier is that it is nonpositive. In view of the interpretation of the multipliers as sensitivities of minimized cost with respect to small negative changes in equality constraints, these conditions are very reasonable since no change in minimized cost results when an inactive inequality constraint is slightly changed, while a decrease in an active constraint amounts to a relaxation for the case of a nonnegativity constraint, and this might possibly lead to a reduction in the minimized cost. Hence the multipliers for active constraints are nonpositive.

The geometric interpretation of the Kuhn-Tucker conditions is interesting to consider. The gradient of L must be a nonnegative linear combination of the gradient vectors of the active constraints. In other words, the direction in which L is decreasing coincides with a direction that leads to violation of one or more of the active constraints. This is a natural intuitive requirement for achieving a constrained local minimum (Fig. 5.7).

Example

In terms of practical applications, the most common form of minimization problem with inequality constraints involves a linear cost function with linear equality and inequality constraints. This is the class of *linear programming problems*. The mathematical description takes the form

$$L(\mathbf{x},\mathbf{u}) = \mathbf{c}_1^T \mathbf{x} + \mathbf{c}_2^T \mathbf{u}$$

$$\mathbf{b} - \mathbf{A}_1 \mathbf{x} - \mathbf{A}_2 \mathbf{u} \geq \mathbf{0}$$

$$\mathbf{d} - \mathbf{B}_1 \mathbf{x} - \mathbf{B}_2 \mathbf{u} = \mathbf{0}$$

where $\mathbf{b}$, $\mathbf{c}_1$, $\mathbf{c}_2$, and $\mathbf{d}$ are constant vectors, and $\mathbf{A}_1$, $\mathbf{A}_2$, $\mathbf{B}_1$, and $\mathbf{B}_2$ are constant matrices. A variety of specialized computational methods for solving linear programming problems is available.

5.2 NUMERICAL OPTIMIZATION TECHNIQUES

The first section of this chapter emphasized the analytical formulation of optimization problems, and now we turn to a more practical aspect: computational solution of optimization problems. Of course, the theory and analysis already discussed is of fundamental importance, but as has already been seen in many places in this book, computational

Sec. 5.2 Numerical Optimization Techniques

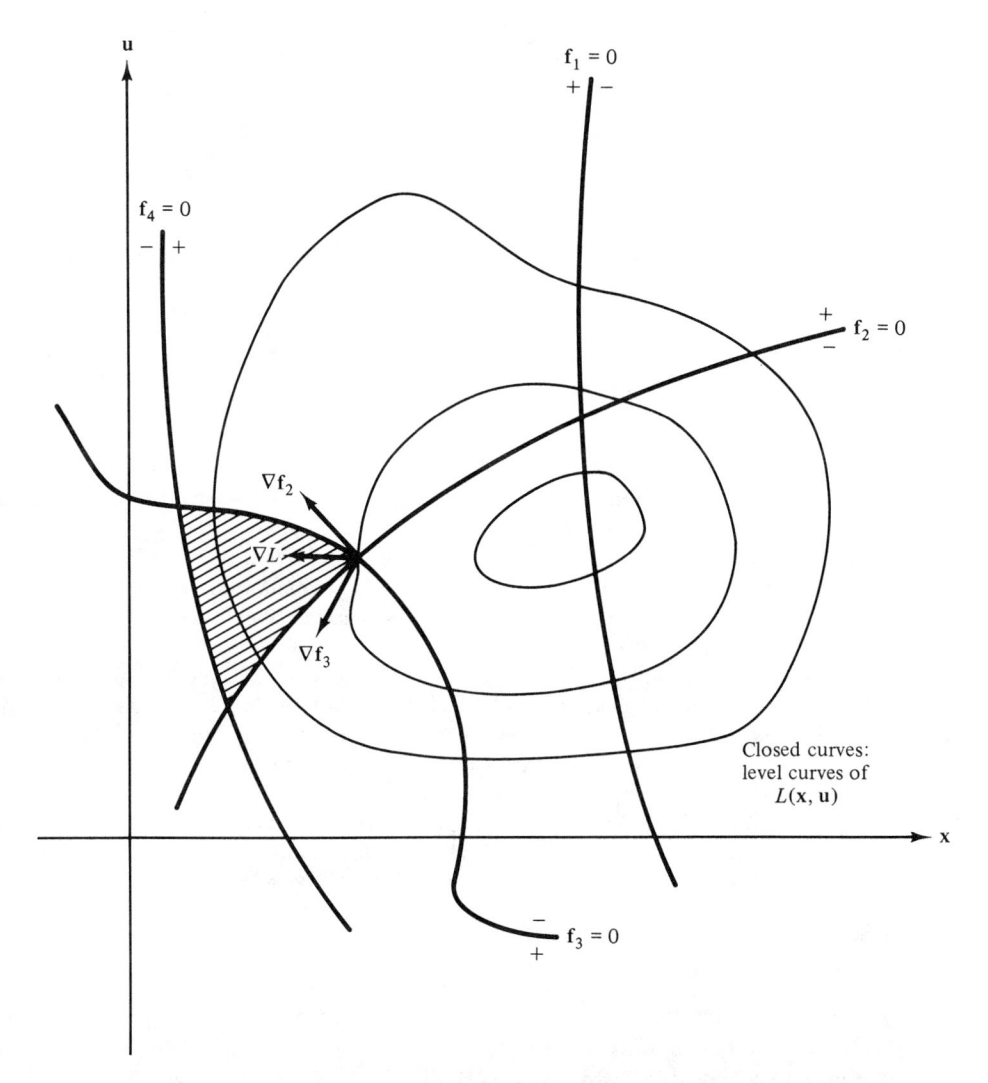

Figure 5.7 Kuhn-Tucker conditions: The gradient vector of $L(x, u)$ is a nonnegative linear combination of the gradient vectors of active constraints. Shaded region shows where all constraints are satisfied. Constraints: f_1, inactive; f_2, active; f_3, active; f_4, inactive. Closed curves: level curves of $L(x, u)$.

considerations are also very important—even assuming a dominant importance in many cases. Our coverage of this topic is intended as an introduction to the ideas and issues that arise rather than as a comprehensive treatment, which would require a book or two itself!

A very simple example serves to illustrate how computational approaches to an optimization problem can be quite different than analytical ones. Consider how you might attack the problem of finding the minimizing value of u, say u^*, for a function

$L(u)$ that is known to be *unimodal* (i.e., that is monotonic on either side of a single local minimum) inside an interval $[u_l, u_r]$ of the real line. If you approach the problem analytically, you might suggest that the problem be solved by finding the zero of the derivative of the function on the given interval. The "problem" with this solution is that it assumes that the derivative is known (or can be found) in a form that is amenable to evaluation. (If you were thinking of using Newton's method for determining the zero, you made some similar assumption about the second derivative.) This assumption is not justified in certain situations; possible reasons are: the function is not differentiable; the function itself is given implicitly, not analytically, and determining the derivative is not possible even though it is known to exist; and the function is differentiable but evaluating the expression for the derivative is a difficult computational task (say, as compared with evaluating the function).

Example

We will describe an interesting problem where it is not possible to express the cost function in analytical form. Let u be the "launch angle" for a batted baseball, the angle at which the ball leaves the bat, and assume that the initial velocity is given. Let $-L(u)$ equal the horizontal distance (range) traveled by the ball after it is hit. An obvious question is how u should be chosen to maximize range (i.e., to minimize $L(u)$). We make the reasonable assumption that the gravitational force is constant. Should we simplify the problem by assuming "ideal" projectile motion, thereby neglecting all forces except gravity, we may express the range analytically and easily determine that 45° is the best angle. However, the range must be determined by (numerical) integration of the relevant differential equations when the Newtonian aerodynamic drag effect (a force proportional to the square of the ball's speed in the direction opposite to the instantaneous velocity vector), the Magnus effect (a lift force perpendicular to the instantaneous velocity vector arising from rotation of the ball), and the drag torque (acting to decrease the initial angular velocity of the ball) are taken into account. For a range of reasonable initial velocities and angular velocities, the optimum choice of u has been determined to be about 35°.*

A Search Algorithm for Minimization

A few quick sketches will provide some intuitive justification of one strategy for locating the minimizing value, u^*, to any desired degree of accuracy. The method uses the idea of a *bounding subinterval*, a subinterval within which we can be sure that the local minimum is achieved. Given just two points in the interval $[u_l, u_r]$ and the corresponding function values at these points, it is clear that the bounding subinterval is something less than the entire interval. With three points chosen to subdivide the original interval into quarters, it is clear that a bounding subinterval of half the original length is achieved. Consider this bounding subinterval as defining the interval for a new minimization problem; we can repeat the process of and achieve another reduction in the length of the bounding subinterval by a factor of 2. Notice that each repetition will require only two additional function evaluations because the value at the midpoint of the bounding subinterval at each step is known from previous steps (Fig. 5.8). In this way, any specified solution accuracy can be achieved; the required number of function

---

*Robert K. Adair, *The Physics of Baseball* (New York: Ballantine Books, 1990), Chapter 2.

Sec. 5.2 Numerical Optimization Techniques

$L(u)$ — unimodal on $[u_l, u_r]$

Figure 5.8 Construction of bounding subintervals. The order of evaluation of points is indicated.

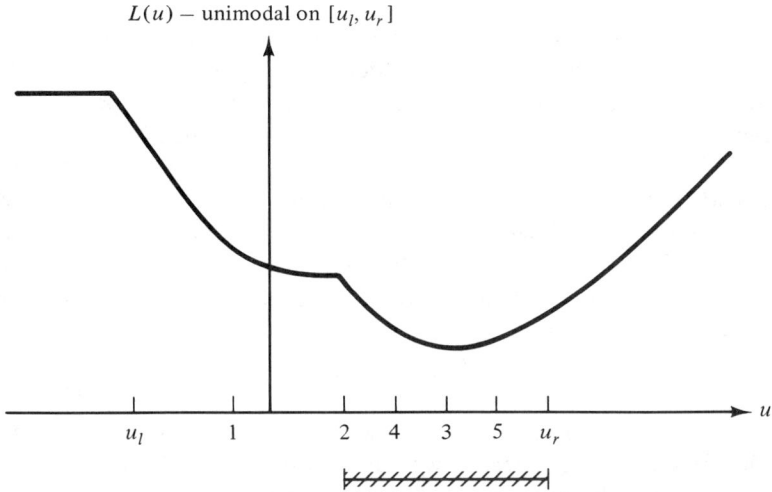

First bounding subinterval, based on points 1, 2, 3.

Second bounding subinterval, based on points 3, 4, 5.

evaluations can even be found without any knowledge of the unimodal function; in other words, the worst-case performance of the interval halving method can be determined.

The simple interval halving method just described can be improved in certain ways, but we will leave them for the interested reader to explore. Our purpose was to use it as an example of how numerical approaches to optimization can differ from methods following directly (more or less) from the analytical theory developed in the first section of this chapter. It is worthwhile to emphasize that even for some problems involving multidimensional variables, finding a shrinking sequence of sets in which an optimizing value lies can be a powerful computational strategy, especially in cases where worst-case performance guarantees are needed. There are problems where the similar idea of *functional bounds*, upper and lower bounds on the minimized value that can be iteratively improved, may be exploited. Also, the fact that the method employs only function evaluations serves to point out that optimization methods based on analytical approaches may have very significant computational requirements connected with the use of derivatives (in evaluating them or approximating them). These requirements result in benefits, of course, but it is always necessary to balance price and performance when matching methods to applications.

The Gradient Method for Minimization

We turn now to a discussion of a variety of numerical optimization methods exploiting the analytical results on optimization discussed in the first section of this chapter. It should come as no surprise that using gradient information is fundamental to a variety of important numerical methods. After all, the gradient of a scalar function of several variables is a vector pointing in the direction of steepest slope (i.e., perpendicular to the level curves (the contour lines of constant function value)).

The most basic of all numerical minimization algorithms, the *gradient method*, or *method of steepest descent*, is obtained simply by selecting a sequence of trial points by proceeding in the direction of the negative gradient. For the parameter optimization problem of minimizing the scalar function $L(\mathbf{u})$, where $\mathbf{u}$ is a vector with m components and $\mathbf{u}_0$ is an initial guess at the minimizing value, the gradient method defines a sequence of vectors by

$$\mathbf{u}_{i+1} = \mathbf{u}_i - \gamma (\nabla L(\mathbf{u}_i))^T \qquad (5.46)$$

where γ is a gain or step size parameter that must be chosen to "tune" the algorithm. (The transpose arises because we have defined the derivative of a scalar function (i.e., its gradient) to be the row vector of its partial derivatives.) The algorithm forms the $(i+1)$st vector in the sequence by moving from the ith vector in the negative gradient direction, the direction of greatest decrease of the function L at the point $\mathbf{u}_i$. The length of the step taken is proportional to the magnitude of the gradient vector; this has the desirable effect of reducing the step size as the derivative approaches $\mathbf{0}$, which will occur at local minima of L. The gain γ, which will be a small positive number, allows the algorithm to be adjusted; the idea is to find as large a value of γ as possible, so that the number of steps required to move to a local minimum is kept as small as possible, while being cautious enough to avoid any instability that would be caused by "overstepping."

The gradient algorithm takes a very simple form, but it is representative of a wide range of more elaborate algorithms, some of which will be discussed later. For this reason it is helpful to look at the analysis that deals with its performance in a fair amount of detail. Two approaches will be described in what follows.

The gradient algorithm will first be viewed as describing a nonlinear discrete-time system by a set of state equations:

$$\mathbf{u}_{i+1} = \mathbf{f}_L(\mathbf{u}_i) = \mathbf{u}_i - \gamma (\nabla L(\mathbf{u}_i))^T \qquad (5.47)$$

When γ is a small positive number, the system is nearly linear, and this suggests that linearization be used to study its behavior, just as was done in Chapter 4 when nonlinear continuous-time systems were considered. Indeed, since the minimizing value of $\mathbf{u}$ occurs at a point where (the possibly nonlinear term) $\nabla L(\mathbf{u}) = \mathbf{0}$, the desired minimizing vector is an equilibrium point of the discrete-time system. The performance analysis of the gradient algorithm is simply a matter of determining the stability of this equilibrium point (or more generally all equilibrium points corresponding to local minima) and, given stability, its domain of attraction.

Linearization of the system involves the derivative of the function $\mathbf{f}_L(\mathbf{u})$, and this requires the derivative of the vector $(\nabla L)^T$. This is just the matrix of second partial derivatives of L introduced earlier in this chapter. To see this, recall that if we expand L up to second-order terms in its Taylor series, we have

Sec. 5.2 Numerical Optimization Techniques

$$L(\mathbf{u}) \approx L(\mathbf{u}_0) + L_\mathbf{u}(\mathbf{u}_0)(\mathbf{u}-\mathbf{u}_0) + \frac{1}{2}(\mathbf{u}-\mathbf{u}_0)^T L_{\mathbf{uu}}(\mathbf{u}_0)(\mathbf{u}-\mathbf{u}_0) \tag{5.48}$$

where $L_\mathbf{u}(\mathbf{u}_0)$ is simply another notation for $\nabla L(\mathbf{u}_0)$ and

$$L_{\mathbf{uu}} = \begin{bmatrix} L_{u_1 u_1} & \cdots & L_{u_1 u_m} \\ \vdots & & \vdots \\ L_{u_m u_1} & \cdots & L_{u_m u_m} \end{bmatrix} = \begin{bmatrix} (L_\mathbf{u})_{u_1} \\ \vdots \\ (L_\mathbf{u})_{u_m} \end{bmatrix} = (L_\mathbf{u}^T)_\mathbf{u} \tag{5.49}$$

The corresponding expansion to first-order terms in the Taylor series for the gradient is thus given by

$$\nabla L(\mathbf{u}) \approx \nabla L(\mathbf{u}_0) + (\mathbf{u}-\mathbf{u}_0)^T L_{\mathbf{uu}}(\mathbf{u}_0) \tag{5.50}$$

Therefore, in the vicinity of a point where $\nabla L(\mathbf{u}^*) = \mathbf{0}$,

$$\nabla L(\mathbf{u}) \approx (\mathbf{u}-\mathbf{u}^*)^T L_{\mathbf{uu}}(\mathbf{u}^*) \tag{5.51}$$

The linearized gradient method system is

$$\begin{aligned} \mathbf{u}_{i+1} &= \nabla(\mathbf{f}_L(\mathbf{u}^*))\mathbf{u}_i \\ &= (\mathbf{I} - \gamma L_{\mathbf{uu}}(\mathbf{u}^*))\mathbf{u}_i \end{aligned} \tag{5.52}$$

From the solution to this linear discrete-time system,

$$\mathbf{u}_i = (\mathbf{I} - \gamma L_{\mathbf{uu}}(\mathbf{u}^*))^i \mathbf{u}_0 \tag{5.53}$$

the following condition for stability is obtained:

$$\gamma < \frac{1}{\lambda_{\max}} \tag{5.54}$$

where the denominator denotes the largest eigenvalue of $L_{\mathbf{uu}}(\mathbf{u}^*)$. (Recall that for a local minimum, $L_{\mathbf{uu}}(\mathbf{u}^*)$ is positive definite and therefore has positive eigenvalues.)

From a practical point of view, this result is not altogether satisfying because the matrix $L_{\mathbf{uu}}$ depends on $\mathbf{u}^*$, so the eigenvalues depend on the equilibrium point, which is not known. However, if it is possible to obtain an upper bound on the maximum eigenvalue, this may be used to guide the choice of γ.

Example

Minimizing a quadratic function makes a useful test case to consider because the solution may be expressed analytically. Furthermore, it represents the situation where the Taylor series approximation up to second order is exact. Thus, with

$$L(\mathbf{u}) = \frac{1}{2}\mathbf{u}^T \mathbf{Q}\mathbf{u} + \mathbf{m}^T \mathbf{u}$$

where $\mathbf{Q}$ is a positive definite matrix and $\mathbf{m}$ is a vector of constants, it is easily determined that

$$\nabla L(\mathbf{u}) = \mathbf{u}^T \mathbf{Q} + \mathbf{m}^T$$

and

$$L_{\mathbf{uu}}(\mathbf{u}) = \mathbf{Q}$$

In this case, the second derivative matrix is independent of **u** and γ can be chosen to be any value smaller than the reciprocal of the largest eigenvalue of **Q**.

Writing out the gradient algorithm explicitly gives

$$\mathbf{u}_{i+1} = \mathbf{u}_i - \gamma(\mathbf{Q}\mathbf{u}_i - \mathbf{m})$$
$$= (\mathbf{I} - \gamma\mathbf{Q})\mathbf{u}_i - \gamma\mathbf{m}$$

As shown in Chapter 2, the solution to this linear discrete-time system takes the form

$$\mathbf{u}_i = (\mathbf{I} - \gamma\mathbf{Q})^i \mathbf{u}_0 - \gamma \sum_{k=0}^{i-1} (\mathbf{I} - \gamma\mathbf{Q})^k \mathbf{m}$$

With γ chosen as described above, the system is asymptotically stable and $\mathbf{u}_i$ approaches a limiting value, $\mathbf{u}_\infty$, satisfying

$$\mathbf{u}_\infty = \mathbf{u}_\infty - \gamma\mathbf{Q}\mathbf{u}_\infty - \gamma\mathbf{m}$$

This may be solved to obtain $\mathbf{u}_\infty$, giving

$$\mathbf{u}_\infty = -\mathbf{Q}^{-1}\mathbf{m}$$

as expected. In this problem, since **Q** is positive definite, there is a unique solution that is a global minimum.

A second approach to analyzing the behavior of a numerical optimization algorithm such as the gradient algorithm is to consider the limiting case for vanishing step size (i.e., to let $\gamma \to 0$). It is easy to recognize that the limiting form of the algorithm is the differential equation

$$\dot{\mathbf{u}}(t) = -(\nabla L(\mathbf{u}(t)))^T \tag{5.55}$$

In fact, the gradient algorithm may be recognized as the Euler discretization of this differential equation using discretization step size γ. Recall from Chapter 3 that the issue of stability of the discretization was examined—mainly for linear systems, but from Chapter 4 we know that asymptotic stability of a linearized system guarantees asymptotic stability of the corresponding nonlinear system. At any local minimum of $L(\mathbf{u})$, the gradient vanishes and thus the local minimum is an equilibrium point for the differential equation. The linearized system around such an equilibrium solution is

$$\frac{d}{dt}\Delta\mathbf{u}(t) = -L_{\mathbf{uu}}\Delta\mathbf{u}(t) \tag{5.56}$$

(By now this should be no surprise!) The stability condition for Euler discretization provides exactly the same result as obtained earlier: $\gamma < 1/\lambda_{max}$.

The strong point of the gradient algorithm is its simple form, which makes for computational efficiency when the gradient vector is available in a computationally convenient form. The algorithm's performance is degraded by a large eigenvalue spread associated with the $L_{\mathbf{uu}}$ matrix (i.e., in the case where the associated differential equation is "stiff"). Problems of this kind arise when minimizing a function $L(\mathbf{u})$ whose level curves are like the contour lines of a very narrow gorge or ravine where, except for the points very near the "valley floor," the direction of steepest descent is nearly perpendicular to the direction in which the minimum value lies.

Example

A challenging numerical problem is to minimize the following function, known as the Rosenbrock function for its "inventor," H. Rosenbrock:

$$L(\mathbf{u}) = L(u_1, u_2) = 100(u_2 - u_1^2)^2 + (1 - u_1)^2$$

The global minimum, at the point $u_1^* = u_2^* = 1$, lies in a ravine that presents the gradient method with considerable difficulty. As an exercise, consider the solution starting at the point $u_1 = -1.2$ and $u_2 = 1$.

A variety of techniques have been developed to improve upon the basic gradient method described above. We will not attempt to cover all of the specifics, but some of the general ideas are worth mentioning. Based on our discussion in Chapter 3 about stability of discretization methods, the reader may think of applying backward difference discretization (or other methods developed to deal with stiff differential equations) to the associated continuous-time system; the drawback to such an approach is that the algorithm becomes an implicit one and generally requires much more computation at each step, but the price might be worth paying in particularly difficult problems.

Example

For minimizing the function $L(\mathbf{u})$ using the backward difference discretization of the associated differential equation, we have the following algorithm:

$$\mathbf{u}_{i+1} = \mathbf{u}_i - \gamma(\nabla L(\mathbf{u}_{i+1}))^T$$

Typically, it will not be possible to solve this implicit equation analytically for $\mathbf{u}_{i+1}$. This drawback must be addressed with a numerical approach of some kind. Since γ is small, we expect that $\mathbf{u}_{i+1} \approx \mathbf{u}_i$ and it is reasonable to use one step of Newton's method to obtain an approximate solution of the implicit equation for $\mathbf{u}_{i+1}$. This amounts to the approximation

$$\nabla L(\mathbf{u}_{i+1}) \approx \nabla L(\mathbf{u}_i) + (\mathbf{u}_{i+1} - \mathbf{u}_i)^T L_{\mathbf{uu}}(\mathbf{u}_i)$$

and it gives the following minimization algorithm:

$$\mathbf{u}_{i+1} = (\mathbf{I} + \gamma L_{\mathbf{uu}}(\mathbf{u}_i))^{-1}(\mathbf{I} - \gamma L_{\mathbf{uu}}(\mathbf{u}_i))\mathbf{u}_i - \gamma \nabla L(\mathbf{u}_i)$$

As this example shows, the use of backward discretization leads, quite naturally, to an algorithm employing second derivative information about the function $L(\mathbf{u})$; Newton's method applied to solving $L(\mathbf{u}) = \mathbf{0}$ generally makes better use of this extra information.

Another approach is quite commonly used in practice. The gradient algorithm is modified slightly by using a variable step size, where the gain γ is adjusted at each step in an optimal, or near-optimal, way. The idea behind this kind of method is that the gradient directions may be used more efficiently by choosing the largest step size possible, consistent with the objective of minimization. Such a strategy would hopefully prevent overstepping such as would be expected to occur in minimizing ravine-like functions when γ is too large in the gradient method.

For differentiable functions (and we have tacitly assumed that $L(\mathbf{u})$ is twice differentiable in our analysis involving $\mathbf{L_{uu}}$) a curve-fitting approach to minimization along a line (as opposed to the interval halving method described at the beginning of this section) is usually very effective. This is applied to the variable step size gradient method by considering the basic iterative step,

$$\mathbf{u}_{i+1} = \mathbf{u}_i - \gamma_i (\nabla L(\mathbf{u}_i))^T \tag{5.57}$$

where we use a subscript on γ to indicate that it will change from iteration to iteration. Consider the value of the function to be minimized at the point $\mathbf{u}_{i+1}$:

$$L(\mathbf{u}_{i+1}) = L(\mathbf{u}_i - \gamma_i (\nabla L(\mathbf{u}_i))^T) \tag{5.58}$$

We view this as a function of γ_i to be minimized; for ease of notation we will denote it by $\lambda(\gamma_i)$. The function represents a cross section taken through the surface defined by $L(\mathbf{u})$ in the direction of $(\nabla L(\mathbf{u}_i))^T$ at the point $\mathbf{u}_i$ (Fig. 5.9). There is no real need to seek the minimizing value of γ_i to high accuracy, and a reasonable approach to take is to select a candidate step size, such as $\gamma_c = \gamma_{i-1}$, the step size used at the previous iteration, and to decide upon the choice of γ_i as follows.

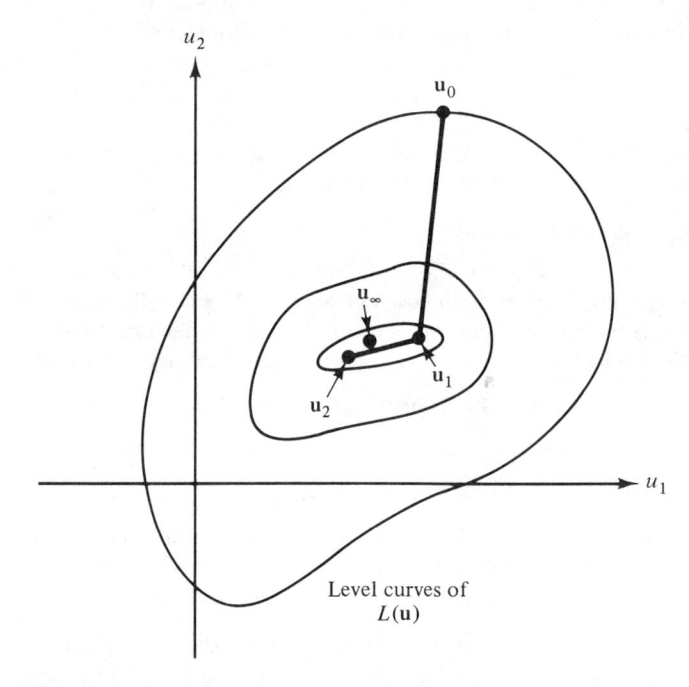

Figure 5.9 Path follows the sequence of gradient vectors; step size selected is for one-dimensional minimization.

Step size adjustment algorithm:

First, $\lambda(\gamma_c)$ is evaluated to see whether or not a value smaller than $\lambda(0)$ is obtained. We first describe the procedure for handling the situation when $\lambda(\gamma_c) < \lambda(0)$. In this case, $\lambda(2\gamma_c)$ is evaluated also. Then, if $\lambda(2\gamma_c) > \lambda(\gamma_c)$, the candidate step size is selected as γ_i. Otherwise, $\lambda(2\gamma_c) < \lambda(\gamma_c)$, and a quadratic function could be fit to the values determined so far: $\lambda(0)$, $\lambda(\gamma_c)$, and $\lambda(2\gamma_c)$. If this quadratic achieves its minimum at a point between $2\gamma_c$ and $4\gamma_c$—it turns out that this does not require that the quadratic function coefficients be determined explicitly because the condition can be conveniently expressed in terms of the three

Sec. 5.2 Numerical Optimization Techniques

function values—then $2\gamma_c$ is selected as γ_i; otherwise, $2\gamma_c$ replaces γ_c as the candidate step size, and the procedure is continued: $\lambda(4\gamma_c)$ is evaluated; this causes $2\gamma_c$ to be selected if an increase in the value is found; if not, the minimum of the quadratic function fit to $\lambda(0)$, $\lambda(2\gamma_c)$, and $\lambda(4\gamma_c)$ is tested to determine if $4\gamma_c$ should be taken as the step size, and so on.

Now we return to case not yet handled, when $\lambda(\gamma_c) > \lambda(0)$ at the beginning of the search process. In this case a half-length step is tried, and if $\lambda(0.5\gamma_c) < \lambda(0)$, this becomes γ_i. If the step size is still too big, a quadratic function is imagined to be fit to the three values $\lambda(0)$, $\lambda(0.5\gamma_c)$, and $\lambda(\gamma_c)$. If this quadratic function has a positive derivative at 0, or if the derivative at 0 is negative but the minimum is achieved at a point less than $0.1\gamma_c$, the candidate step size is reduced to $0.1\gamma_c$ and the search starts over. Otherwise, the point at which the quadratic function achieves its minimum is taken as the new candidate step size and the search is restarted.

Many variations on this general theme are possible, such as the use of derivative information as well as function values in the curve-fitting step, or the fitting of cubic polynomials or other functions. Line search methods are employed in the quasi-Newton optimization algorithms to be described later.

Modifying the choice of directions, instead of using only the gradient direction at each step, is another variation on the gradient method that provides some important practical alternatives. In the simplest case this is accomplished by selecting a direction in a way that depends on the gradient and on directions selected previously. One popular method of this type is described in the following example.

Example

A simple variation on the gradient algorithm is obtained by adding an extra term, called a momentum term, giving

$$\mathbf{u}_{i+1} = \mathbf{u}_i - \gamma (\nabla L(\mathbf{u}_i))^T + \eta (\mathbf{u}_i - \mathbf{u}_{i-1})$$

The momentum coefficient, η, determines how strongly the most recent direction affects the choice of a new direction. This algorithm may be viewed as employing a multipoint discrete approximation to the derivative term in the associated differential equation.

Another way of obtaining variations on the gradient method is to exploit the fact that minimizing a quadratic function is accomplished by solving a linear equation; the choice of **u** minimizing

$$Q(\mathbf{u}) = \tfrac{1}{2}\mathbf{u}^T \mathbf{Q}\mathbf{u} + \mathbf{m}^T \mathbf{u} \qquad (5.59)$$

is the solution to

$$\mathbf{Q}\mathbf{u} = -\mathbf{m} \qquad (5.60)$$

Only a slight change in perspective is required to conclude, on the basis of the example already carried out, that the gradient algorithm provides an iterative method of solving the linear equation. In fact, the gradient algorithm is precisely Richardson's method, introduced in Chapter 4 as an example of the general method of successive approximation. Other iterative solution methods for linear equations, such as the Gauss-Seidel and successive overrelaxation methods introduced in Chapter 1, may also be adapted to provide numerical minimization procedures by identifying $\mathbf{Q}$ as the matrix of second partial derivatives, $L_{\mathbf{uu}}$, for the quadratic case.

Quasi-Newton Methods for Minimization

With more elaborate numerical methods it is possible to exploit the analytical theory for minimization even further. As mentioned earlier, finding a local extremum of $L(\mathbf{u})$ requires finding values of $\mathbf{u}$ where the gradient, $\nabla L(\mathbf{u})$, is zero. Newton's method, possibly using a set of different starting points if many local extrema are present, offers one approach to solving for such values of $\mathbf{u}$. In the notation of this chapter, Newton's method is the iteration

$$\mathbf{u}_{i+1} = \mathbf{u}_i - L_{\mathbf{uu}}^{-1}(\mathbf{u}_i)(\nabla L(\mathbf{u}_i))^{\mathrm{T}} \quad (5.61)$$

A modified Newton method is commonly used in practice. The iteration is

$$\mathbf{u}_{i+1} = \mathbf{u}_i - h_k L_{\mathbf{uu}}^{-1}(\mathbf{u}_i)(\nabla L(\mathbf{u}_i))^{\mathrm{T}} \quad (5.62)$$

where h_k is a step size adjustment parameter selected by using a line search along the "Newton direction" to perform an approximate minimization of L.

In the equation describing Newton's method we have "solved" for $\mathbf{u}_{i+1}$ using a matrix inverse in order to put it in a form similar to the gradient algorithm. Indeed, notice that we could interpret the gradient algorithm as an approximation to Newton's method obtained by using the scaled identity matrix $\gamma \mathbf{I}$ in place of $L_{\mathbf{uu}}^{-1}$. The variable step size gradient method provides the same approximation to the modified Newton method. (Using a diagonal matrix consisting of the reciprocals of the diagonal elements of $L_{\mathbf{uu}}$ amounts to a Jacobi-type method or, with immediate use of partial results of updating $\mathbf{u}_{i+1}$, to a Gauss-Seidel-type method.)

The point we wish to emphasize is that using Newton's method requires the solution of a set of linear equations involving the derivative of ∇L (i.e., the matrix $L_{\mathbf{uu}}$ of second partial derivatives of $L(\mathbf{u})$), at every step of the algorithm. In practice, even the evaluation of this matrix can be a formidable task when the function L is complicated or the number of components of $\mathbf{u}$ is large. A class of methods known as *quasi-Newton* methods have been developed for the purpose of reducing the computational costs of Newton's method without losing its excellent convergence properties. Our view is that the key idea behind the quasi-Newton methods is that the crude but trivial-to-compute approximation leading to the gradient method can be modified to give an approximation with better properties which imposes only a modest increase in computational costs.

We use $\mathbf{H}_i$ to denote an approximation to the inverse of the matrix $L_{\mathbf{uu}}(\mathbf{u}_i)$; the form of the quasi-Newton iteration is

$$\mathbf{u}_{i+1} = \mathbf{u}_i - h_i \mathbf{H}_i (\nabla L(\mathbf{u}_i))^{\mathrm{T}} \quad (5.63)$$

where we have included the step size adjustment parameter h_i which is to be determined by a line search minimization. In order to view such an iteration as an approximation to the modified Newton method, $\mathbf{H}_i$ needs to be suitably constructed to approximate $L_{\mathbf{uu}}(\mathbf{u}_i)$. Three popular choices for $\mathbf{H}_i$ lead to quasi-Newton methods known by the names *Davidon-Fletcher-Powell*, *symmetric rank-one*, and *Broyden-Fletcher-Goldfarb-Shanno*. To give a concise description of these methods, we use the following notation:

$$s_i = \mathbf{u}_{i+1} - \mathbf{u}_i \tag{5.64}$$

$$y_i = (\nabla L(\mathbf{u}_{i+1}))^\mathrm{T} - (\nabla L(\mathbf{u}_i))^\mathrm{T} \tag{5.65}$$

Then we have the following:

Davidon-Fletcher-Powell:

$$\mathbf{H}_{i+1} = \mathbf{H}_i + \frac{s_i s_i^\mathrm{T}}{y_i^\mathrm{T} s_i} - \frac{(\mathbf{H}_i y_i)(\mathbf{H}_i y_i)^\mathrm{T}}{y_i^\mathrm{T} \mathbf{H}_i y_i} \tag{5.66}$$

Symmetric rank-one:

$$\mathbf{H}_{i+1} = \mathbf{H}_i + \frac{(s_i - \mathbf{H}_i y_i)(s_i - \mathbf{H}_i y_i)^\mathrm{T}}{y_i^\mathrm{T}(s_i - \mathbf{H}_i y_i)} \tag{5.67}$$

Broyden-Fletcher-Goldfarb-Shanno:

$$\mathbf{H}_{i+1} = \mathbf{H}_i + \frac{y_i^\mathrm{T}(\mathbf{H}_i y_i + s_i) s_i s_i^\mathrm{T}}{(y_i^\mathrm{T} s_i)^2} - \frac{s_i y_i^\mathrm{T} \mathbf{H}_i + \mathbf{H}_i y_i s_i^\mathrm{T}}{y_i^\mathrm{T} s_i} \tag{5.68}$$

Various details of implementation and further discussion of these methods may be found in the references. We simply note that the quasi-Newton methods require an initial approximation, $\mathbf{H}_0$, which is commonly taken to be $\mathbf{I}$. Also, other quasi-Newton methods can be obtained by directly approximating $L_{\mathbf{uu}}$ instead of its inverse.

Methods for Constrained Minimization Problems

As a final topic to be considered in this section, we turn to constrained optimization problems. More often than not, engineering design problems involve several quantities that may not be chosen arbitrarily and that may be interdependent. Some examples: Physical quantities such as mass and area are nonnegative; energy is conserved by the motion of a frictionless pendulum; mass balance is preserved in chemical reactions. The discussion here will be limited to minimization problems with equality constraints. There are various "tricks" available for handling other kinds of constraints within the more restricted setting. For example, variable transformations such as $w = \arctan u$ may be used to map a finite interval, in this case $[-\pi/2, \pi/2]$, onto the real line, thereby allowing "box constraints" to be handled. For single inequality constraints such as $u \geq 0$, taking $u = v^2$ produces an unconstrained variable v. These tricks are not without their costs, however, since the resulting problems are likely to be difficult to solve; after all, there is no such thing as a free lunch.

Penalty function approaches are popular for turning certain equality constrained minimization problems into unconstrained ones so that methods like those described earlier in this section may be applied. The general idea is to replace a problem such as

$$\min L(\mathbf{x}, \mathbf{u}) \tag{5.69}$$

subject to the constraints

$$\mathbf{f}(\mathbf{x},\mathbf{u}) = \mathbf{0} \tag{5.70}$$

with the unconstrained problem

$$\min L(\mathbf{x},\mathbf{u}) + r\,\mathbf{f}^T(\mathbf{x},\mathbf{u})\mathbf{f}(\mathbf{x},\mathbf{u}) \tag{5.71}$$

The idea is that when the parameter r is very large, the second term severely penalizes potential solutions unless the constraints are nearly satisfied; in the limit as $r \to \infty$, a finite minimum can only be achieved when the constraints are exactly satisfied. More elaborate problems, where the added penalty term takes the form $\mathbf{f}^T \mathbf{R} \mathbf{f}$, with $\mathbf{R}$ a diagonal matrix with different, large entries, can be handled in a similar way.

For the simple case, numerical solutions for a sequence of unconstrained problems, corresponding to a sequence of parameters $1 \ll r_1 < r_2 < \cdots$, are computed. When there is little change in the resulting solutions, the process is halted and the last solution is used as the solution to the constrained problem. This is justified by theoretical results that guarantee that the solutions converge for well-behaved problems. It should also be noted that the kind of compromise solution obtained by using a penalty function might be even more appropriate than a solution to a constrained problem when there are modeling uncertainties involved in the specification of $\mathbf{f}(\mathbf{x},\mathbf{u})$.

There is a potential drawback to the penalty function approach arising from the fact that the sequence of unconstrained problems may become more and more ill-conditioned as r grows. A deep local minimum may dominate and prevent fast convergence toward the constrained solution, or a problem with many shallow local minima may produce solutions dominated by the penalty term. Design of the penalty term (it need not be quadratic) and the selection of a "good" parameter sequence are largely matters of experience and judgment.

The reader might wonder why the penalty function method is introduced to find solutions of a problem where there are analytical results available (namely the Lagrange multiplier theory developed earlier) which apparently provide a basis for numerical computation. Of course, the answer is that the theory does not always lead to a well-conditioned problem with a computationally effective solution. For example, the assumption that the constrained minimum occurs at a regular point of the constraint function can fail to hold, or "nearly fail" to hold. Another difficulty can arise from the fact that the second-order conditions for a local constrained minimum do not require that the matrix of second partial derivatives with respect to $(\mathbf{x},\mathbf{u})$ be positive definite, so the application of quasi-Newton methods to solving constrained minimization problems must be done with some care.

Even the application of the gradient algorithm involves a small subtlety. Recall that the necessary conditions satisfied by a local minimum of $L(\mathbf{x},\mathbf{u})$ subject to the constraints $\mathbf{f}(\mathbf{x},\mathbf{u}) = \mathbf{0}$ are conveniently written in terms of the function $H(\mathbf{x},\mathbf{u},\lambda)$, where

$$H(\mathbf{x},\mathbf{u},\lambda) = \lambda^T \mathbf{f}(\mathbf{x},\mathbf{u}) + L(\mathbf{x},\mathbf{u}) \tag{5.72}$$

The resulting equations are

$$H_\mathbf{x} = \lambda^T \mathbf{f}_\mathbf{x} + L_\mathbf{x} = \mathbf{0} \tag{5.73}$$

$$H_\mathbf{u} = \lambda^T \mathbf{f}_\mathbf{u} + L_\mathbf{u} = \mathbf{0} \tag{5.74}$$

$$H_\lambda = \mathbf{f}^T = \mathbf{0} \tag{5.75}$$

While the solution to these equations is a critical point of the H function, it is typically a saddle point and not a local minimum!

To remedy this situation, the gradient algorithm can be modified to perform steepest ascent on the λ variables in order to obtain a convergent algorithm. (An alternative would be to use steepest descent on $-\lambda$.) Thus the algorithm takes the form

$$\mathbf{u}_{i+1} = \mathbf{u}_i - \gamma \lambda_i^T \mathbf{f}_\mathbf{u}(\mathbf{x}_i, \mathbf{u}_i) - \gamma L_\mathbf{u}(\mathbf{x}_i, \mathbf{u}_i) \tag{5.76}$$

$$\mathbf{x}_{i+1} = \mathbf{x}_i - \gamma \lambda_i^T \mathbf{f}_\mathbf{x}(\mathbf{x}_i, \mathbf{u}_i) - \gamma L_\mathbf{x}(\mathbf{x}_i, \mathbf{u}_i) \tag{5.77}$$

$$\lambda_{i+1} = \lambda_i + \gamma \mathbf{f}^T(\mathbf{x}_i, \mathbf{u}_i) \tag{5.78}$$

It is possible to combine the penalty function and Lagrange multiplier approaches. This amounts to using the Lagrange multiplier approach with

$$H(\mathbf{x}, \mathbf{u}, \lambda) = \lambda^T \mathbf{f}(\mathbf{x}, \mathbf{u}) + L(\mathbf{x}, \mathbf{u}) + r \mathbf{f}^T(\mathbf{x}, \mathbf{u}) \mathbf{f}(\mathbf{x}, \mathbf{u}) \tag{5.79}$$

The advantage of the combined approach is that convergence to the solution of the original constrained minimization problem is achieved for some finite value of r, rather than in the limit as $r \to \infty$. Intuitively, once r is large enough, the penalized, constrained problem admits neighboring optimal solutions, and methods like the gradient method are assured to converge to the solution.

5.3 PATH OPTIMIZATION PROBLEMS AND THE PRINCIPLE OF OPTIMALITY

Now we turn to a discussion of what we have called path optimization problems. We include in this category those optimization problems that involve the design of signals (discrete-time or continuous time) for the purpose of optimizing some measure of system performance. As a simple example you can think of determining the throttle input, as a function of time, required to drive a car from location A to location B in the shortest time but using no more than some fixed amount of fuel.

We want to emphasize a particular framework for thinking about path optimization problems: the *principle of optimality*. To do this we will think in terms of three kinds of problems. The first kind is typified by the problem of finding a shortest path between two points in a graph whose branches are assigned (positive) lengths. The length of a path between two nodes is the sum of the lengths of the branches making up the path. To find a shortest path is a combinatorial optimization problem that is solved in a sequence of (discrete) steps, each step involving a discrete set of possible choices. The key to solving this problem efficiently (at least improving on the brute force attack of simply enumerating all possible paths and their lengths and then choosing a shortest one) is the principle of optimality:

subpaths of a shortest path are themselves shortest paths.

More specifically, if a node k is on the shortest path between nodes i and j, then the subpath from node i to node k and the subpath from node k to node j must also be shortest paths themselves (Fig. 5.10).

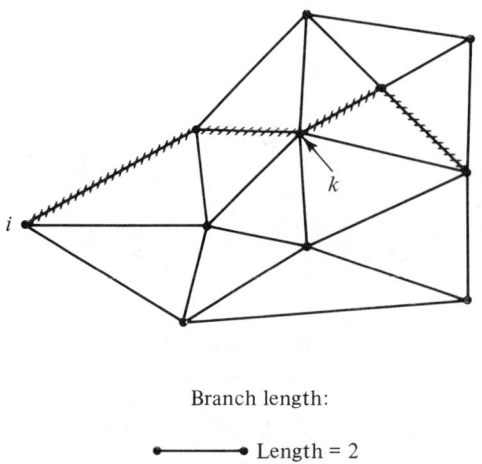

Branch length:

•——————• Length = 2

•⁓⁓⁓⁓⁓• Length = 1

Figure 5.10 Graph with shortest path between nodes i and j passing through k.

The proof of this statement is quite simple, so we give it for the sake of completeness. If path $P_{i \to j}$ is a shortest path from i to j, and if it passes through an intermediate node k, then two subpaths are formed, $P_{i \to k}$ and $P_{k \to j}$. Suppose there is another path from i to k, say $\bar{P}_{i \to k}$. The length of the path $\bar{P}_{i \to j}$ that consists of $\bar{P}_{i \to k}$ followed by $P_{k \to j}$ is no less than the length of the path $P_{i \to j}$, since the latter is a shortest path; it follows that the length of path $\bar{P}_{i \to k}$ is no less than the length of the path $P_{i \to k}$. Hence path $P_{i \to k}$ is a shortest path. A similar argument shows that $P_{k \to j}$ is a shortest path.

A second kind of path optimization problem involves a generalization of the first in the following way. Again, the problem is posed in terms of a sequence of discrete steps (e.g., a time sequence corresponding to the time evolution of a discrete-time system). Now, however, at each step instead of there being a finite number of possible choices to be made, a parameter optimization problem must be solved (i.e., there are an infinite number of alternative choices for a set of parameters that must be resolved by a suitable parameter optimization procedure). We will give an example of this kind of problem after a bit more discussion. For these problems we may again appeal to the principle of optimality to determine optimum solution paths: Subpaths of an optimal path must themselves be optimal paths.

The third kind of path optimization problem generalizes the second. Instead of having a sequence of discrete steps at which parameter optimization problems must be solved, there is a continuum of such steps. For example, a system described by a differential equation is to have its performance optimized by choosing an input function; the automobile throttle control problem mentioned earlier falls into this class. It turns out that the principle of optimality again may be used to describe (and in principle to

Sec. 5.3 Path Optimization Problems and the Principle of Optimality

find) optimum solution paths for such problems. The mathematical formulation of this kind of path optimization problem involves considerable mathematical sophistication, and we will content ourselves with a brief description. One intuitive way to think about such problems is in terms of the limiting case of discrete stage problems as the sampling times become closer and closer together; this will be illustrated later. It should not be too surprising that the optimal solution paths are described in terms of certain differential equations analogous to setting derivatives to zero in parameter optimization problems. To put this approach into rigorous mathematical terms requires the use of *variational calculus*; our earlier discussion of function-space derivatives provides a suitable framework, but our discussion will not include details of the development.

Discrete-Stage Path Optimization and the Principle of Optimality

A representative problem of the second kind mentioned above will serve as an illustrative example. For the scalar discrete-time system

$$x_{n+1} = ax_n + u_n, \quad x_0 = x \tag{5.80}$$

it is desired to choose the control sequence $\{u_n\}$ to achieve $x_N = 0$. Unless $N = 1$, there are many possible solutions to this problem, so we introduce a cost function as a criterion for choosing the "best" solution. We seek to minimize the function

$$\sum_{n=1}^{N}(x_n^2 + u_{n-1}^2) \tag{5.81}$$

Note that the cost is a sum of nonnegative terms, one corresponding to each of the control values that is to be chosen. If $N = 1$, there is only one choice to be made, and furthermore, there is only one value of u_0 that will make $x_1 = 0$, namely $u_0 = -ax_0$. When $N > 1$, for any choice of $u_0, \ldots, u_{N-2}$ there is a single choice of u_{N-1} that produces $x_N = 0$, namely $u_{N-1} = -ax_{N-1}$. Thus our problem is in determining how to best choose the control values $u_0, \ldots, u_{N-2}$ according to the cost function given above.

We will apply the principle of optimality to determine how the control values should be chosen. To formulate the problem in mathematical terms, we introduce the *value function* associated with this problem: Let $J_N(x)$ denote the value of the minimized cost function; as indicated, it is a function of two variables, the number of stages, N, and the initial condition, x. From the fact that there is a unique u_0 leading to $x_1 = 0$, the one-step problem has associated with it the minimum cost $J_1(x) = a^2x^2$, and this provides a boundary condition for the value function.

To go further, we use the principle of optimality to obtain an equation that relates the value function for successive stages, say stages $N-1$ and N. Notice that with the initial condition $x_0 = x$, if the first input takes the value u, then $x_1 = ax + u$. For an optimizing sequence of control values, the principle of optimality assures that an optimizing sequence of control values is chosen for all time instants after 0. For the $(N-1)$-stage problem starting at the value $x_1 = ax + u$, the minimum cost is $J_{N-1}(ax + u)$. Thus the cost of the N-stage problem must obey the equation

$$J_N(x) = \min_{u}[(ax + u)^2 + u^2 + J_{N-1}(ax + u)] \tag{5.82}$$

where $N > 1$, and where the minimum is taken over u, with the minimizing choice being u_0.

It is quite easy to obtain a similar equation for the value function of any problem from a broad class that includes this example. The difference equation defining x_{k+1} may depend nonlinearly on x_k and u_k, and the cost function may be expressed as a sum of general nonnegative terms instead of being quadratic in form. Hence, for

$$\mathbf{x}_{n+1} = \mathbf{F}(\mathbf{x}_n, \mathbf{u}_n), \qquad \mathbf{x}_0 = \mathbf{x} \tag{5.83}$$

where the goal is to choose the input sequence so that $\mathbf{x}_N = \mathbf{0}$ while minimizing the cost function

$$\sum_{n=1}^{N} l(\mathbf{x}_n, \mathbf{u}_{n-1}) \tag{5.84}$$

we find that the value function satisfies the equation

$$J_N(\mathbf{x}) = \min_{\mathbf{u}} [l(\mathbf{F}(\mathbf{x}, \mathbf{u}), \mathbf{u}) + J_{N-1}(\mathbf{F}(\mathbf{x}, \mathbf{u}))] \tag{5.85}$$

For such a general problem, determination of the boundary condition cannot be done analytically; it is required to find $\mathbf{u}^*$ as a function of $\mathbf{x}$ so that $\mathbf{F}(\mathbf{x}, \mathbf{u}^*) = \mathbf{0}$, and then $J_1(\mathbf{x}) = l(\mathbf{0}, \mathbf{u}^*(\mathbf{x}))$.

Even for the simple case chosen as an example, it is not clear how to go about solving the equation to obtain an analytical expression for the value function. About all that can be done in general is to compute a tabular partial description or some other approximation for successive value functions. Or one can be lucky enough to guess the correct functional form for the value functions and to verify that it is indeed correct.

For the example problem, a good guess is available, and we will carry out a more detailed analysis of the problem to illustrate some important ideas. Notice that the known boundary condition, $J_1(x) = a^2 x^2$, is a quadratic function; this makes plausible the guess that a quadratic function will solve the difference equation, so we assume the form

$$J_N(x) = p_N x^2 \tag{5.86}$$

where the sequence of coefficients, $\{p_N\}$, is to be determined subject to the initial condition $p_1 = a^2$. Plugging this form into the equation, we find that for $N > 1$

$$p_N x^2 = \min_{u} [(ax + u)^2 + u^2 + p_{N-1}(ax + u)^2] \tag{5.87}$$

This minimization may be performed by setting the derivative of the bracketed quantity with respect to u equal to zero, with the minimizing choice of u giving the optimal control value at time 0:

$$u_0 = -\frac{(1 + p_{N-1})ax}{2 + p_{N-1}} \tag{5.88}$$

Plugging this back into equation (5.87) leads to the following equation for the unknown multipliers in the value function:

$$p_N = \frac{a^2(1 + p_{N-1})}{2 + p_{N-1}} \tag{5.89}$$

Sec. 5.3 Path Optimization Problems and the Principle of Optimality

This may be solved successively, starting with the initial condition $p_1 = a^2$. (This is a *nonlinear* difference equation and cannot be solved analytically using z-transform methods.)

We notice that the equation for the optimal control may be written in the form

$$u_0 = -\frac{(1+p_{N-1})ax_0}{2+p_{N-1}} \tag{5.90}$$

So the first control for an N-stage problem is given by the product of the initial condition, x_0, and a coefficient depending on the number of stages. Again appealing to the principle of optimality, we deduce that this must be the form of the optimal control at each stage, namely that for $0 \leq n < N-1$,

$$u_n = -\frac{(1+p_{N-1-n})ax_n}{2+p_{N-1-n}} \tag{5.91}$$

(Of course we already know that $u_{N-1} = -ax_{N-1}$, since we require that $x_N = 0$.) This provides a complete solution for this path optimization example in the form of a *feedback control law*, which means that we have specified the optimal control as a function of the current value of the "state."

It is interesting to pursue this example a little further. We can view the "time horizon" N as a design parameter whose selection will determine the feedback control gains used to implement optimal control. (Notice that the gains depend only on quantities that can be calculated without knowledge of what initial state x_0 is to be driven to 0. The gains can be precomputed and stored for use "on line" or the quantity p_N can be computed and then the required gains can be computed "on line" by reversing the difference equation.) In the limit of large N, there will be an important practical simplification in that the feedback control gains become constant, independent of the time step, greatly simplifying implementation of the optimal control.

In the limit of large N, our problem amounts to specifying the natural stability condition $\lim_{n \to \infty} x_n = 0$, and some simple reasoning shows that good limiting behavior can be expected. First, we see that $J_{n+1}(x) \leq J_n(x)$ for every x because it is always possible to extend an n-step solution to $n+1$ steps by taking $u_n = 0$. Thus for an optimal solution, the choice of control values for an $(n+1)$-step problem must produce a cost no greater than the cost of using the best n-step control sequence with a final 0 control value appended; this gives the inequality.

Since the value functions are quadratic, we conclude that $p_{n+1} \leq p_n$ (i.e., the $\{p_n\}$ sequence is nonincreasing). Since the cost of any solution is always nonnegative, the $\{p_n\}$ sequence is bounded below by 0. Thus the $\{p_n\}$ sequence has a limit. (Notice that it is not so simple to deduce this "stability" result directly from the difference equation specifying $\{p_n\}$ without appealing to the connection with the value functions.) Taking the limit as $N \to \infty$ on both sides of the difference equation for p_N, using p_∞ to denote the limiting value, gives

$$p_\infty = \frac{a^2(1+p_\infty)}{2+p_\infty} \tag{5.92}$$

This equation may be solved by cross-multiplying to reduce it to a quadratic equation and using the fact that $p_\infty \geq 0$ to choose the correct solution. This gives

$$p_\infty = \frac{(a^2 - 2) + \sqrt{4 + a^4}}{2} \tag{5.93}$$

For the infinite-step control problem, the optimal control thus simplifies to the time-invariant feedback control law:

$$u_n = -\frac{(1 + p_\infty)ax_n}{2 + p_\infty} \tag{5.94}$$

Furthermore, since this control law assures that $x_n \to 0$ as $n \to \infty$, we may conclude from the equation for the controlled system,

$$x_{n+1} = ax_n - \frac{(1 + p_\infty)ax_n}{2 + p_\infty} = \frac{a}{2 + p_\infty} x_n \tag{5.95}$$

that the quantity $a/(2 + p_\infty)$ has magnitude less than 1. This follows because the solution of the controlled equation is given by

$$x_n = \left(\frac{a}{2 + p_\infty}\right)^n x_0 \tag{5.96}$$

which tends to zero exactly under this condition.

Discrete-Stage Path Optimization: Lagrange Multiplier Approach

The principle of optimality provides a characterization of global minimizing solutions, but the generality of the formulation is matched by the difficulty of solving the resulting equation for the value function. Necessary conditions based on the Lagrange multiplier approach are commonly applied to solving discrete-time path optimization problems because they turn out to be more tractable.

We will sketch the Lagrange multiplier formulation in the case of the basic path optimization problem described earlier for a system described by discrete-time state equations,

$$\mathbf{x}_{n+1} = \mathbf{F}(\mathbf{x}_n, \mathbf{u}_n) \tag{5.97}$$

with $\mathbf{x}_0$ given. The object is to choose the input sequence, $\{\mathbf{u}_k : 0 \leq k \leq N - 1\}$ to minimize

$$L(\mathbf{x}, \mathbf{u}) = \sum_{n=1}^{N} l(\mathbf{x}_n, \mathbf{u}_{n-1}) \tag{5.98}$$

where we use $\mathbf{x}$ and $\mathbf{u}$ to denote "stacked" state and input vectors:

$$\mathbf{x} = \begin{bmatrix} \mathbf{x}_1 \\ \mathbf{x}_2 \\ \cdot \\ \cdot \\ \cdot \\ \mathbf{x}_N \end{bmatrix}, \quad \mathbf{u} = \begin{bmatrix} \mathbf{u}_0 \\ \mathbf{u}_1 \\ \cdot \\ \cdot \\ \cdot \\ \mathbf{u}_{N-1} \end{bmatrix} \tag{5.99}$$

Sec. 5.3 Path Optimization Problems and the Principle of Optimality

Notice that the state equations may be regarded as a set of equality constraints of the form $\mathbf{f}(\mathbf{x},\mathbf{u}) = \mathbf{0}$, by stacking the N constraint equations,

$$\mathbf{f}_i = \mathbf{x}_i - \mathbf{F}(\mathbf{x}_{i-1},\mathbf{u}_{i-1}) = \mathbf{0} \quad \text{for } 1 \leq i \leq N \tag{5.100}$$

Posed in this way, the path optimization problem is simply a parameter optimization problem involving equality constraints, and the Lagrange multiplier method may be applied directly to obtain a solution. Notice that it is a conceptually easy step to generalize this problem by including pointwise inequality constraints on the individual control and state vectors, using the Kuhn-Tucker conditions for obtaining a solution.

Continuous Path Optimization and the Principle of Optimality

Rather than presenting details of the discrete-time path optimization problem just described, we will turn to continuous-time problems. As we did in the discrete-time case, we will use a simple example to introduce the methodology. It is an optimal control problem for a continuous-time system described by the differential equation

$$\dot{x}(t) = \alpha x(t) + u(t) \tag{5.101}$$

where $x(t_0) = x$ and where the control function $u(t)$ is to be chosen so that $x(T) = 0$ while minimizing the cost function

$$\int_{t_0}^{T} x^2(\tau) + u^2(\tau) \, d\tau \tag{5.102}$$

for $t_0 = 0$. (An analogy with the previous discrete-time example is obvious.) Let $J(x,t)$ denote the value function, the minimized value of the cost function when $t_0 = t$ and $x(t_0) = x$. By the Mean Value Theorem for integrals, we write

$$\int_{t}^{T} x^2(\tau) + u^2(\tau) \, d\tau = \int_{t}^{t+\Delta t} x^2(\tau) + u^2(\tau) \, d\tau + \int_{t+\Delta t}^{T} x^2(\tau) + u^2(\tau) \, d\tau$$

$$\approx (x^2(t_*) + u^2(t_*))\Delta t + \int_{t+\Delta t}^{T} x^2(\tau) + u^2(\tau) \, d\tau \tag{5.103}$$

for some t_* with $t \leq t_* \leq t + \Delta t$. This allows us to write the following equation for the value function by using the principle of optimality:

$$J(x,t) = \min_{u}[(x^2(t_*) + u^2(t_*))\Delta t + J(x + \Delta x, t + \Delta t)] \tag{5.104}$$

where Δx is the change $x(t+\Delta t) - x(t)$. (Recall that $x(t) = x$.) The minimum is taken over functions on the interval from t to $t + \Delta t$. With Δt small, and with sufficient continuity assumptions, an approximation of the last term in the bracketed expression may be obtained by using derivatives, namely

$$J(x + \Delta x, t + \Delta t) \approx J(x,t) + J_x(x,t)\Delta x + J_t(x,t)\Delta t \tag{5.105}$$

where the subscripts denote partial derivatives with respect to the two arguments of $J(x,t)$ and the approximation is accurate to first-order in Δx and Δt. Similarly, using the differential equation for $x(t)$, an approximation for Δx is

$$\Delta x \approx (\alpha x + u(t))\Delta t \tag{5.106}$$

Thus the value function equation becomes

$$J(x,t) = \min_{u} [\,(x^2(t_*) + u^2(t_*))\Delta t + J(x,t) + J_x(x,t)(\alpha x + u(t))\Delta t$$
$$+ J_t(x,t)\Delta t\,] \qquad (5.107)$$

(Actually, the quantity inside the brackets is accurate to first-order in Δt.) Blithely assuming that the operation of taking the limit as $\Delta t \to 0$ can be interchanged with the operation of taking the minimum over u leads to the following partial differential equation for the value function:

$$-J_t(x,t) = \min_{u}[x^2 + u^2 + J_x(x,t)(\alpha x + u)] \qquad (5.108)$$

where now the minimum is over the real number u and the minimizing value gives $u(t)$. We have the obvious boundary condition $J(0,t) = 0$ for $0 \le t \le T$.

For the example at hand a solution may be guessed on the basis of the discrete-time problem already solved:

$$J(x,t) = q(t)x^2 \qquad (5.109)$$

and the function $q(t)$, through which the optimal control function is expressed, may be specified as the solution to a nonlinear differential equation. (Simply plug in the proposed solution and carry out the minimization by setting the derivative of the quantity in brackets to zero to find the minimizing u.) We will not pursue this example any further, since our main purpose has only been to suggest that the principle of optimality may be applied to optimal control problems for continuous-time systems. We conclude by sketching the formulation of general problems of the same type.

For the problem where the state equations are

$$\dot{\mathbf{x}}(t) = \mathbf{f}(\mathbf{x}(t),\mathbf{u}(t)), \quad \mathbf{x}(0) = \mathbf{x} \qquad (5.110)$$

and it is desired to select the control $\mathbf{u}(t)$ so that $\mathbf{x}(T) = \mathbf{0}$ while minimizing the integral cost function

$$\int_0^T L(\mathbf{x}(\tau),\mathbf{u}(\tau))\,d\tau \qquad (5.111)$$

the principle of optimality may be applied to show that the value function $J(\mathbf{x},t)$ obeys the so-called Hamilton-Jacobi-Bellman equation:

$$-J_t(\mathbf{x},t) = \min_{\mathbf{u}} [\,L(\mathbf{x},\mathbf{u}) + J_{\mathbf{x}}(\mathbf{x},t)\mathbf{f}(\mathbf{x},\mathbf{u})\,] \qquad (5.112)$$

As is probably clear from our discussion, putting this approach on a rigorous mathematical foundation requires substantial technical detail. And just as in the discrete-time case, solving for the value function is a very difficult problem in general.

Continuous Path Optimization: Lagrange Multiplier Approach

To conclude this discussion, we will illustrate the Lagrange multiplier approach to the continuous-time path optimization problem of (5.110) and (5.111). The roots of this

Sec. 5.3 Path Optimization Problems and the Principle of Optimality

approach may be found in the discussion of derivatives in Chapter 4, where the idea that derivatives are linear mappings was exploited to generalize Newton's method from problems involving vectors of variables to problems involving functions. Thinking of functions as vectors indexed by a continuous variable was the key to an intuitive understanding of the resulting waveform relaxation technique for solving differential equations. The same concept underlies the Lagrange multiplier approach to path optimization problems; the resulting equations represent an appropriate "continuous limit" of discrete-time problems.

As discovered in the analysis of waveform relaxation, the integral equation formulation that is used to pose the generalized version of Newton's method gives way in the end to a differential equation description as a "practical" implementation. The same turns out to be the case for path optimization problems, and we will leave the details for interested readers to find in more advanced books. Since the end result is easy to comprehend and to apply in a variety of important problems, we will simply give the equations and two examples. The important issues involved in numerical solution of path optimization problems, where ideas such as gradient descent methods are combined with discretization, for example, will not be covered.

The necessary conditions are specified in terms of the augmented cost integrand (known as the *Hamiltonian function*)

$$H(\mathbf{x},\mathbf{u},\boldsymbol{\lambda}) = \boldsymbol{\lambda}^T \mathbf{f}(\mathbf{x},\mathbf{u}) + L(\mathbf{x},\mathbf{u}) \tag{5.113}$$

The resulting necessary conditions for a constrained local extremum are known as the *Euler-Lagrange equations*:

$$H_{\mathbf{x}}(\mathbf{x}(t),\mathbf{u}(t),\boldsymbol{\lambda}(t)) = -\dot{\boldsymbol{\lambda}}^T(t) \tag{5.114}$$

$$H_{\mathbf{u}}(\mathbf{x}(t),\mathbf{u}(t),\boldsymbol{\lambda}(t)) = \mathbf{0} \tag{5.115}$$

$$H_{\boldsymbol{\lambda}}(\mathbf{x}(t),\mathbf{u}(t),\boldsymbol{\lambda}(t)) = \dot{\mathbf{x}}^T(t) = \mathbf{f}^T(\mathbf{x}(t),\mathbf{u}(t)) \tag{5.116}$$

The boundary conditions for these differential equations reflect the conditions specified in the problem description: $\mathbf{x}(0) = \mathbf{x}_0$ is a given initial condition and $\mathbf{x}(T) = \mathbf{0}$ is the desired terminal condition. (Some variations on the problem description can be handled by selecting the boundary conditions for the Euler-Lagrange equations differently. One such case arises when a penalty term $\phi(\mathbf{x}(T))$ is added to the cost function and the requirement that $\mathbf{x}(T) = \mathbf{0}$ is removed.)

Examples

For the optimal control problem involving a linear system and a quadratic cost function, the Euler-Lagrange equations can be transformed to a form that is particularly amenable to numerical solution. To keep the notation simple, this example will concern the case of time-invariant system and costs. The state equations for the system are

$$\dot{\mathbf{x}}(t) = \mathbf{A}\mathbf{x}(t) + \mathbf{B}\mathbf{u}(t), \quad \mathbf{x}(0) = \mathbf{x}_0$$

where the input $\mathbf{u}(t)$ is to be chosen to achieve $\mathbf{x}(T) = \mathbf{0}$ while minimizing an integral cost function

$$\tfrac{1}{2}\int_0^T \mathbf{x}^T(\tau)\mathbf{Q}\mathbf{x}(\tau) + \mathbf{u}^T(\tau)\mathbf{R}\mathbf{u}(\tau)\,d\tau$$

where $\mathbf{Q}$ and $\mathbf{R}$ are positive definite symmetric matrices. The corresponding Hamiltonian function is

$$H(\mathbf{x}(t),\mathbf{u}(t),\lambda(t)) = \lambda^T(t)\mathbf{A}\mathbf{x}(t) + \lambda^T(t)\mathbf{B}\mathbf{u}(t) + \tfrac{1}{2}\mathbf{x}^T(t)\mathbf{Q}\mathbf{x}(t) + \tfrac{1}{2}\mathbf{u}^T(t)\mathbf{R}\mathbf{u}(t)$$

and the Euler-Lagrange equations take the form

$$\dot{\lambda}(t) = -H_{\mathbf{x}}^T = -\mathbf{A}^T\lambda(t) - \mathbf{Q}\mathbf{x}(t)$$

$$\dot{\mathbf{x}}(t) = H_{\lambda}^T = \mathbf{A}\mathbf{x}(t) + \mathbf{B}\mathbf{u}(t)$$

$$\mathbf{0} = H_{\mathbf{u}} = \lambda^T(t)\mathbf{B} + \mathbf{u}^T(t)\mathbf{R}$$

The last equation may be solved to give $\mathbf{u}$ in terms of λ,

$$\mathbf{u}(t) = -\mathbf{R}^{-1}\mathbf{B}^T\lambda(t)$$

and $\mathbf{u}(t)$ may be eliminated, leaving two coupled linear differential equations for $\mathbf{x}(t)$ and $\lambda(t)$:

$$\begin{bmatrix}\dot{\mathbf{x}}(t)\\ \dot{\lambda}(t)\end{bmatrix} = \begin{bmatrix}\mathbf{A} & -\mathbf{B}\mathbf{R}^{-1}\mathbf{B}^T\\ -\mathbf{Q} & -\mathbf{A}^T\end{bmatrix}\begin{bmatrix}\mathbf{x}(t)\\ \lambda(t)\end{bmatrix}$$

These equations comprise a *two-point boundary value problem*, so named because the set of boundary conditions for the differential equation are specified at two different instants of time.

One approach to obtaining a solution is to "guess" that the variables are linearly related, say

$$\mathbf{x}(t) = \mathbf{S}(t)\lambda(t)$$

so that

$$\dot{\mathbf{x}}(t) = \dot{\mathbf{S}}(t)\lambda(t) + \mathbf{S}(t)\dot{\lambda}(t)$$

Substituting into the equation for $\dot{\mathbf{x}}$, replacing $\mathbf{x}(t)$ by $\mathbf{S}(t)\lambda(t)$ wherever it occurs, gives

$$\dot{\mathbf{S}}(t)\lambda(t) - \mathbf{S}(t)(-\mathbf{A}^T - \mathbf{Q}\mathbf{S}(t))\lambda(t) = \mathbf{A}\mathbf{S}(t)\lambda(t) - \mathbf{B}\mathbf{R}^{-1}\mathbf{B}^T\lambda(t)$$

If $\mathbf{S}(t)$ is taken to be the solution of the nonlinear matrix differential equation,

$$\dot{\mathbf{S}} = \mathbf{A}\mathbf{S} + \mathbf{S}\mathbf{A}^T + \mathbf{S}\mathbf{Q}\mathbf{S} - \mathbf{B}\mathbf{R}^{-1}\mathbf{B}^T$$

with (terminal) boundary condition $\mathbf{S}(T) = \mathbf{0}$, then the linear relation between $\mathbf{x}$ and λ holds for all t. Thus the solution of the Euler-Lagrange two-point boundary value problem is determined from the solution of a nonlinear matrix differential equation (known as the *matrix Riccati equation*). A key feature of the solution obtained this way is that the optimal control is expressed in the form of a time-varying linear state feedback control law:

$$\mathbf{u}^{\text{opt}}(t) = -\mathbf{R}^{-1}\mathbf{B}^T\mathbf{S}^{-1}(t)\mathbf{x}(t)$$

As a second example, consider the problem of designing a "time-optimal slide" that will allow the fastest descent from height y_0 meters over a horizontal distance of d meters. The slide is assumed to be frictionless and the vertical drop is assumed to be small enough so that the gravitational force on the sliding body, modeled as a point mass, is constant. Denoting the horizontal distance by z and the vertical distance by y, the desired solution is a curve, $y = y(z)$, with endpoints $y(0) = y_0$ and $y(d) = 0$ (Fig. 5.11). The curve may be regarded as the solution to the differential equation

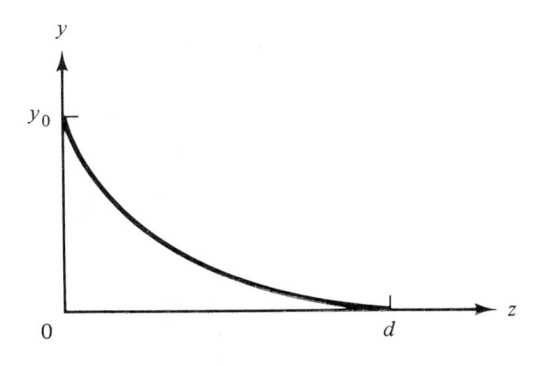

Sec. 5.3 Path Optimization Problems and the Principle of Optimality

Figure 5.11 One possible solution to the time-optimal slide problem.

$$\frac{dy}{dz} = u(z)$$

where $u(z)$ is a function that gives the pointwise slope of the slide, which is to be determined. Since the sliding body moves with constant total energy, E, we have

$$E = \tfrac{1}{2}m\dot{y}^2 + \tfrac{1}{2}m\dot{z}^2 + mgy(z)$$

where the dots indicate time derivatives and m is the mass of the sliding body. The value of E may be obtained by assuming that the initial velocity is zero, giving

$$E = mgy_0$$

The vertical component of the velocity may be expressed in terms of the horizontal component of the velocity by using the chain rule,

$$\dot{y} = \frac{dy}{dz}\dot{z} = u(z)\dot{z}$$

and the total energy equation may be written as

$$mgy_0 = \tfrac{1}{2}m(1+u^2(z))\dot{z}^2 + mgy(z)$$

This may be solved to obtain the horizontal velocity

$$\frac{dz}{dt} = \dot{z} = \frac{(2g(y_0-y(z)))^{1/2}}{(1+u^2(z))^{1/2}}$$

Separating variables and integrating gives

$$T = \int_0^d \frac{(1+u^2(z))^{1/2}}{(2g(y_0-y(z)))^{1/2}}\,dz$$

This is the quantity to be minimized, and the differential equation that describes the curve $y(z)$ in terms of its slope $u(z)$ is the "state equation" constraint. (In this problem z, not t, is the independent variable.) Thus the Hamiltonian function for the problem is

$$H(y,u,\lambda) = \lambda(z)\mathbf{f}(y(z),u(z)) + L(y(z),u(z))$$

$$= \lambda(z)u(z) + \frac{(1+u^2(z))^{1/2}}{(2g(y_0-y(z)))^{1/2}}$$

and the Euler-Lagrange equations are

$$\frac{dy}{dz} = H_\lambda = u(z)$$

$$\frac{d\lambda}{dz} = -H_y = -L_y = -\frac{(1+u^2(z))^{1/2}}{(8g(y_0 - y(z))^3)^{1/2}}$$

$$0 = H_u = \lambda(z) + L_u = \lambda(z) + \frac{u^2(z)}{(2g(y_0 - y(z))(1+u^2(z)))^{1/2}}$$

To solve for $y(z)$, it is convenient to use a general property of any optimal control problem whose cost integrand, L, has no explicit dependence on the independent variable (in this example, z): The Hamiltonian function for such a problem is constant along the optimizing trajectory. This fact is easy to show for the optimal slide problem, and it will simplify the determination of $y(z)$ considerably. The following calculation shows that H is constant, thanks to the Euler-Lagrange equations (we use a prime to denote differentiation with respect to the independent variable z):

$$H'(y,u,\lambda) = H_y y'(z) + H_u u'(z) + H_\lambda \lambda'(z)$$
$$= -\lambda'(z)u(z) + 0 + u(z)\lambda'(z) = 0$$

Let C be the constant value of H, a quantity to be determined. We have

$$C = H(y,u,\lambda) = \lambda(z)u(z) + L(y(z),u(z))$$
$$= -L_u(y(z),u(z))u(z) + L(y(z),u(z))$$
$$= -\frac{u^2}{(2g(y-y_0)(1+u^2))^{1/2}} + \frac{(1+u^2)^{1/2}}{(2g(y-y_0))^{1/2}}$$

This equation is easily solved to give $u(z)$ as a function of $y(z)$:

$$u(z) = \frac{(A - (y(z) - y_0))^{1/2}}{(y(z) - y_0)^{1/2}}$$

where $A = 1/(2gC^2)$. Using this expression, the state equation $y'(z) = u(z)$ may be solved for $y(z)$ (e.g., using separation of variables and an appropriate substitution of variables). The resulting curve, $y(z)$, is a segment of a *cycloid*, which is the curve traced out by a point on the rim of a wheel rolling in a straight line, without slipping, on a flat surface. In other words, the curve $y(z)$ is expressed in parametric form as

$$z = A(\theta - \sin\theta)$$
$$y = y_0 - A(1 - \cos\theta)$$

where $0 \le \theta \le \theta_d$ and the values of A and θ_d are chosen so that $y = 0$ when $z = d$.

5.4 NOTES AND REFERENCES

Parameter optimization is an important tool in computer-aided design. The underlying theory and the computational methods are discussed in many books, including Brayton and Spence [4], Cuthbert [6], Gill, Murray, and Wright [7], Luenberger [8], Mastascusa [10], and Stengel [13]. Dynamic programming was pioneered by Richard Bellman [1];

the book by Bertsekas [2] provides a more up-to-date account. Path optimization problems for continuous-time systems are covered in Luenberger's books, [8] [9], by Sontag [11], and by Stengel [13]. For the classical approach to problems like the slide example, see Weinstock [15]. The book by Boyd and Barratt [3] develops a comprehensive optimization-based approach to linear control design problems.

Linear programming is a subject unto itself, as is combinatorial optimization. The book by Chvátal [5] covers the "classical" linear programming theory and methods; "interior methods," recently introduced by Karmarkar, are discussed by Strang [14]. The book by Steiglitz [12] introduces combinatorial optimization in the context of some simple graph problems; it is interesting to see how a local optimality condition, analogous to conditions derived in Section 6.1 for parameter optimization problems, plays a central role in solving the *shortest tree problem*, whereas a global condition (i.e., the principle of optimality) is the key to solving the *shortest path problem*.

BIBLIOGRAPHY

[1] R. Bellman, *Dynamic Programming*, Princeton University Press, Princeton, NJ, 1957.

[2] D.P. Bertsekas, *Dynamic Programming: Deterministic and Stochastic Models*, Prentice-Hall, Englewood Cliffs, NJ, 1987.

[3] S.P. Boyd and C.H. Barratt, *Linear Controller Design: Limits of Performance*, Prentice-Hall, Englewood Cliffs, NJ, 1991.

[4] R.K. Brayton and R. Spence, *Sensitivity and Optimization*, Elsevier, Amsterdam, 1980.

[5] V. Chvátal, *Linear Programming*, W.H. Freeman, New York, 1983.

[6] T.R Cuthbert, Jr., *Optimization Using Personal Computers*, John Wiley & Sons, New York, 1987.

[7] P.E. Gill, W. Murray, and M.H. Wright, *Practical Optimization*, Academic Press, New York, 1981.

[8] D.G. Luenberger, *Optimization by Vector Space Methods*, John Wiley & Sons, New York, 1969.

[9] D.G. Luenberger, *Introduction to Dynamic Systems*, John Wiley & Sons, New York, 1979.

[10] E.J. Mastascusa, *Computer-Assisted Network and System Analysis*, John Wiley & Sons, New York, 1988.

[11] E.D. Sontag, *Mathematical Control Theory: Deterministic Finite Dimensional Systems*, Springer-Verlag, New York, 1990.

[12] K. Steiglitz, *An Introduction to Discrete Systems*, John Wiley & Sons, New York, 1974.

[13] R.F. Stengel, *Stochastic Optimal Control: Theory and Application*, John Wiley & Sons, New York, 1986.

[14] G. Strang, *Introduction to Applied Mathematics*, Wellesley-Cambridge Press, Wellesley, MA, 1986.

[15] R. Weinstock, *Calculus of Variations*, Dover, New York, 1974; original edition: McGraw-Hill, New York, 1952.

PROBLEMS

1. Find the minimum of the function
$$L(x,u) = x^2 + 2u^2$$
over the variable u, subject to the constraint
$$f(x,u) = x + 3u + 5 = 0$$
(Check both the necessary and sufficient conditions.) Also find the minimizing choice of u.

2. Determine the choice of u which minimizes the function
$$L(x,u) = x^2 + cu^2$$
subject to the constraint
$$f(x,u) = x + 2u + 4 = 0$$
Here c is a positive constant. The minimizing u will obviously depend on c. What is the minimized value of $L(x,u)$? Discuss the limiting cases: $c \to 0$ and $c \to \infty$ and interpret the results.

3. Use the Lagrange multiplier method for solving constrained parameter optimization problems to determine the triangle of maximum area that may be inscribed in a circle of radius r. Determine how the value of the Lagrange multiplier is related to the rate of change of the maximized area with respect to changes in r^2.

4. Compare, analytically and experimentally, the performance of Newton's method for solving the nonlinear equation $\mathbf{f}(\mathbf{x}) = \mathbf{0}$ with the gradient method for minimizing $\|\mathbf{f}(\mathbf{x})\|^2$. Choose a quadratic function of three variables and a fifth-degree polynomial of one variable as test cases.

5. Given the season totals for a baseball team's runs scored R_s and runs allowed R_a, a plausible form of the estimated winning percentage is
$$\frac{1}{1 + (R_a/R_s)^\alpha}$$
for some positive number α. This is a reasonable form because the estimated winning percentage is 50% when $R_s = R_a$, and the percentages add to 100% for two teams whose R_s/R_a ratios are inversely related. Look up data from a recent major league season and find a least-squares fit for the parameter α. For obvious reasons of simplicity, an integral value of α would be desirable. Round off the value obtained by least squares and find the largest change in estimated winning percentage for the data analyzed. (This estimate was introduced by Bill James, a well-known sabermetrician; for further discussion, see *The Bill James Baseball Abstract*, Ballantine Books, New York, 1982, or one of the editions from the period 1983–1988.)

6. For a second-order bandpass filter with transfer function
$$H(s) = \frac{1}{s^2 + 2\zeta\omega_n s + \omega_n^2}$$
determine the values of the damping ratio and natural frequency, ζ and ω_n, corresponding to a Bode plot whose peak occurs at 1.21 MHz and whose half-power bandwidth is 10 kHz.

Problems

7. Compare the performance of several numerical methods for minimizing the Rosenbrock cost function

$$L(\mathbf{u}) = 100(u_2 - u_1^2)^2 + (1 - u_1)^2$$

8. Investigate numerical methods for minimizing the following cost function (the Fletcher-Powell helical valley):

$$L(\mathbf{u}) = 100((u_3 - 10\theta)^2 + (r - 1)^2) + u_3$$

where r and θ are polar coordinates given by $u_1 = r \cos 2\pi\theta$, $u_2 = r \sin 2\pi\theta$, and θ lies in the interval $-0.25 \leq \theta \leq 0.75$.

9. A discrete-time optimal control problem takes the form

$$\min_{\{u_n\}} \sum_{n=1}^{3} (x_n^2 + |u_n|)$$

over the variables u_1, u_2, and u_3 subject to the constraints $|u_1| \leq 1$, $|u_2| \leq 1$, and $|u_3| \leq 1$, and the state equations

$$x_{k+1} = x_k + u_k; \quad x_0 = 5$$

Use any method you can come up with to show that the optimum choice of inputs is $u_1 = u_2 = u_3 = -1$. Show, if you can, how the principle of optimality may be applied to this problem.

10. For the path optimization problem

$$\min_{\{u_k\}} \sum_{k=0}^{N-1} q(x_k, u_k) + x_N^2$$

subject to

$$x_{n+1} = f(x_n, u_n); \quad x_0 = c$$

let $J_N(c)$ be the minimized cost as a function of c for the N-step problem. The minimization is carried out with respect to the inputs $u_0, \ldots, u_{N-1}$. Use the principle of optimality to obtain an equation for $J_N(c)$ in terms of J_{N-1}.

Index

A

A-stable discretization method, 134
AM demodulator, 161–62
Accuracy of discretization methods, 132
    (*See also* particular method)
Active inequality constraints, 253
Airplane pitch dynamics, 70
    Euler discretization, 135
    sample-and-hold discretization, 126
    stability, 94
Amplitude density function, 214
Angle between vectors, 56–57
Asymptotic stability
    discrete-time linear system, 116
    equilibrium solution, 180
    linear system, 90

B

BIBO stability, 95
Backward difference discretization, 127, 128
    accuracy, 132
    stability, 134
    transfer function relation, 133
Banded matrix, 25
Banded systems of linear equations, 25

Bandwidth
    Butterworth filters, 108
    first order system, 101
    second order system, 104
Basis for a vector space, 51
Basis transformation, 4
Batch learning, 206–7
Bendixson test, 193–94
Bilinear transformation method, 135
Binding inequality constraints, 253
Block diagrams, 85–87
    state space equations, 88, 89
Bode plot, 99–100
    first order system, 101–2
    second order system, 103–6
Bounded-input, bounded-output (BIBO) stability, 95
Butterworth filters, 105–8

C

$\mathbb{C}^n$, definition, 50
Cart and stick system, 164–65
    linearization, 170
Catastrophe, 223
Cayley-Hamilton theorem, 8
    proof, 13
Center, 190

Chaos, 199–200, 225
Characteristic polynomial, 7–8
Column space of a matrix, 60
Companion and diagonal form relationships, 10–11
Companion matrix, 9, 83
Complementary orthogonal projection, 21
Completing the square, 27
Condition number, 24, 31, 56
Constrained local minimum
　necessary conditions, 244
　sufficient conditions, 250
Content-addressable memory, 204–10
Continuous path optimization, 273–78
Contraction mapping, 178
Convolution integral, 79
Convolution sum, 114
Coulomb friction-stiction function, 215
Cramer's rule, 48
Critical points, 235
Curve fitting, 21–22
Cusp catastrophe, 223

D

D, 81
Damped nonlinear pendulum, 184
Damping ratio, 103
Dead zone function, 215
Deflation, 36
Derivative
　definition, 166
　for integral operator, 175
Describing functions, 213–19
Determinant of a matrix, 47–48
　Laplace expansion, 47
　properties, 48–49
　two-by-two matrix, 47
Diagonal and companion form relationships, 10–11
Diagonal dominance, 34
Diagonalizable matrix, 5–6
Diagonalization by similarity transformation, 5
Differentiation operator, **D**, 81
Digital filter design, 141–45, 239–40
　using impulse invariance, 143–45
　using optimization, 239–40
　using Tustin discretization, 142–43
Dimension of a vector space, 51
Discrete-time linear system
　frequency response, 115–16

input-output model, 114
solution to state equations, 113
stability, 116
state equations, 112
transfer function, 115–16
unit pulse response, 114
zero-state output response, 114
Discrete-time nonlinear system
　linearization, 168
　stability based on linearization, 182
Discrete-time phasor, 115
Discretization methods
　A-stable, 134
　accuracy, 132
　backward difference, 127, 128
　Euler, 127, 128
　explicit, 129
　forward difference, 127, 128
　implicit, 129
　midpoint method, 179
　nonlinear systems, 128, 173–74, 179
　sample-and-hold, 125–27
　second order Ralston-Runge-Kutta, 130
　second order Runge-Kutta, 130
　stability, 132–34
　trapezoidal, 129
　Tustin, 135
Distributed parameter system, 145
Duffing's equation, 223–25

E

Ebers-Moll transistor model, 162
Eigenvalues, 4
　computation
　　inverse power method, 36
　　power method, 35–36
　　QR factorization, 30
　distinct, 5
　multiplicity of, 7
　repeated, 7
Eigenvectors, 4
　computation
　　inverse power method, 36
　　power method, 35–36
　linearly independent, 6
　orthogonal, 6
Eigenvector-eigenvalue factorization, 26–27
Electric circuit components, 64–65
Electric motor, 69
Electrostatic potential
　finite difference discretization, 149–51

Equilibrium point, 186
Equilibrium solution, 167
　stability, 180
Euclidian norm of a vector, 53
Euler discretization method, 127, 128
　accuracy, 132
　stability, 133
　transfer function relation, 133
Euler-Lagrange equations, 275
Explicit discretization methods, 129

F

FM demodulator, 219–21
Finite difference discretization, 147
Finite element discretization, 153–55
Fixed-point problem, 178
Forced nonlinear pendulum, 225
Forward difference discretization method, 127, 128
　accuracy, 132
　stability, 133
　transfer function relation, 133
Fourth order Runge-Kutta discretization, 131
Frequency response function, 98
　discrete-time system, 115–16
Frequency warping, 139
Frobenius norm, 55
Fundamental Theorem of Algebra, 6–7

G

Gauss-Seidel method, 34
Gaussian elimination, 25
Gradient, definition, 166
Gradient method, 258–60
　for constrained optimization problems, 267
　step size selection, 261–63
Gradient vector, 236
Gram matrix, 15
Gram-Schmidt procedure, 27–28, 57–58

H

Harmonic balancing, 221
Heat equation
　Crank-Nicolson discretization, 151–53
　semi-discretization, 151–53

Hebb learning, 209–10
Hermitian symmetric matrix, 49–50
Hermitian transpose, 49
Hessian matrix, 236
Homogeneous linear equations, 16–17
Homogeneous solution for linear system, 71, 72, 76
Householder transformation, 28
Hyperplane, 24

I

Ill-conditioned linear equations, 24
Ill-conditioned matrix, 56
Implicit discretization methods, 129
Inconsistent linear equations, 15
Inconsistent linear equations
　minimum length approximate solutions, 18, 21–22, 29
Induced matrix norm, 55
　1-norm, 56
　Euclidean norm, 55
　uniform norm, 55
Initial conditions for state space equation, 84
Inner product of vectors, 53
Input-output difference equation, 114
Input-output differential equation, 80
Input-output model
　discrete-time linear system, 114
　for linear state-space equations, 81
　linear system, 79–81
　nonlinear system, 212–13
Input-output stability, 95–96
Interconnected systems, 85–88
　feedback loop, 87
　parallel combination, 87
　series combination, 86
Inverse function theorem, 171
Inverse matrix, 46
　Cramer's rule, 48
　properties, 48–49
　two-by-two matrix, 47
Inverse power method, 36
Inverted pendulum balancer, 164–65
　linearization, 170
Invertible matrix, 46
Iterative calculation of eigenvectors
　inverse power method, 36
　power method, 35–36
Iterative solution of linear equations, 32–35
　Gauss-Seidel method, 34
　Jacobi method, 34

Iterative solution of linear equations (*cont.*)
   Richardson method, 178–79
   successive overrelaxation method, 35
   splitting methods, 33–35

J

Jacobi method, 34
Jump resonance, 223

K

Kernel of a linear function, 60
Kirchhoff's Laws, 65
Kuhn-Tucker conditions, 254

L

L_2 stability, 95
LC transmission line, 67–68
LU factorization, 25
Lagrange multipliers, 245–48
   as cost sensitivities, 251
   for continuous path optimization, 274–75
   for discrete-stage path optimization, 272–73
Laplace expansion of a determinant, 47
Least squares curve fitting, 21–22
Length of a vector, 53
Limit of a sequence of matrices, 30
Limit cycle, 191
   Bendixson test, 193–94
   in nonlinear feedback systems, 222
   Poincaré index theorem, 193
   Poincaré-Bendixson theorem, 194
   quenching, 225
Linear equations, 15–24
   banded matrix, 25
   computation of solutions, 24–25, 27–30, 32–35, 178–79
   existence of solutions, 15
   homogeneous, 16–17
   ill-conditioned, 24
   inconsistent, 15
   nontrivial solution, 16
   overdetermined, 15
   solution set, 24
   sparse matrix, 25
   triangular matrix, 24
   underdetermined, 15
   uniqueness of solutions, 15
Linear functions, 59–60
   matrix representations, 59
Linear independence, 51
Linear mapping (*See* Linear functions)
Linear operator (*See* Linear functions)
Linear prediction filters, 238–39
Linear programming problem, 254
Linear system (*See also* Discrete-time linear system)
   external description, 80
   frequency response, 97–105
   homogeneous solution, 71, 72, 76
   input-output model, 79–81
   input-output stability, 95–96
   internal description, 78
   output equation, 78
   output solution, 78
   particular solution, 71, 76
   Routh-Hurwitz stability test, 93–94
   solution, 71
   stability, 89
   state space equations, 71
   step response, 108
   superposition principle, 80
   time constant, 109
   time-invariant, 71
   total solution, 77
   transfer function, 98
   transient response, 108–10
   zero-state output response, 79
Linear transformation (*See* Linear functions)
Linearization, 167
   for analysis of stability, 181–82
   systems with inputs, 169
Linearized system, 167
Linearly dependent vectors, 51
Linearly independent vectors, 51
Local minimum
   necessary conditions, 235–37
   sufficient conditions, 237
Lorenz attractor, 199–200
Lower triangular matrix, 24, 47
Lyapunov function, 183–84
   for damped pendulum, 184
   for neural network, 205, 208–9
Lyapunov's theorem, 181

M

Marginal stability
   linear system, 90, 94–95

Index

discrete-time linear system, 116
Matrix
  banded, 25
  companion, 9, 83
  definition, 42
  degenerate, 7
  diagonalizable, 7
  Gram, 15
  Hermitian symmetric, 49
  Hessian, 236
  Householder, 28
  identity, 46
  ill-conditioned, 56
  inverse, 46
  nonnegative definite, 27, 237
  nonsingular, 46
  orthogonal, 26–27, 50, 58–59
  permutation, 24–25
  positive definite, 27, 237
  pseudo-inverse, 20
  sparse, 25, 32
  symmetric, 49
  triangular, 24, 47
  unitary, 50
  zero, 46
Matrix calculus, 36–41
Matrix condition number, 56
Matrix differential equations, 41
Matrix exponential function, 73–76
  for diagonal matrix, 74
  for diagonalizable matrix, 74–75
  properties, 75–76
  Taylor series definition, 75
Matrix function, 37
Matrix inversion identities, 61
Matrix multiplication, 44–45
  composition of linear functions, 59
Matrix norm, 55
Matrix representation of linear functions, 59
Matrix transposition, 49
Midpoint discretization method, 179
Monic polynomial, 6

N

Natural frequency, 103
Neighboring minima, 248
Neural networks, 202–11
  discrete-time model, 208
Newton's method, 172–73
Noninvertible matrix, 46
Nonlinear feedback system
  limit cycles, 222
  stability, 218–19
Nonlinear pendulum, 163
  forced, 225
  linearization, 167–68
  period of oscillation, 188
  phase plane analysis, 187–88
  solution using successive approximation, 177–78
  solution using Newton's method, 176
  stability analysis, 184
Nonlinear system
  chaotic solution, 199–200
  discretization, 128, 173–74
  frequency response, 213
  input-output model, 212
  integral equation form of solution, 175
  jump resonance, 223
  limit cycles, 191–94
  linearization, 167
  phase plane analysis, 185–98
  solution using successive approximation, 176–77
  solution using Newton's method, 175–76
  stability analysis, 181–84
  state equations, 159
  strange attractor, 200
  subharmonic response, 224
Nonnegative definite matrix, 27, 237
Nonsingular matrix, 46
Norm
  function, 55
  matrix, 55
  vector, 53–54
Normed vector space, 54
Nullspace of a linear function, 60
Nullspace of a matrix, 17
Numerical optimization, 254–67
  bounding subinterval search, 256–57
  gradient method, 258–60
  penalty function methods, 265–66
  quasi-Newton methods, 264–65
  unimodal cost function, 256

O

On/off function, 215
Orbital motion of satellite, 164
  linearization, 169–70
Orthogonal basis, 57–58
Orthogonal matrix, 26–27, 50, 58–59

Orthogonal projection, 20
Orthogonal vectors, 57
Orthogonality, 57
Orthogonalization, 27–28, 57–58
Orthonormal basis, 57–58
Overdetermined linear equations, 15
    minimum length approximate solutions, 18, 21–22

P

p-norm, 54
PLU factorization, 24–25
Parametric sensitivity, 251
Particular solution, 71, 76
Path optimization, 267–78
    discrete-stage, 269–73
Penalty function methods, 265–66
Pendulum (*See* Nonlinear pendulum)
Phase-locked loop, 219–21
Phase plane analysis, 185–98
Phasor, 97
Picard iteration, 177
Piecewise linear systems, 226–29
Pivoting, 25
Poincaré index theorem, 193
Poincaré-Bendixson theorem, 194
Poisson equation
    finite difference discretization, 149–51
    finite element discretization, 153
Poles and zeros, 99
Positive definite matrix, 27, 237
Power method, 35–36
Power series, 38
Predator-prey equations, 195–98
Principal axis vectors, 52
Principle of optimality, 267
    continuous path optimization, 274
    discrete-state path optimization, 270
Pseudo-inverse matrix
    computation
        QR factorization, 29
        SVD, 31
    defining equations, 20
    geometric properties, 20

Q

QR factorization, 27–30
Quadratic convergence, 173
Quadratic forms, 26–27

Quasi-Newton methods, 264–65
Quasi-static solution, 170
Quasilinearization, 219–25

R

$\mathbb{R}^n$, definition, 50
Range of a linear function, 60
Range space, 15
Rank of a matrix, 31
Regular matrix, 46
Richardson method, 178–79
Routh array, 93–94
Routh-Hurwitz stability test, 93–94

S

SOR, 35
SVD, 30–32
Saddle point, 190
Sample-and-hold discretization, 125–27
Sampled phasor, 115
Sampling interval, 112
Sampling Theorem, 115
Satellite motion (*See* Orbital motion of satellite)
Saturating amplifier, 215
Schur-Cohn stability test, 117–18
Schwarz inequality, 56
Second order Ralston-Runge-Kutta discretization, 130
Second order Runge-Kutta discretization, 130
    stability, 138
Sensitivity of minimized, 251
Sequence of matrices, 30
Series RLC circuit, 66
    initial conditions for states, 85
    solution, 77
    state space equations, 82–83
    zero-state output response, 79
Similar matrices, 4
Similarity transformation, 4
    companion and diagonal matrices, 11
Singular Value Decomposition (SVD), 30–32
Singular matrix, 46
Singular points, 186
Singular points, classification, 189
Singular values, 31

Index

Solution of nonlinear equations
  Newton's method, 172–73
  modified Newton's method, 175
  successive approximation, 178
Spanning set of vectors, 51
Sparse matrix, 25, 32
Spectral decomposition of a vector, 11
Spectral representation of a matrix, 12
Splitting methods, 33–35
Stability of discretization methods
  backward difference discretization, 134
  Euler discretization, 133
  second order Runge-Kutta, 138
  trapezoidal discretization, 136
  Tustin discretization, 136
Stable
  discrete-time linear system, 116
  equilibrium, 180
  linear system, 89
Stable focus, 189
Stable node, 189
Standard basis, 52
Steepest descent, 258
Step response, 108
Stick balancer, 164–65
  linearization, 170
Stiff system, 133
Straight line fitting, 21–22
Subharmonic response, 224
Subspace of a vector space, 50
Successive Overrelaxation (SOR) method, 35
Successive approximation, 176–77
Superposition principle, 80, 98
Symmetric matrix, 6, 49

T

Taylor series, 38
Telegraph equation, 147–49
Time constant
  first order system, 109
  second order system, 111
Time-invariance, 71
Time-invariant system, 71
Transfer function, 98
Transfer function for discrete-time system, 115–16
Transfer function, poles and zeros, 99
Transformation of bases, 4
Transient response, 108
Transistor
  Ebers-Moll model, 162
  hybrid pi model, 66–67
Transmission line
  finite difference discretization, 147–49
Transpose of a matrix, 49–50
Trapezoidal discretization method, 129
  accuracy, 137–38
  stability, 136
  transfer function relation, 135–37
Triangle inequality, 54
Triangular (PLU) factorization, 24–25
Triangular matrix, 24, 47
Triangular systems of linear equations, 24
Tustin discretization method, 135
  accuracy, 137–38
  frequency prewarping, 141
  frequency warping, 139–40
  stability, 136
  transfer function relation, 135–37

U

Underdetermined linear equations, 15
  minimum length solutions, 18–19, 23
Uniform norm, 54
Unit pulse response, 114
Unit step function, 79, 108
Unit vectors, 52
Unitary matrix, 50
Unstable equilibrium, 180
Unstable focus, 189
Unstable node, 190
Upper triangular matrix, 24, 47

V

Value function, 269, 273
Van der Pol equation, 193, 225
Variable step size, 261–63
Vector norm, 53–54
Vector space
  definition, 50
  examples, 51
Volterra series, 212
Volterra-Lotka equations, 195–98
Vortex, 190

W

Waveform relaxation, 176
Weighting function, 79